Environmental Performance Measurement:
The Global Report 2001–2002

An Initiative of:

World Economic Forum Global Leaders for Tomorrow Environment Task Force

Yale University's Yale Center for Environmental Law and Policy (YCELP)

Columbia University's Center for International Earth Science Information Network (CIESIN)

Daniel C. Esty and Peter K. Cornelius, editors

The Environmental Performance Measurement: The Global Report 2001–2002 is published by the Yale Center for Environmental Law and Policy (YCELP), Columbia University's Center for International Earth Science Information Network (CIESIN), and the World Economic Forum, where it is a special project within the framework of The Global Competitiveness Programme.

At the Yale Center for Environmental Law and Policy:

Professor Daniel C. Esty
Director

Ilmi M. E. Granoff
Project Director

Barbara Ruth
Administrative Associate

Marguerite Camera
Manuscript Coordinator

At Columbia University's Center for International Earth Science Information Network:

Marc A. Levy
Associate Director for Science Applications

Kobi Abayomi
Assistant Research Associate

Robert S. Chen
Deputy Director

Alex de Sherbinin
Research Associate

Francesca Pozzi
Research Associate

Maarten Tromp
Geographic Information Systems Specialist

Antoinette Wannebo
Research Associate

At the World Economic Forum:

Professor Klaus Schwab
President

Dr. Peter K. Cornelius
Director

The term *country* as used in this report does not in all cases refer to a territorial entity that is a state as understood by international law and practice. The term covers well-defined, geographically self-contained economic areas that are not states but for which statistical data are maintained on a separate and independent basis.

Oxford University Press

Oxford   New York
Athens   Auckland   Bangkok   Bogotá   Buenos Aires
Cape Town   Chennai   Dar es Salaam   Delhi   Florence
Hong Kong   Istanbul   Karachi   Kolkata   Kuala Lumpur
Madrid   Melbourne   Mexico City   Mumbai   Nairobi
Paris   São Paulo   Shanghai   Singapore   Taipei   Tokyo
Toronto   Warsaw

associated companies in
Berlin   Ibadan

Published by
Oxford University Press, Inc.
198 Madison Avenue
New York, New York 10016
http://www.oup-usa.org

Oxford is a registered trademark of
Oxford University Press

ISBN 0–19–515255–7

Printing (last digit): 9 8 7 6 5 4 3 2 1

Printed in the United States of America
on acid-free paper.

Environmental Performance Measurement:
The Global Report 2001–2002

An Initiative of:

World Economic Forum Global Leaders for Tomorrow Environment Task Force

Yale University's Yale Center for Environmental Law and Policy (YCELP)

Columbia University's Center for International Earth Science Information Network (CIESIN)

Daniel C. Esty and Peter K. Cornelius, editors

Table of Contents

iii

World Economic Forum Global Leaders for Tomorrow Environment Task Force

Kim Samuel-Johnson
Samuel Group of Companies
Canada

PROJECT DIRECTOR:

Daniel C. Esty
Yale Center for Environmental Law and Policy
United States

MEMBERS:

Manny Amadi
Cause and Effect Marketing Limited
United Kingdom

Alicia Barcena Ibara
Economic Commission for Latin America
and the Caribbean
Chile

Ugar Bayar
Privatization Administration
Turkey

Matthew Cadbury
Cadbury Schweppes PLC
United Kingdom

Carlos E. Cisneros
Cisneros Television Group
Venezuela

Craig A. Cohon
GlobaLegacy
United Kingdom

Colin Coleman
Goldman Sachs
South Africa

Dominiqe-Henri Freiche
Pinault Printemps-Redoute
France

Thomas Ganswindt
Siemens
Germany

Francisco Gutierrez-Campos
Paraguay

Guy Hands
Nomura International PLC
United Kingdom

Molly Harriss-Olson
Eco Futures Pty. Ltd.
Australia

George M. Kailis
MG Kailis Group
Australia

Shiv Vikram Khemka
Sun Group of Companies
India

Loren Legarda
Senator
The Philippines

Christopher B. Leptos
Southrock Corporation
Australia

Philippa Malmgren
Malmgren and Company
United Kingdom

John Manzoni
BP AMOCO PLC
United Kingdom

Liavan Mallin
Onemade.com
United States

Jonathan Mills
Melbourne Festival
Australia

Rodrigo Navarro Banzer
Corporacion Andina de Fomento
Venezuela

Patrick Odier
Lombard, Odier et Cie
Switzerland

Paul L. Saffo
Institute for the Future
United States

Kiyomi Tsujimoto
Member of the House of Representatives
Japan

The World Economic Forum, the Yale Center for Environmental Law and Policy, and the Center for International Earth Science Information Network would like to acknowledge the support of The Samuel Family Foundation in launching the Environmental Sustainability Index.

v

Contributors

Peter K. Cornelius is Director of the Global Competitiveness Programme at the World Economic Forum. Previously, he was the Head of International Economic Research at Deutsche Bank and a Senior Economist of the International Monetary Fund, where he also served as the IMF's resident representative in Lithuania from 1993 to 1995. A former Staff Economist at the German Council of Economic Advisors and a former consultant to the United Nations Industrial Development Organization and the European Union, Cornelius has also been an advisor to Deutsche Asset Management, an Adjunct Professor at Brandeis University, a Visiting Lecturer at Wissenschaftliche Hochschule für Unternehmensführung, School of Management, Koblenz and a Visiting Scholar at the Harvard Institute for International Development at Harvard University. Cornelius studied Economics and Philosophy at the London School of Economics and Political Science, and received his doctorate in Economics from the University of Göttingen.

Frank Dixon is Managing Director of Research and Development at Innovest Strategic Value Advisors. His work at Innovest includes developing methodologies for assessing the relative financial impacts of corporate environmental and social strategies. Dixon also oversees the analysis of nearly 1,500 companies around the world, then helps financial sector clients use this research to develop socially responsible investment products. Prior to joining Innovest, he worked as a management consultant, specializing in the energy and manufacturing sectors, where his work included advising companies on improving competitive position through enhanced environmental and social performance. Earlier, he worked in the financial area, arranging debt and equity financing for major energy facilities and venture financing for early-stage manufacturing companies. He has an MBA from the Harvard Business School.

Daniel C. Esty is Professor of Environmental Law and Policy at Yale University. He teaches in both the Law School and the Environment School, where he also serves as Associate Dean. He is Director of the Yale Center for Environmental Law and Policy and of the Yale World Fellows Program. Prior to coming to Yale in 1994, Esty was a Senior Fellow at the Institute for International Economics, a Washington, D.C. think tank. From 1989 to 1993, he served in a variety of senior positions in the U.S. Environmental Protection Agency. He is the author or editor of six books and numerous articles on environmental policy issues and the relationships between the environment and trade, security, global governance, competitiveness, international institutions, and development. Esty holds degrees from Harvard College, Oxford University, and the Yale Law School.

Alois M. Flatz is Head of Research at Sustainable Asset Management (SAM), Zurich, where he is responsible for Sustainability Investment Methodology. His work at SAM has included the development of the Dow Jones Sustainability Index. Previously he was an advisor to the Austrian Minister of Environment. He also worked for the Federation of Austrian Industry, where he led several projects including work on the Austrian Packaging Ordinance. Flatz holds an MBA from the Vienna University of Economics, Austria, studied at Haute Ecole de Commerce (HEC) in Paris, and received a Ph.D. from the Institute of Management at the University of St. Gallen, Switzerland.

Marc A. Levy is Associate Director for Science Applications at CIESIN, the Center for International Earth Science Information Network in the Columbia University Earth Institute, where he oversees programs concerning indicators of environmental sustainability, measures of state capacity, information tools for international environmental agreements, and other work aimed at integrating natural and social science information on the environment. He is a Project Scientist for CIESIN's Socioeconomic Data and Applications Center, one of the primary liaisons between NASA's Earth Observing System and the social science and policymaking communities. Levy has taught political science and international environmental policy since 1987, with appointments at Princeton, Williams, and Columbia. He has published on the effectiveness of international environmental institutions, on social learning and environmental policymaking, and on environment-security connections. Levy is coeditor (with Robert O. Keohane and Peter M. Haas) of *Institutions for the Earth* (MIT Press, 1993) and coeditor (with Keohane) of *Institutions for Environmental Aid* (MIT Press, 1996).

Mondher Mimouni is a market analyst with the International Trade Centre UNCTAD/WTO. As part of his work on international trade he has built a bilateral database on market access, and developed the Trade Performance Index for assessing and monitoring the multifaceted dimensions of the export performance and competitiveness of countries. He holds an M.Phil. equivalent in Development Economics from the Economic University of Montpellier, an

MS in Agricultural Policy and Development Administration from the International Centre for Advanced Mediterranean Agronomic Studies (CIHEAM-Montpellier) and is currently completing a Ph.D. at the University of Paris I (Panthéon-Sorbonne).

Michael E. Porter is the Bishop William Lawrence University Professor at Harvard University. He directs the Institute for Strategy and Competitiveness based at the Harvard Business School. He is a leading authority on competitive strategy and international competitiveness. The author of 16 books and over 75 articles, his ideas have guided economic policy throughout the world. Porter has led competitiveness initiatives in nations and states such as Canada, India, New Zealand, and Connecticut, and guides regional projects in Central America and the Middle East. He is cochairman of the annual *Global Competitiveness Report*. In 1994, he founded the Initiative for a Competitive Inner City, a nonprofit private sector initiative formed to catalyze business development in distressed inner cities across the United States. The holder of eight honorary doctorates, Porter has won numerous awards for his books, articles, public service, and influence on several fields.

Forest Reinhardt is a Professor at Harvard Business School, where he teaches courses on strategy, business-government relations, and business and the environment. He is the author of *Down to Earth: Applying Business Principles to Environmental Management* (Harvard Business School Press, 2000). He received an MBA from Harvard Business School and a Ph.D. in Business Economics from Harvard University.

Friedrich von Kirchbach is Chief of the Market Analysis Section in the Geneva-based International Trade Centre United Nations Conference On Trade and Development/World Trade Organization. His team has developed a variety of Web-based tools for strategic market research, which are now being used around the world. He has been associated with several national and international institutions, including the United Nations Economic and Social Commission for Asia and the Pacific, INSEAD, the International Labor Organization, Organization for Economic Cooperation and Development, and the Volkswagen Foundation. Von Kirchbach holds a Ph.D. in economics focusing on international investment and has written and lectured widely on trade-related issues.

Foreword

Klaus Schwab
President
World Economic Forum

Over the last few decades a broad consensus has emerged that our standards of living are driven not just by economic success, but are also determined by the quality of the environment we live in. Today, it is widely accepted that traditional measures of economic welfare, such as GDP or per capita incomes, can provide only an incomplete picture of how well we are doing. Yet, few issues have remained so controversial as the relative importance of economic versus environmental objectives. For many, more stringent environmental regulation is indispensable if our natural resources and the global commons are to be protected for future generations. Others, however, have rejected such calls, pointing to the potentially adverse effects on competitiveness and economic growth.

Unfortunately, the debate has not always led to rational policy outcomes. In many countries, inadequate regulatory regimes continue to prevent economic and environmental objectives from being reconciled. Subsidies continue to result in the overuse of certain resources, polluters remain undercharged, undermining incentives for innovators to develop cleaner technologies, and capital markets shy away from long-term risks because of regulatory uncertainty.

A key problem lies in the lack of critical information and the substantial amount of uncertainty regarding the causes and effects of environmental decision-making. In the absence of reliable data and sound analyses, choices have often been based on generalized observations, best guesses, and, all too often, on rhetoric and emotion. Not surprisingly, therefore, the use of natural resources has remained suboptimal, resulting in lower standards of living than would otherwise be possible.

To be sure, the need for better information and analyses concerns all levels of decision-making—individuals, companies, and governments. While this need is particularly acute in developing countries, it is by no means limited to them. Taking into account that many environmental issues are not confined to national borders, it becomes clear that "measurement matters." We have to redouble our efforts to improve our knowledge about environmental sustainability and the multidimensional factors determining it.

The present book represents a major step forward in this direction. It collects several research papers, which have one common objective: to help facilitate data-driven environmental decision-making.

The centerpiece of this collection consists of the first serious attempt to measure environmental sustainability in one summary indicator and to rank a large number of countries on the basis of this index. I am very proud that the results presented here are based on a research project that was initiated three years ago by a task force of the Global Leaders for Tomorrow of the World Economic Forum. The Environmental Sustainability Index (ESI), whose pilot approach was first presented at the Forum's annual meeting in Davos in January 2000, represents an outstanding example of commitment to the Forum's overall mission: business and society in partnership to improve the state of the world.

Since its first presentation in Davos, the ESI has received considerable attention in many different circles, including governments, investors, academia, and civil society. In several countries, the ESI has triggered a lively debate and has already begun to help shape environmental policies. With the ESI now being made available to a broad audience, I have no doubt that its relevance and impact will increase even further.

I would like to extend my sincere thanks to the members of the ESI task force and, in particular, to Mrs. Kim Samuel-Johnson whose commitment to this project has been truly exceptional. Without the very generous financial support of the Samuel Foundation, the ESI would not yet exist.

I would also like to thank Professor Daniel Esty of Yale University who has served as the task force's project director. Indeed, his scientific leadership has proved instrumental in this important endeavor. My thanks go as well to Marc Levy and his team of data experts and researchers at Columbia University. The list of those involved in this project has increased substantially over the last two years, and my gratitude extends to all who have contributed to the success of the project.

The presentation of the ESI is supplemented by a number of studies, which discuss other important aspects of evaluating and measuring environmental performance at different levels of economic activity. Written by both academics and practitioners, these studies are analytically rigorous and highly relevant from a policy and business perspective. Together, they form the first *Environmental Performance Measurement: The Global Report 2001–2002*, which undoubtedly will have a substantial impact on how environmental decisions will be made. With the World Summit on Sustainable Development being held in mid-2002 in Johannesburg, South Africa, the publication of the *Report* could not have been more timely.

In addition to the aforementioned, I wish to thank all other authors for their important contributions to the *Report*, and, in particular, Frank Dixon of Innovest Strategic Value Advisors; Dr. Alois Flatz of Sustainable Asset Management; Dr. Friedrich von Kirchbach of the International Trade Centre; Professor Michael E. Porter of the Harvard Business School and Director of the Institute of Strategy and Competitiveness; and Professor Forest Reinhardt of the Harvard Business School.

The *Environmental Performance Measurement: The Global Report 2001–2002* is being published under the Forum's *Global Competitiveness Program* directed by Dr. Peter K. Cornelius. It represents an integral part of this program, reflecting our conviction that economic prosperity and our standards of living are inextricably tied to the quality of our environment.

Preface

Michael E. Porter

While in most nations there is widespread consensus about the importance of protecting the environment, the environmental field remains mired in controversy. Environmentalists advocate policies based on broad principles, but marshal scant evidence of their effectiveness. Affected communities and companies oppose costly requirements as excessive, but offer no credible alternatives. Sustainability is accepted as a principle, but it has become a catchphrase whose practical meaning is far from clear. Environmental debates are often polarized and degenerate into little more than rhetoric on all sides. The unfortunate result is that progress on environmental issues is halting.

This book is based on the premise that the whole approach to environmental policymaking needs to move to the next level. First and foremost, we need the facts about environmental performance in a wide sample of countries. Objective, comparative data on rates of pollution and energy usage have been almost absent. As shown in Chapter 2, there are huge differences in environmental performance even for countries with similar levels of GDP per capita. Consistent data will lead to inevitable comparisons and spur competition that can only encourage innovation and more rapid progress.

Second, there is a pressing need for comparative data across countries on environmental policy choices. Simply knowing what countries are doing, and comparing one country to another, will pay huge dividends. There are tremendous differences in practices among countries, demonstrating much room for learning and improvement.

Third, there is a pressing need to employ advanced statistical techniques to link policies and environmental outcomes. Today, there is much advocacy of policies without any evidence of impact. A hard-nosed approach to testing which environmental policies and programs are working and which are not is sure to speed up the diffusion of best practices and better utilize society's inevitably limited resources.

Other fields have benefited from a move to more data-intensive approaches and more rigorous analytical methods, even in the social sciences. Educational policy, for example, is now heavily informed by the facts and careful empirical study. Making environmental choices without rigorous data and methods is like choosing medical treatments without studying the link between treatments and actual health outcomes.

This book demonstrates the opportunity for the environmental field to progress rapidly in these areas. It describes the promising efforts at measurement already underway at both the policy and corporate scales. It discusses the learning that is resulting, providing answers to some questions, and raising a number of new issues. The chapters in this book will be far from the last word on the subject, but they represent a significant step forward. Hopefully, they will motivate a stream of further research that transforms the environmental debate during the coming decade.

Chapters

Chapter 1

Why Measurement Matters

Daniel C. Esty

Environmental decision-making has long been plagued by uncertainties and a lack of critical information. The data and analyses needed—by governments, companies, and individuals—for thoughtful and systematic action to minimize pollution harms and to optimize the use of natural resources are often unavailable or seem too costly to obtain. As a result, choices are made on the basis of generalized observations and best guesses, or worse yet, rhetoric and emotion. We stand, however, on the verge of an opportunity to transform our approach to pollution control and natural resource management through deployment of digital technologies in support of a more careful, quantitative, empirically grounded, and systematic environmentalism. This volume explores how better data and greater emphasis on statistical analysis might strengthen environmental problem solving in the policy and corporate worlds.

At the center of this potential revolution in the environmental arena lies a commitment to performance measurement and data-driven decision-making. It has long been understood that better analytic underpinnings generate better envi-

ronmental outcomes. Information has a cost, however, so there are limits to how much investment in environmental data and knowledge makes sense. In an ideal world, environmental decisions would be made on the basis of many factors, including levels of emissions from every relevant pollution source, who and what is being affected, how much ecological or epidemiological harm each "receptor" is suffering, how much value to place on these injuries, what options exist for mitigating the harms, and the costs and benefits of the alternative harm-reducing interventions available. But, historically, gathering and analyzing all this information has come at a very high price. Therefore trade-offs are made to reduce "administrative" costs, even at the expense of imprecision in environmental policy or actions.[1]

Some elements of the decision process are inescapably matters of judgment or values, making a purely quantitative decision process not just impossible but unwise (Sagoff, 1982; Wagner, 1995). But the inappropriateness of rigid numerical algorithms does nothing to diminish the value of more data and analysis as a decision tool. Indeed, narrowing the reducible zone of technical uncertainty would be of enormous value in the environmental domain (Esty, 1999). In addition, over the last few years, collecting, storing, tabulating, dissecting, and sharing information—including data on key environmental parameters—has become dramatically easier and cheaper as a result of advances in computer and telecommunications technologies (McRae, 1996). As the cost of gathering information falls, data-intensive, fine-grained analysis becomes relatively more cost-effective and sensible. *Environmental Performance Measurement* builds on this Information Age opportunity and explains how a shift toward more analytically rigorous underpinnings can and should unfold in the environmental realm.

An emphasis on developing and tracking a core set of environmental "indicators" could restructure our understanding of environmental problems and redefine our thinking about which response strategies work and how best to deploy them. Specifically, broader access to better data on air and water quality, emissions levels, toxic and solid waste, and energy use, as well as stocks and flows of other critical natural resources, would make it easier to identify issues, spot trends, evaluate risks, set priorities, establish policy options, test solutions, target technology development, and refine policies. With a clearer picture of the environmental impacts for which they are responsible, nations, states, communities, corporations, factories and other facilities, and even households and individuals would be positioned to reconfigure their behavior to reduce the level of harm.

Not only does a more "measured" approach to environmental problem solving promise enhanced analysis and decision-making, it makes it possible to evaluate policy and program performance, track on-the-ground progress in addressing pollution control and natural resource management challenges, and identify successful (and unsuccessful) efforts and approaches. As the pages that follow explain and the chapters in this volume demonstrate, data—particularly comparative data—facilitate performance benchmarking, support quantitative goal setting, and trigger competitive pressures to identify and implement environmental "best practices." In a world where environmental shortcomings are often a function of implementation deficiencies, a data-oriented approach to environmental action offers a new, harder-edged strategy for achieving better results.

To the extent that significant environmental shortcomings are a function not simply of "uninternalized externalities" but rather of waste and inefficiency, data-driven environmentalism provides a firmer foundation for improving resource productivity among both producers and consumers. Quantification also offers a basis for moving governments toward refined environmental programs and the use of more efficient policy instruments. A shift of emphasis toward information and data analysis thus represents, to some extent, a new environmental policy paradigm, emphasizing solutions to information failures rather than narrowly focusing on cost internalization.

Better Decision-Making
Reducing Uncertainty

Better data and information can help to address some of the most pervasive shortcomings in environmental decision-making. A central element of the environmental challenge is the fact that pollution and natural resource management problems are often hard to see and, therefore, easily overlooked or underestimated.[2] In some cases, such as automobile exhaust, the harm arises from numerous sources of emissions that are individually infinitesimally small but cumulatively very significant. Likewise, the impact of a single fisherman or fishing boat on fish stocks seems minute, but collectively, fishing fleets can deplete entire fisheries. Good data provide the perspective needed to see such aggregate effects and to spot "tragedies of the commons" in the making.

In other cases, emissions mix in ways that are difficult to sort out and make sense of. Air pollution in any major metropolitan area is a complex "soup" derived not only from millions of cars and trucks emitting particulates, oxides of nitrogen (NO_x), carbon monoxide, volatile organic chemicals (VOCs), and other by-products of combustion in their exhaust, but also releases from hundreds of thousands of households and tens of thousands of factories and other facilities discharging a range of toxics and other harmful substances. Sorting out responsibility for environmental damage across these many types and sources of harm can seem daunting. But sophisticated monitors, sensors, and data tracking systems can help to identify the separate "ingredients" and pollution sources.

In other circumstances, emissions spread spatially in ways that make analysis difficult. The effects of sulfur dioxide and other acid rain precursors spewed from coal-burning power plants, for instance, would be quite noticeable if concentrated. But tall smokestacks spread the emissions widely, making the harm hard to monitor and control. Data on downwind impacts (cases of respiratory distress, acidification of lakes, etc.) can help to sharpen the focus on such diffuse emissions. Perhaps the most difficult categories of pollutants are those that are dispersed not only across space but also over time. Greenhouse gas emissions, for example, persist in the atmosphere for as long as several centuries. The nearly impossible-to-observe buildup of these gases creates the risk of climate change. But numbers permit patterns to be spotted and trends to be traced. And with modern analytic techniques, such as regression analysis, correlations can be established, causal linkages identified, models built, and future effects forecast.

Simply put, data can make the invisible visible, the intangible tangible, and the complex manageable. The "realization" effect of numbers can be transformative. For example, while none of us can see the ozone layer, credible measurements of the thinning of this shield against the sun's ultraviolet rays convinced the public and politicians of the need for action (Benedick, 1991). Computer-supported data collection and analysis allows us to "see" many more environmental problems and to begin to disentangle the full range of risk factors implicated. Information Age gains in other fields (such as statistics, epidemiology, and meteorology) promise, moreover, to strengthen further our capacity to track ecological and public health threats and to identify how best to reduce these harms.

Perfect information, of course, will never be achieved. Significant environmental uncertainties are likely to persist for a long time to come. Some questions are inherently difficult to answer, and new problems are constantly emerging. Unintended consequences and countervailing risks constantly threaten to undermine environmental efforts (Graham and Wiener, 1995). Identifying all of the relevant variables and elements of a comprehensive analytic framework thus represents an enormous undertaking.

One must also remember that "environment" is not a narrow category or a single issue but rather a vast rubric covering an array of pollution control and natural resource management questions. Thus, environmental progress cannot be measured with reference to a small number of variables; it must be understood as a multidimensional concept demanding attention to a panoply of issues.

To add to the complexity, environmental decision-making involves a reducible but inescapable dimension of political judgment. Differences in values and assumptions can be diminished over time with good analysis and shared data, but ultimate agreement on how much weight to put on various goals will never be achieved. Individuals and societies differ in how much value they place on a pretty view or a life saved, and policy judgments on these matters will evolve as countries develop (Esty, 1999). Richer countries can afford higher environmental standards than poor ones can. Fundamentally, the multicriteria nature of environmental goals ensures that pollution control and natural resource management decision-making can never be reduced to a narrowly numerical quantitative risk assessment.

Complexity will remain a hallmark of the environmental realm, but it can be managed. Computers, in particular, make complexity much easier to cope with, permitting much more information-rich and nuanced analytic processes. With better data and derivative knowledge, sloganeering and guesswork can be supplanted by a hard-nosed focus on key problems and the search for effective and efficient solutions. An enhanced information base thus promises to solidify the foundations of environmental decision-making, which have too often been shaky, leading to the entire policy domain being dismissed as "soft" (Esty and Porter, 2000).

Enhancing Comparative Analysis

Environmental decisions almost always turn on comparisons and trade-offs. Many choices require the decision maker to identify the costs and benefits of investments in pollution prevention or abatement, and decide whether the risk reduction to be obtained justifies the expenditure entailed. Are the public health gains from reducing arsenic in water worth the filtration and other control costs? Would a tighter standard for particulate emissions be justified by lower incidence of respiratory disease? Does it make sense to invest in a new smokestack scrubber or to switch from coal to natural gas to reduce SO_2 emissions? Numbers facilitate such analyses, and measurement of key parameters is therefore critical.

Comparative analysis also makes it possible to target environmental spending. With data on risks, their relative significance, and alternative ways of dealing with the most pressing issues, decision makers can set priorities and evaluate competing policy options. As funds for investment in pollution control and natural resource management inevitably are limited, efficient use of the available resources is essential to sound environmental management in every sphere.

Comparisons, furthermore, spur competition. And competition, in turn, unleashes innovation (Porter, 1990). Everyone loves rankings, and no one likes to be revealed as a lagging performer. Just as knowledge that a competitor in the marketplace has higher profits or faster-growing sales drives executives to redouble their efforts, evidence that others are outperforming one's country, community, or company on environmental criteria can sharpen the focus on opportunities for improved pollution control and resource use efficiencies.

Finding Points of Leverage

Beyond providing a snapshot of current circumstances and a basis for systematic environmental decision-making, data can be used to identify the "drivers" of environmental outcomes. With data that are relevant, valid, and reliable, statistical analysis permits the correlates of good performance to be identified.[3] Empirical evidence should be used much more widely by both governments and business as a foundation for their environmental decisions. As time series data become available, causal relationships will increasingly emerge, making it easier to identify the determinants of top-tier environmental performance at the policy and corporate levels. Esty and Porter demonstrate the potential in this regard in Chapter 3, providing a preliminary empirical analysis of the variables affecting national environmental policy success.

Such data-driven decision-making is firmly established in other fields. Corporations spend a great deal of time and money on accounting to get a vantage point on their various activities and to understand better the strengths and weaknesses of their business strategies. Numbers permit options—for capital expenditures, choices of product lines, marketing and advertising, etc.—to be analyzed methodically and results to be systematically tracked. Data also allow goals to be set based on both internal targets and comparisons (for example, derived from observed results within the company at other facilities or in other product lines) and with reference to external benchmarks, such as industry-wide financial returns. Investments that pay off are continued or augmented; those that do not are discontinued.

Numerical analyses thus permit a degree of clarity and specificity that cannot be achieved otherwise. Success or failure can be quantified. Decisions can be made on an objective basis. Verified comparative data not only allow companies to gauge their own performance but also enable those in the capital markets to make independent judgments about who is doing well.[4]

Empirically based decision-making is critical for another reason: intuition is often wrong. Robyn Dawes (2001) and others have demonstrated in a number of fields—from the diagnoses of emergency room doctors to the ability of parole officers to identify likely recidivists—that good statistical analysis beats "expert" judgment nearly every time. Cass Sunstein (2001) has similarly analyzed why people are prone to significant errors in making risk assessments, building on the work of Paul Slovic (2000) in understanding the limits of human cognition. It is becoming increasingly clear that intuition can "top up" data and analysis, but it cannot replace it. The advantages of quantification and statistical analysis are now recognized in many disciplines. It is high time that the same logic is brought to bear in the environmental realm.

Improved Performance
Measuring Progress

Greater emphasis on data also can help to make environmental decision-making more "output" rather than "input" oriented. Too often in the past, environmental performance has been assessed based on how much money has been spent or how many inspections have been completed—or, worse yet, how many laws or rules have been adopted. These input measures may or may not be indicative of progress. Actual environmental success can be judged only "on the ground" as a matter of reduced public health or ecological impacts. Results are what matters—improved air and water quality, reduced waste, and more sustainably managed natural resources.

In a world where good intentions are not enough and implementation is key, quantitative policy evaluation is essential. In business, not every new product sells. In government, not every program works. Finding the failing efforts is thus an important element of good environmental management. But historically, the environmental community has not been supportive of rigorous evaluation and the weeding out of unsuccessful policies, strategies, and approaches, perhaps fearful that negative reviews would result in lower overall environmental spending. A tough-minded environmentalist should insist on having all programs monitored continuously against empirically defined benchmarks—and on redeploying the resources of those initiatives that do not measure up.

Comparative data and a focus on output measurement can improve corporate environmental performance as well. Facility-by-facility results allow executives to track pollution control and resource management practices within their own companies. Such data can be used to identify top-tier performance, establish targets, and build programs to move all of a corporation's operations toward leading-edge standards. Industry-wide performance data provide another basis for goal setting. Results identified by international bodies or the scientific community can also generate guidelines for what environmental action is possible.

Similar opportunities are also available at the household level. Electric bills in most places, for example, show how much energy was consumed in the month before. They may even provide a comparison with the last few months' electricity use or with the same month last year. But they do not say how much an average household of comparable size in the same geographic locale consumes or, better yet, what the most energy-efficient families are able to achieve under similar circumstances. Such targets, practically determined and easily understood, would provide a real spur to action with society-wide potential for reduced pollution, especially in combination with information on *how* the top performers have been able to reduce their use of electricity.

Benchmarking and Best Practices

To the extent that many environmental efforts fall short in implementation, a more data-intensive approach to policy-making offers special opportunities. Measurement facilitates policy evaluation, performance comparisons, and identification of superior regulatory approaches. In fact, enormous environmental gains can be obtained simply by moving lagging jurisdictions toward the "best practices" of those with top-ranked results. Quantitative measures also provide a basis for judging which specific regulatory tools, technologies, or strategies are succeeding and which need to be rethought (Eccles et al., 2001).

The potential to use benchmarking to drive progress applies at many scales and across a diverse set of environmental issues and actors. Comparing results across environmental challenges (air versus water versus waste) and jurisdictions (California versus Texas versus Connecticut, or the United States versus France versus Germany) allows conspicuous achievements or difficulties to be spotlighted, facilitating movement toward better results over time. Moreover, in the Information Age, both benchmarking and dissemination of information on best practices, strategies, and technologies promises to become ever cheaper as computer and telecommunications technologies advance.

As the Environmental Sustainability Index (ESI) discussed in Chapter 2 makes clear, countries have much to gain by learning how they compare across the spectrum of pollution control and natural resource issues they manage on behalf of their citizens. Every country ranked in the ESI study lags its peer group (defined as those at a comparable level of development) on some issues. Spotting these opportunities for improvement, and having access to the information on what leading jurisdictions are doing, sharply clarifies the policy challenge. States and communities would benefit from the same opportunity to see how they rank against their counterparts and to have best practices illuminated.

The availability of information on how others are doing in reducing pollution and improving resource productivity tends to stimulate comparisons, benchmarking, and a push to improve performance in the corporate sector as well. While some critical information is kept confidential because of its strategic value or because it creates exposure vis-à-vis regulatory authorities, a great deal of data is now available. And the Internet puts the answers to many pollution control or natural resource management question just a few clicks away. It appears, moreover, that the competitive pressures are mounting. As Frank Dixon (Chapter 5) and Alois Flatz (Chapter 6) explain, financial analysts and others who track the capital markets have a growing appetite for information on corporate environmental results as an element to fold into their forecasting of future corporate profitability. Such interest has sharpened the environmental focus of many businesses, triggering competition among companies to optimize their handling of pollution and natural resource challenges. Performance data and identification of best practices thus promote an action orientation and stronger environmental performance in several ways. First, clear numerical measures highlight what is possible, as a matter of fact, in improved environmental results. In many cases, governments, corporations, and households do not have a clear picture of what might be obtainable in pollution control or resource management gains. Data on the results others are achieving can help to clarify what constitutes an appropriate target or goal.

Second, benchmarking can be used to reveal best practices and to provide a road map for laggards to follow in moving to adopt better strategies, technologies, or policies. Modern telecommunications make dissemination of information on best practices much easier and cheaper. The excuses for continuing sub-par performance thus become ever more limited.

Third, as noted earlier, comparative data often stimulate competitive pressures and, therefore, innovation that can lead to improved results. The benefits of competition have been demonstrated repeatedly (Porter, 1990). Absent data and appropriate benchmarks from comparable jurisdictions, it is hard to spot lagging performance. Complacency and inertia are hard to overcome, but when citizens (or environmental groups or the media) find out that other cities, states, or countries are delivering much better environmental results than their own government, they have a basis for complaint. Indeed, Belgium's poor showing in the 2001 Environmental Sustainability Index (ranking 79th, just below Albania) caused a huge uproar in Brussels and has led to a significant focus on the country's pollution problems. The environmental facts before and after the publication of the ESI remained exactly the same. The ESI simply gave the Belgians (and especially the Belgian media) a context for understanding their pollution numbers and a benchmark for judging their government's relative performance.

Comparative information on government activities and results is also useful for business. It gives companies operating in jurisdictions with inefficient regulatory systems an independent (and not purely self-interested) basis on which to press for better government performance. The prospect of focused oversight by citizens, NGOs, the media, and the regulated community tends to induce better government performance. In particular, the presence of data and benchmarks from other jurisdictions can trigger a process of "regulatory competition" among governments that yields more effective and cost-conscious environmental controls (Esty and Geradin, 2001).

Better corporate-scale environmental data can generate a similar competitive dynamic. Numbers make vivid the fact that others are achieving superior environmental results. Just as environmental indicators pressure governmental authorities to sharpen their policies, so can comparative data drive lagging companies to improve their environmental performance.

Greater Efficiency

Many environmental problems arise from market and regulatory failures, but other issues can be traced to inefficiency and waste, reflecting ignorance or mistakes on the part of polluters and natural resource users. In fact, a significant percentage of pollution arises not from emissions intentionally sent up the smokestack or out the effluent pipe to avoid control costs (uninternalized externalities) but from materials or energy that are unwittingly underutilized in fabrication or elsewhere in a product's life cycle. Such "inadvertent" pollution can be traced to poorly designed goods, outdated technologies, unnecessarily wasteful packaging, and general inattention to the dictates of environmental management. Data on the losses attributable to such mistakes, as well as easy access to information about alternative production or consumption practices, promise to improve resource productivity, enhance consumer welfare, and improve corporate competitiveness (Porter and van der Linde, 1995; Esty and Porter, 1998). The U.S. Environmental Protection Agency's Toxic Release Inventory (TRI) highlights the potential in this regard. Merely being required to tabulate the amount of chemicals flowing out of a facility into the land, air, and water pushed many companies to adopt waste minimization programs (Karkannian, 2001).

Where excess emissions can be attributed to inefficiency, polluters generally can be induced to shift to less harmful production or consumption practices if they are told about better alternatives. No government mandate is needed. The opportunity for individual gain or competitive advantage provides all the incentive needed. Without any regulatory mandate, for instance, the U.S. EPA's "Green Lights" initiative convinced thousands of enterprises to substitute high-efficiency fluorescent lighting for traditional incandescent bulbs—reducing electricity use (and thus the emissions from power generation) and lowering company costs.

There has been much policy discussion and academic debate over the extent of opportunities for "win-win" outcomes that yield both environmental benefits and economic gains (Walley and Whitehead, 1994; Porter and van der Linde, 1995; Jaffe et al., 1995). This issue has been viewed in far too static terms, however. Without any doubt, as the cost of performance data, and information more generally, falls, new opportunities to reduce environmental impacts while simultaneously improving competitiveness emerge. Information technologies lower the cost of access to data and knowledge and, in doing so, transform the trade-off between the precision permitted by a detailed analytic foundation for decision-making and the administrative cost of assembling all the relevant facts and figures. Thus, the "policy possibility frontier" for both governments and corporations changes.

In the corporate setting, environmental metrics can shed light on opportunities to eliminate waste and squeeze out inefficiencies in production and product use.

In fact, substituting information for scarce resources or polluting materials represents a major move toward improved environmental results. More precise data, careful engineering calculations, and resource utilization benchmarking can reduce material and energy throughput in production, by substantial amounts in some cases. Computer-guided saws are helping forest products companies to reduce scrap and increase lumber yields per tree by 30 percent or more. In manufacturing, digital monitoring and management of equipment has generated an array of process refinements that reduce scrap, eliminate emissions, and improve production efficiency. Likewise, farmers are increasingly using portable soil test kits, global positioning systems, and on-board tractor computers to refine their fertilizer and pesticide applications, reducing costs and chemical runoff.

Data mining by corporations eager to understand their markets and data-enabled mass customization by businesses trying to meet the precise needs of their customers promise further opportunities to reduce waste. Catalogue companies used to send out mass mailings based on generalized assumptions about the buying habits of people living in certain zip codes. Today, they can select and solicit customers on a much more refined basis, saving literally tons of paper. In a similar vein, using a state-of-the-art information system, Dell builds computers to each customer's requirements, which translates into reduced material inputs, less waste, and lower pollution.

Data links may also facilitate efficiency gains up and down the value chain and even beyond. Online connections between suppliers and customers have helped a number of companies, from General Electric to Seven-Eleven, to shrink inventories, limit spoilage, and cut waste. While the full potential of e-commerce has not yet been realized, and there may be downsides as well as upsides, the potential for Internet-driven efficiency gains is already visible (Esty, 2001). Companies increasingly are looking upstream and downstream to find ways to reduce costs and increase value. Customer-supplier data exchanges and networks, for example, are deepening commercial relationships, and interconnected companies often find it easier to identify the "least cost avoider" from a pollution perspective (Esty and Porter, 1998).

Beyond cutting costs and improving resource productivity, companies are finding a great many other ways to use environmental information to advantage, as Forest Reinhardt demonstrates in Chapter 4. Good data can facilitate environmental product differentiation, "managing" of competitors through regulatory interventions, and more sophisticated risk reduction. A data-backed environmental lens may even help companies to redefine their "market space" by leading them to new products or services to sell (Kim and Mauborgne, 1999). A number of "environmental information" companies (or environmental divisions of established consulting firms) have sprung up in the last few years to help businesses exploit the wealth of data that is now available to improve their eco-efficiency and to strengthen their environmental strategic positioning. Whether the entities in this niche will survive the dot.com shakeout remains to be seen, but the opportunity to use data to sharpen corporate environmental performance is clearly understood.

Policy Efficiency and Regulatory Reform

In the government realm, quantitative analysis can support efforts to improve regulatory efficiency. Notably, a more data-rich policy process facilitates efforts to shift away from "command and control" regulation toward more sophisticated approaches to managing shared natural resources and controlling pollution. In particular, numbers make possible the use of "market mechanisms," which put a price on emissions, thereby "monetizing" environmental harms.

In some instances, actual markets—the allocation of shares in a scarce resource and the protection of these environmental property rights—can be the basis for pollution control or natural resource management schemes (Demsetz, 1967). In the last few years, for example, New Zealand's fisheries have revived under a system of tradable quotas, with fish landings carefully measured and tracked (Pearse and Walters, 1992). In the United States, acid rain emissions have been cut in half since 1990 under an SO_2 control regime based on tradable pollution allowances (Stavins and Whitehead, 1997). The acid rain reduction program depends heavily on smokestack monitors that transmit data on power plant emissions on a real-time 24-hour-a-day basis.

Fundamentally, market-based approaches to environmental protection can be effective if the transaction costs involved in exchanging (and enforcing) environmental property rights are low (Coase, 1960; Williamson, 1989). But in many circumstances, high transaction costs lead to market failures. Better data, however, can help to bring down these costs by making property boundaries easier to delineate, lowering the cost of vindicating environmental rights, and allowing environmental property rights markets to work more smoothly (Esty, 1996). Just as barbed wire made possible fencing off individual ranches in the American West, thus diminishing the risk of overexploitation of range land (Rose, 1998), data can serve as virtual barbed wire, enabling property rights in various shared resources to be demarcated and protected.

More generally, data makes regulatory approaches that seek to harness a range of economic incentives easier to implement. Specifically, information on the level of harm from particular pollutants represents a critical first step toward internalizing these externalities. Measures of emissions and impacts can be used to set pollution taxes, facilitate tradable permit schemes, or simply to signal to consumers (for example, through ecolabels) which products are environmentally preferable.

Even where economic-incentive-based regulation is not feasible, better data promise to provide a clearer starting point for "command and control" mandates. More precise pollution information allows regulators to address problems at the scale of the harm, avoid overly broad uniform rules, and tailor control strategies to individual circumstances. In effect, low-cost and easily accessible data make it easier to refine regulations and, thus, to accommodate the diversity across the regulated community. As noted earlier, data and analysis enable more regular and careful policy evaluations leading to continuous improvement in regulatory design. When substandard results are observed more readily, special interest lobbying and other manipulations of the regulatory process become more difficult. Thus, a data-driven policy process may be less susceptible to "public choice" failures.

Better and cheaper data also tend to increase "transparency." As discussed earlier, the increased intensity of information available to opposition leaders, the media, business critics, and NGOs means governments are now subject to greater scrutiny than in the past. Similarly, government regulators, environmental groups, community activists, and the media all have extraordinary access—via the Internet, email, etc.—to facts and figures about corporate environmental activities. Bad acts and poor results are now almost impossible to hide. Although it may be uncomfortable for some companies (as it has been for some governments), this new world of instantaneous connections, open access, and transparency seems likely to intensify focus on pollution problems and natural resource management, speed up feedback loops, and increase the pace of environmental progress.

Bumps in the Road

More data does not necessarily translate into more knowledge in either the policy or the corporate domain. More information could mean more *mis*information and *dis*information, and it could mean that decision processes are overwhelmed. The risk of overloaded systems and data being contorted to serve narrow interests suggests that mechanisms will be needed to ensure quality control and some degree of standardization.

Similarly, data can enrich environmental debates and facilitate "triangulation" on answers in the face of uncertainties. But more information can also diminish the quality of policy dialogues, translate into a battle of numbers, fuel chaos, and lead to breakdown in the decision-making process. What will be determinative are the relevance, validity, and reliability of the new data flows—and the emergence of institutions to promote quality assurance of both the raw information and the analyses that flow from it.

Numbers, moreover, are not value neutral. What is measured and how things are measured builds on presumptions about what is important, which in turn reflect the values of those engaged in the data exercise (Wagner, 1995; Kahan and Braman, 2001). But the "political" nature of statistics, and of science more generally, can be overplayed. Even if precise quantitative results cannot be developed or cardinal rankings are susceptible to challenge, comparative data often will give a relatively clear general picture of the scale and importance of environment challenges or generate an ordinal ranking of policy options.

It is nonetheless important that the assumptions that underlie any particular data set, performance ranking, statistical methodology, or line of analysis be laid bare and exposed to peer review and critical comment. Data should always be tested for quality, consistency, and robustness. Moreover, tools, such as sensitivity analysis, should be used to help data users understand when underlying assumptions (and which ones) drive results.

The Path Forward

As the business community has long understood, measurement matters. Or, more precisely, what matters is what gets measured. Thus, if pollution control and natural resource management efforts are to become more serious, environmental decision-making must become more systematic, data-driven, and analytically rigorous. The chapters that follow illustrate the possibilities that lie ahead. Taken together, they show the revolutionary potential of a commitment to environmental performance measurement.

In Chapter 2, Marc Levy of the Center for International Earth Science Information Network (CIESIN) at Columbia University lays out the structure of the Environmental Sustainability Index (ESI) developed by the World Economic Forum's Global Leaders for Tomorrow Environmental Task Force with the Yale Center for Environmental Law and Policy and CIESIN. He explains the methodology of the analysis (more details can be found in Annex 1). He further spells out how the ESI initiative builds on data on 22 core environmental indicators (see Annex 2) for 122 countries (see Annex 3 for country-by-country results) based on 67 underlying variables (spelled out in Annex 4). The ranking generated can be seen as an environmental quality of life forecast for a generation or two from now. Levy also outlines the difficulties encountered in constructing the ESI and identifies some of the methodological challenges to a more data-driven environmental future. He argues that these challenges have workable solutions, but that the lack of good data represents a serious obstacle to more rigorous environmental decision-making.

In Chapter 3, Dan Esty of Yale University and Mike Porter of Harvard Business School attempt to identify empirically the drivers of good environmental performance. Focusing on three "output" measures of environmental results—particulate levels, energy efficiency, and SO_2 levels—they find that the breadth and rigor of environmental policies matter. But, perhaps more surprisingly, they also find a strong correlation between environmental performance and a society's underlying legal and economic structure.

Forest Reinhardt of Harvard Business School offers in Chapter 4 a theoretical overview of how a more data-rich environmental domain might affect corporate pollution control and natural resource management. Reinhardt identifies five core strategies by which companies might improve their environmental performance, and simultaneously their competitiveness. He also analyzes the role of information and measurement in making such gains possible.

In Chapter 5, Frank Dixon of the New York–based environmentally oriented investment group, Innovest, applies the logic of environmental performance measurement to the corporate sector. Dixon finds rapidly expanding interest in environmental results as an element of financial analysis with the capital markets. He also highlights the growing evidence that superior environmental performance translates into superior financial returns.

Alois Flatz of Zurich-based Sustainable Asset Management expands on this line of argument in Chapter 6, demonstrating why a growing number of investors are seeking data on corporate sustainability. He also explains how this data is being used to develop indexes tracking the financial performance of leading companies.

In Chapter 7, Peter Cornelius of the World Economic Forum and Friedrich von Kirchbach and Mondher Mimouni of the International Trade Center based in Geneva analyze whether stringent environmental regulations promote or undermine national competitiveness. Using trade data from a range of environmental product sectors, they conclude that, at least in some cases, a strict regulatory regime translates into greater market strength.

Overall, this volume demonstrates how more data-driven analysis might be folded into the environmental realm. The results presented suggest that a more empirical approach to environmental decision-making is possible—and likely to be advantageous. But the contributions to this study also make clear the need for much greater investment in pollution control and natural resource management data collection and generation. The need for more sophisticated analytic methods and statistical techniques is also evident. Despite these gaps and existing limitations, the promise of a more empirical environmentalism looms large.

References

Benedick, Richard, *Ozone Diplomacy: New Directions in Safeguarding the Planet*, Cambridge, Mass.: Harvard University Press, 1991.

Coase, Ronald H., "The Problem of Social Cost," *Journal of Law and Economics* 3 (1960).

Dawes, Robyn M., *Everyday Irrationality: How Pseudo-Scientists, Lunatics, and the Rest of Us Systematically Fail to Think Rationally.* Boulder, Colo.: Westview Press, 2001.

Demsetz, Harold, "Toward a Theory of Property Rights," *American Economic Review* 57 (1967).

Diver, Colin, S., "The Optimal Precision of Administrative Rules," *Yale Law Journal* 93 (1983).

Eccles, Robert, Robert H. Herz, Mary E. Keegan, and David M. H. Phillips, *The Value Reporting Revolution: Moving Beyond the Earnings Game*, New York: John Wiley and Sons, 2001.

Esty, Daniel C., "Toward Optimal Environmental Governance," *New York University Law Review* 74 (1999), p. 1495.

Esty, Daniel, C., "Revitalizing Environmental Federalism," *Michigan Law Review* 95 (1996).

Esty, Daniel, "Measuring National Environmental Performance and Its Determinants," in Michael Porter and Jeffrey Sachs et al. (eds.), *The Global Competitiveness Report 2000.* New York: Oxford University Press, 2000.

Esty, Daniel C., "Digital Earth: Saving the Environment," *OECD Observer* (May 2001).

Esty, Daniel C., and Damien Geradin, "Regulatory Co-opetition," in Daniel C. Esty and Damien Geradin (eds.), *Regulatory Competition and Economic Integration: Comparative Perspectives.* New York: Oxford University Press, 2001.

Esty, Daniel C., and Michael E. Porter, "Industrial Ecology and Competitiveness," *Journal of Industrial Ecology* 2 (1998).

Esty, Daniel C., and Michael E. Porter, "Measuring National Environmental Performance and its Determinants," in Michael E. Porter, Jeffrey Sachs, et al. (eds), *The Global Competitiveness Report 2000.* New York: Oxford University Press.

Graham, John D., and Jonathan Baert Wiener, "Confronting Risk Tradeoffs." in John D. Graham and Jonathan Baert Wiener (eds.), *Risk vs. Risk: Tradeoffs in Protecting Health and the Environment.* Cambridge, Mass.: Harvard University Press, 1995.

Hahn, Robert W., and Robert E. Litan, *Improving Regulatory Accountability*, Washington: American Enterprise Institute and the Brookings Institution, 1997.

Jaffe, Adam, Steven R. Peterson, Paul R. Portney, and Robert Stavins, "Environmental Regulation and the Competitiveness of US Manufacturing: What Does the Evidence Tell Us?" *Journal of Economic Literature* 33 (1995), p. 32.

Kahan, Dan, and Donald Braman, "More Statistics, Less Persuasion: Cultural Theory of Gun-Risk Perception" (unpublished manuscript), 2001.

Karkannian, Bradley, "Information as Environmental Regulation: TRI and Performance Benchmarking, Precursor to a New Paradigm?" *Georgetown Law Journal* 89 (2001), p. 257.

Kim, Chan W., and Renee Mauborgne, "Creating New Market Space," *Harvard Business Review* 11 (1999), p. 83.

McRae, Hamish, *The World in 2020: Power, Culture and Prosperity*, Cambridge, Mass.: Harvard Business School, 1996.

Pearse, Peter H., and Carl J. Walters, "Harvesting Regulation under Quota Management Systems for Ocean Fisheries," *Marine Policy* 16 (1992), p. 167.

Porter, Michael E., *Competitive Advantage of Nations*, New York: Free Press, 1990.

Porter, Michael, and Claas van der Linde, "Toward a New Conception of Environment-Competitiveness Relationship," *Journal of Economic Perspectives* 9 (1995).

Posner, Richard A., *An Economic Analysis of Law*, Boston: Little and Brown, 1992.

Rose, Carol M., "The Several Futures of Property: Of Cyberspace and Folktales, Emissions Trades and Ecosystem," *Minnesota Law Review* 83 (1998), p. 129.

Sagoff, Mark, "We Have Met the Enemy and He Is Us or Conflict and Contradiction in Environmental Law," *Environmental Law*, 12 (1982), p. 283.

Slovic, Paul, *The Perception of Risk*, Sterling, Va.: Earthscan Publications, 2000.

Stavins, Robert, and Bradley Whitehead, "Market-Based Environmental Policies," in Marian Chertow and Daniel Esty (eds.), *Thinking Ecologically: The Next Generation of Environmental Policy*, New Haven: Yale University Press, 1997.

Sunstein, Cass R., "Cognition and Cost-Benefit Analysis," in Mathew Adler and Eric Posner (eds.), *Cost-Benefit Analysis: Legal, Economic, and Philosophical Perspectives*, Chicago: University of Chicago Press, 2001.

Wagner, Wendy, "The Science Charade in Toxic Risk Regulation," *Columbia Law Review* 95 (1995)

Walley, Noah, and Bradley Whitehead, "It's Not Easy Being Green," *Harvard Business Review* 72 (1994).

Williamson, Oliver E., "Transaction Cost Economics," in R. Schmalensee and R. D. Willig (eds.), *Handbook of Industrial Organization*, New York: Elsevier Science Publishers, 1989, p. 136.

Endnotes

1   Indeed, a whole body of academic literature has grown out of the question of "optimal specificity" in regulation (Diver, 1983; Posner, 1992).

2   In some cases, the obscurity results in problems being overestimated or overstated (Hahn and Litan, 1997).

3   Unfortunately, as the Environmental Sustainability Index discussed in Chapter 2 highlights, the existing data sets in the environmental domain are woefully incomplete and lacking in quality assurance. One of the most important conclusions from this study is that the prospect of a more analytically rigorous approach to environmental problem solving depends fundamentally on investments in better data.

4   The same emphasis on data and statistical analysis dominates economics and is increasingly important in other social science disciplines, such as political science and sociology. The environmental field curiously lags in the shift toward empirical underpinnings.

Chapter 2
Measuring Nations' Environmental Sustainability

Marc A. Levy

Sustainability as an environmental policy objective has been embraced by a growing number of governments and organizations. Yet the ability to measure environmental sustainability—and to give concrete meaning to the term (Esty, 2001)—has not kept pace with the spread of this goal. This chapter reports on efforts to create one of the first such measures: the Environmental Sustainability Index (ESI). The ESI seeks to measure overall progress toward, and capacity to achieve, environmental sustainability for 122 countries. It covers the full spectrum of issues that contribute to long-term sustainability: baseline conditions and natural resource endowments, current pollution flows and resource stresses, human vulnerability, social and institutional capacity to respond to environmental challenges, and national contributions to global stewardship. The ESI can be seen as an attempt to assess empirically which nations are most likely to possess favorable environmental conditions several generations into the future. The challenges faced, the choices and assumptions made, and the lessons learned in the course of creating the ESI illuminate the nature of measuring national environmental performance more broadly.[1]

The core conclusions of the essay can be summarized as follows:

- it is possible to construct a single index measuring environmental sustainability, generating results that appear to be both plausible and useful. Such an index can serve a helpful role in gauging progress in achieving environmental sustainability. The ESI makes use of a breadth of available information while generating a wealth of comparative data, strengthening the capacity to benchmark national performance in meeting pollution control and natural resource management challenges and facilitating efforts to identify "best practices" in the environmental domain.

- calculating concrete measures of environmental sustainability makes it possible to explore important policy debates in greater depth and with more analytic rigor than was previously possible. In particular, our results suggest that economic and environmental performance objectives, broadly speaking, are not in conflict, but rather reflect discrete policy choices.

- while a number of serious challenges face any effort to measure environmental sustainability, the most critical by far is the striking underinvestment in data generation and collection at both the national and international levels.

The Motivation behind the ESI

The ESI was designed to fill a very specific measurement gap in the context of numerous environmental indicator efforts. The World Bank issues an annual series of indicators that includes a number of useful environmental and natural resource indicators. The Organization of Economic Cooperation and Development (OECD) issues a similar set of measures. The World Resources Institute collects a valuable range of environmental indicators and publishes them in its *World Resources Indicators*. Many other global and regional bodies perform such activities. What none of these efforts does, however, is relate these separate indicator measurements to each other, or relate them as a group to the broad goal of environmental sustainability.

The lack of good data, particularly comparative data, represents a serious policy problem in the environmental realm. This shortcoming limits the capacity for data-driven decision-making at the global, national, regional, and community scales. And it makes it hard to implement the concept of "sustainable development," which galvanized the world community at the 1992 Earth Summit. The lack of solid factual underpinnings complicates international negotiations and has slowed progress in finding ways to respond to a range of global environmental issues. And although environmental sustainability has become an important goal in many countries, and dozens of international environmental agreements refer explicitly to sustainability in their articulation of goals, the concept has not been subjected to sustained measurement efforts.

One of the most concrete expressions of the environmental sustainability goal, for example, was in the "International Development Targets" developed in 1996 by the world's major aid donor countries. The International Development Targets identify seven core objectives for development assistance. Most of the targets focus on social and economic objectives, but the seventh objective targets the environment, aiming at "the implementation of national strategies for sustainable development in all countries by 2005, so as to ensure that current trends in the loss of environmental resources are effectively reversed at both global and national levels by 2015" (OECD, 1996). Most donor countries that have evaluated these goals seriously have been struck by the extremely poor ability to measure progress on this environmental target (for example, U.K. Department for International Development, 2000).

Against this backdrop a coalition of environmental activists, business leaders and academic analysts decided in 1999 to create the ESI. The ESI initiative was sponsored by the World Economic Forum, which had previously devoted considerable effort to measuring economic competitiveness (the most recent results are found in Porter, Sachs et al., 2001). Its Global Leaders for Tomorrow Environment Task Force sought to extend the Forum's reach into the environmental realm by creating an index of environmental sustainability. Yale University's Center for Environmental Law and Policy and Columbia University's Center for International Earth Science Information Network designed the methodology and carried out the data and analytical work. The following sections outline the approach pursued in creating the ESI, including its structure, methods, and input data.

Creating the ESI

The first challenge in measuring environmental sustainability is to define the scope in conceptual terms. What are we trying to measure? For many decades, sustainability in an environmental context meant something quite simple: carrying capacity. Commonly, the sustainability concept was applied in terms of "sustainable yields," and the earliest formulations of environmental objectives framed as sustainability goals were in extraction regimes governing fisheries, forests, and other renewable resources. The policy goal in this sense was to limit extraction rates to within the limits imposed by carrying capacities (a goal that was often unsuccessful, especially on international issues). Measurement efforts inspired by this framework were hampered continually by the complicated dynamics that affect carrying capacity and the inability of science to provide timely, believable measures.

Beginning in the 1980s, this connection between sustainability and carrying capacity weakened as the emergence of the sustainable development framework focused attention on the

environment-development connection as the central pillar of sustainability (World Commission on Environment and Development, 1987). Sustainability in this framework had more to do with coping with poverty-environment interactions than with living within carrying capacity limits. But measurement efforts that took this framework to heart have had a hard time combining such disparate phenomena under a single rubric. More important from our perspective, they inevitably have given short shrift to environmental concerns. One recent such effort, for example, contained only five environmental indicators (Jackson, forthcoming).

The enduring feature of the new sustainability framework, embraced in the ESI, is *multidimensionality*. Indeed, it could be argued that multidimensionality is the dominant theme of the recent National Research Council (1999) study on sustainable development transitions. It is now recognized that sustainability cannot be reduced to carrying capacity, but rather requires metrics relevant to the broad sweep of environmental goals that society has embraced, from clean air and water to biodiversity protection to climate change prevention to limiting vector-borne diseases. That is the challenge the ESI was designed to help face.[2]

At the most basic level, environmental sustainability can be presented as a function of five phenomena (see Table 1): (1) the state of the environmental *systems*, such as air, soil, ecosystems, and water; (2) the *stresses* on those systems, in the form of pollution and exploitation levels; (3) the *human vulnerability* to environmental change in the form of loss of food resources or exposure to environmental diseases; (4) the *social and institutional capacity* to cope with environmental challenges; and, finally, (5) the ability to respond to the demands of *global stewardship* by cooperating in collective efforts to conserve international environmental resources such as the atmosphere. We define environmental sustainability as the ability to produce high levels of performance on each of these dimensions in a lasting manner, and we refer to these five dimensions as the core "components" of environmental sustainability.

The current state of scientific knowledge does not permit us to specify precisely what levels of performance are high enough to be truly sustainable, especially at a worldwide scale. Nor are we able to identify in advance whether any given level of performance is capable of being carried out in a lasting manner. Therefore, we have built our index to be primarily comparative. Establishing thresholds of sustainability remains an important endeavor, albeit one that is complicated by the inescapably critical value judgments and assumptions (for example, sustainable at what level of quality of life [Cohen, 1995]) and the dynamic nature of such economic factors as changes in technology over time.

Table 1

Components of Environmental Sustainability

Component	Logic
Environmental Systems	A country is environmentally sustainable to the extent that its vital environmental systems are maintained at healthy levels, and to the extent to which levels are improving rather than deteriorating.
Reducing Environmental Stresses	A country is environmentally sustainable if the levels of anthropogenic stress are low enough to engender no demonstrable harm to its environmental systems.
Reducing Human Vulnerability	A country is environmentally sustainable to the extent that people and social systems are not vulnerable (in the way of basic needs such as health and nutrition) to environmental disturbances; becoming less vulnerable is a sign that a society is on a track to greater sustainability.
Social and Institutional Capacity	A country is environmentally sustainable to the extent that it has in place institutions and underlying social patterns of skills, attitudes, and networks that foster effective responses to environmental challenges.
Global Stewardship	A country is environmentally sustainable if it cooperates with other countries to manage common environmental problems, and if it reduces negative extraterritorial environmental impacts on other countries to levels that cause no serious harm.

Table 2

ESI Structure

Component	Indicator	Variable
Environmental Systems	Air quality	Urban SO_2 concentration
		Urban NO_2 concentration
		Urban TSP concentration
	Water quantity	Internal renewable water per capita
		Per capita water inflow from other countries
	Water quality	Dissolved oxygen concentration
		Phosphorus concentration
		Suspended solids
		Electrical conductivity
	Biodiversity	Percentage of mammals threatened
		Percentage of breeding birds threatened
	Land	Severity of human-induced soil degradation
		Land area impacted by human activities as a percent of total land area
Reducing Stresses	Reducing air pollution	NO_x emissions per populated land area
		SO_2 emissions per populated land area
		VOC emissions per populated land area
		Coal consumption per populated land area
		Vehicles per populated land area
	Reducing water stress	Fertilizer consumption per hectare of arable land
		Pesticide use per hectare of crop land
		Industrial organic pollutants per available fresh water
		Percentage of country's territory under severe water stress
	Reducing ecosystem stresses	Percentage change in forest cover, 1990–1995
		Percentage of country with acidification exceedence
	Reducing waste and consumption pressures	Consumption pressure per capita
		Spent nuclear fuel arisings per capita
	Reducing population growth	Total fertility rate
		Percentage change in projected population between 2000 and 2050
Reducing Human Vulnerability	Basic human sustenance	Daily per capita calorie supply as a percent of total requirements
		Percent of population with access to improved drinking water supply
	Environmental health	Child death rate from respiratory diseases
		Death rate from intestinal infectious diseases
		Under-5 mortality rate

The ESI thus centers on a core set of 22 environmental sustainability *indicators* identified on the basis of a review of the environmental literature and substantiated by statistical analysis. These indicators were deemed the most relevant constitutive elements of the five core components, and therefore are the central element of analysis. In turn, each of the indicators has associated with it a number of *variables* that are measured empirically. The relationship among these ESI building blocks is specified in Table 2.

The choice of variables was driven by a combination of focus on the theoretical logic and relevance of the indicator in question with data quality, and with country coverage. In general, we sought variables with extensive country coverage, but, in some cases, chose to use variables with narrow coverage if they measured critical aspects of environmental sustainability that otherwise would be lost. Air quality and water quality, for example, were especially disappointing in their poor country coverage, but they were included anyway because of their central role in environmental sustainability.

Results and Analysis

For the overall ESI, the top three countries were Finland (80.5), Norway (78.2), and Canada (78.1). The bottom three were Haiti (24.5), Saudi Arabia (29.8), and Burundi (29.8). The complete ranking appears in Table 3. The component and indicator data are reported in Annex 2 of this volume; full country-by-country data can be found in Annex 3, and data for each of the 67 variables are reported in Annex 4.

One of the driving motivations behind the creation of the ESI was to facilitate investigation into trade-offs between economic growth and environmental protection. We explored the

Table 2 (continued)

ESI Structure

Component	Indicator	Variable
Social and Institutional Capacity	Science/technology	Scientists and engineers per million population
		Expenditure for research and development as a percent of GNP
		Scientific and technical articles per million population
	Capacity for debate	World Conservation Union member organizations per million population
		Civil and political liberties
	Regulation, management	Stringency and consistency of environmental regulations
		Degree to which environmental regulations promote innovation
		Percent of land area under protected status
		Number of sectoral Environmental Impact Assessment guidelines
	Private-sector responsiveness	Dow Jones sustainability group index
		Average Innovest EcoValue 21 rating of firms
		World Business Council for Sustainable Development Members
		Environmental competitiveness
	Environmental information	Availability of sustainable development information at the national level
		Environmental strategies and action plans
		Percent of ESI variables in publicly available data sets
	Eco-efficiency	Energy efficiency (total energy consumption per unit GDP)
		Renewable energy production as a percent of total energy consumption
	Reducing public choice distortions	Price of premium gasoline
		Subsidies for energy or materials use
		Reducing corruption
Global Stewardship	International commitment	Number of memberships in environmental intergovernmental organizations
		Percent of reporting requirements met under Convention on International Trade in Endangered Species
		Levels of participation in the Vienna Convention/Montreal Protocol
		Compliance with environmental agreements
	Global-scale funding/participation	Montreal protocol multilateral fund participation
		Global environmental facility participation
	Protecting international commons	Forest Stewardship Council accredited forest area as a percent of total forest area
		Ecological footprint "deficit"
		CO_2 emissions (total times per capita)
		Historic cumulative CO_2 emissions
		CFC consumption (total times per capita)
		SO_2 exports

Table 3

2001 Environmental Sustainability Index

Country	ESI	Country	ESI
Finland	80.5	Papua New Guinea	47.1
Norway	78.2	Ghana	46.9
Canada	78.1	Honduras	46.9
Sweden	77.1	Singapore	46.9
Switzerland	74.6	Azerbaijan	46.5
New Zealand	71.3	Nepal	46.5
Australia	70.9	Egypt	46.4
Austria	68.2	Trinidad and Tobago	46.4
Iceland	67.3	Bhutan	46.3
Denmark	67.0	Turkey	46.3
United States	66.1	Mali	46.1
Netherlands	66.0	Dominican Republic	45.3
France	65.8	Mexico	45.3
Uruguay	64.6	Thailand	45.2
Germany	64.2	Albania	45.1
United Kingdom	64.1	Cameroon	44.7
Ireland	64.0	Belgium	44.4
Slovak Republic	63.2	Romania	44.1
Argentina	62.9	Mozambique	43.8
Portugal	61.4	Uganda	43.8
Hungary	61.0	El Salvador	43.7
Japan	60.6	Kenya	43.7
Lithuania	60.3	Tunisia	43.7
Slovenia	59.9	Pakistan	43.4
Spain	59.5	Indonesia	42.5
Costa Rica	58.8	Jamaica	42.3
Bolivia	58.2	Senegal	42.3
Estonia	57.7	Morocco	41.8
Brazil	57.4	Uzbekistan	41.6
Czech Republic	57.2	Kazakhstan	41.5
Chile	56.6	Malawi	41.0
Latvia	56.3	India	40.7
Russian Federation	56.2	Algeria	40.6
Panama	55.9	Bangladesh	40.4
Cuba	54.9	South Korea	40.3
Colombia	54.8	Jordan	40.1
Italy	54.3	Tanzania	40.1
Peru	54.3	Kyrgyz Republic	39.6
Croatia	54.1	Zambia	39.5
Botswana	53.5	Benin	39.2
Greece	53.1	Macedonia	39.2
Nicaragua	51.9	Togo	38.9
Zimbabwe	51.9	Iran	38.4
Ecuador	51.8	Burkina Faso	38.3
Mauritius	51.2	Syria	37.9
South Africa	51.2	Sudan	37.6
Venezuela	50.8	China	37.5
Armenia	50.7	Lebanon	37.5
Gabon	50.2	Ukraine	36.8
Mongolia	50.1	Niger	36.7
Malaysia	49.8	Philippines	35.6
Sri Lanka	49.8	Madagascar	35.1
Israel	49.6	Vietnam	34.2
Paraguay	48.8	Rwanda	33.2
Belarus	48.1	Kuwait	31.9
Fiji	48.0	Nigeria	31.5
Central African Republic	47.7	Libya	31.3
Poland	47.6	Ethiopia	31.0
Bulgaria	47.4	Burundi	29.8
Moldova	47.4	Saudi Arabia	29.8
Guatemala	47.2	Haiti	24.5

relationship between environmental and economic performance in a number of ways. We found that although per capita income is correlated significantly with the ESI, for countries situated at similar levels of per capita income, variations in environmental sustainability have no significant connection to levels of economic performance.

At a broad level, the ESI is strongly correlated with per capita income (see Figure 1); the correlation coefficient is 0.76. What this means is that, taking the entire set of 122 countries into account, per capita income is a good predictor of the ESI. A similar relationship holds between the ESI and economic competitiveness, as shown in Figure 2. Two important caveats deserve attention, however.

First, wealthy countries are not better across the board. They tend to perform slightly below average in reducing environmental stresses, for example, and for some indicators (such as reducing waste and greenhouse gas emissions), they perform seriously below average. Therefore, it would be a mistake to conclude from the overall ESI-income correlation that environmental sustainability improves across the board with increases in per capita income. To be more specific, while the ESI supports a conclusion that wealthy countries generally have superior environmental sustainability conditions, it does not support the hypothesis that the patterns found in the wealthy countries are in any way sustainable on their present trajectory over the long run. The ESI does not address all the long-term causal interactions among its components that determine sustainability over the long run. For example, the high levels of environmental stress—high pollution levels—in wealthy countries ultimately may erode the currently relatively favorable environmental systems in those countries over the long run.

Nor does the ESI tell us whether a world in which each country exhibited the patterns found in the wealthiest countries would be the most environmentally sustainable over time, because it doesn't tell us what influence aggregate changes in one set of indicators would have on other indicators. Testing those propositions is beyond the reach of the ESI and requires a combination of richer time series data and greater scientific understanding of the causal connections across countries and indicators. The utility of the ESI is in providing an empirically based comparative gauge of where environmental conditions are likely to be most favorable over the next generation or two.

The second caveat is that, for purposes of policy relevance and causal analysis, the broad ESI-income correlation is somewhat beside the point. Suggesting that environmental conditions in Haiti might improve if economic conditions there became more like those in Finland is probably true, but definitely not helpful. Therefore, to explore the ESI-per capita

Figure 1

Relationship between the ESI and per Capita Income

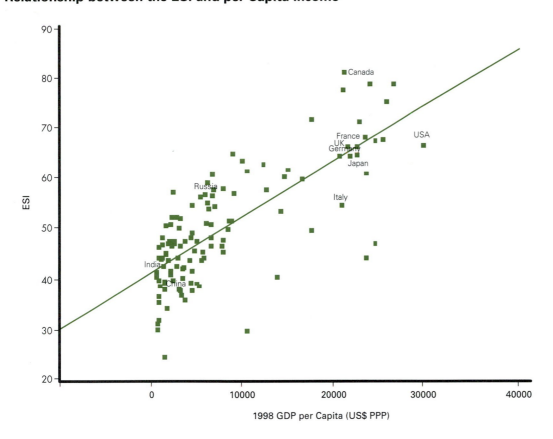

Figure 2

Relationship between the ESI and World Economic Forum Current Competitiveness Rank

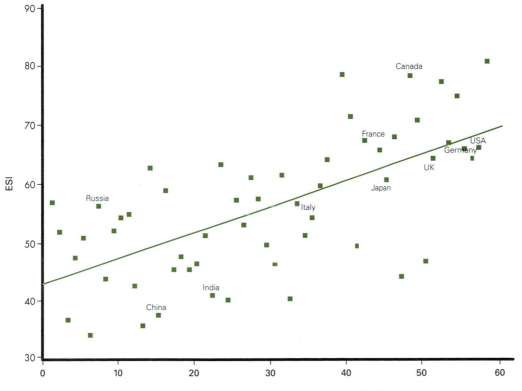

income connection in a more relevant manner, we divided the 122 countries into quintiles based on per capita income.

Within each quintile, the pattern was quite different from the pattern for the world overall. Only in the middle quintile was there a significant correlation, and this correlation was lower than for the overall set (0.59). For the remaining 80 percent of the countries, there was no significant correlation between ESI and per capita income within their respective income quintiles. Belgium, the lowest-scoring country in the high-income group, has virtually the same score as Cameroon, one of the highest-scoring countries in the low-income group. The country with the highest per capita income, the United States, has an ESI score lower than New Zealand's, one of the least wealthy countries in the top quintile.

This suggests that there are most likely policy-relevant drivers of environmental sustainability that transcend the impact of per capita income. This is consistent with the "Porter hypothesis," which suggests that high levels of environmental protection are compatible with high levels of economic growth and may even encourage the innovation that supports growth (Porter and van der Linde, 1995; Esty and Porter, 2000).

What are the most important drivers? Unfortunately, available data limit the search for answers.[3] For example, natural resource subsidies are thought to exert a major negative impact on environmental outcomes, yet we were unable to identify comparable data on such subsidies that covered more than a handful of countries. We can say, however, that there is strong reason to believe that social and institutional factors exert a strong effect on the health of environmental systems, the levels of stress inflicted on those systems, and the levels of human vulnerability to environmental change.

For example, the ESI data suggest that corruption has a profound negative impact on environmental sustainability. Although it is one of just 67 variables that make up the ESI, the reducing corruption measure had the highest correlation with the overall ESI, with a correlation coefficient of 0.75. What is especially striking is that this measure was correlated strongly with so many other more specific measures within the ESI, including such factors as air quality, water quality, population growth, environmental health, availability of environmental information, and energy efficiency (see Table 4).[4]

The strong connection to corruption was unexpected, although some level of correlation with the ESI is expected because of the governance indicators it includes. Indeed, corruption can be seen as a proxy for the presence of a functioning market economy, a public-minded government, and a commitment to the rule of law. We did not, however, expect the correlation to be as high as it was, nor did we expect it to have significant correlations with so many measures of direct environmental conditions. Eleven indicators out of 21 had significant correlations with the reducing corruption measure (not counting the indicator that included corruption as a measure).

The possibility that corruption is an important driver of factors that affect environmental sustainability is reinforced by looking at outliers in the ESI-per capita income relationship. Finland and Sweden, for example, score significantly higher on the ESI than would be predicted from their per capita incomes alone, and they have among the lowest corruption measures. Haiti and Saudi Arabia, on the other hand, score significantly lower on the ESI than would be predicted from their per capita incomes, and they have among the highest corruption measures.

But clearly the ESI by itself raises more questions about the role of corruption than it can answer. What are the mechanisms by which corruption contributes to poor environmental outcomes? Does corruption stifle innovation in the public and private sector? Does it generate inappropriate policy choices? Does it limit basic information on environmental conditions? Does it contribute to poor management across the board? Does it increase incentive to engage in unsustainable use of natural resources? What policy interventions can break the connection most effectively between corruption and environmental mismanagement? Is it necessary to target the root sources of corruption within a political system, or are there intermediate strategies that can help reduce corruption's impact on the environment? Our data by themselves do not answer these questions, but they provide a basis for testing propositions using empirical measures.

Another unexpected result linking social and institutional capacity measures to broader environmental concerns had to do with private-sector innovation. One of the variables with the highest correlation with the overall ESI was Innovest's EcoValue'21™ rating, which measures the quality of environmental management within firms (see Frank Dixon's essay in this volume for more information on this measure). For the ESI, we constructed an index for each country based on the EcoValue'21 ratings of firms with headquarters located within the country, weighted by market capitalization. Because there are only 20 countries with headquarters of companies found in the EcoValue'21 rating system, these results are far more suggestive than the corruption findings. But they are striking for the breadth of potentially significant connections they show between environmental innovation at the firm level and environmental outcomes at the national level. The EcoValue'21 variable was significantly correlated with indicators such as water quality and eco-efficiency, as well as with more specific variables in the environmental stress and system categories.

Table 4

Correlations between Reducing Corruption and Other ESI Indicators

Indicator	Correlation
Social and institutional capacity: science and technology	0.73
Global stewardship: international commitment	0.67
Capacity for debate	0.62
Air quality	0.53
Basic human sustenance	0.53
Environmental health	0.52
Environmental information	0.51
Reducing population pressure	0.49
Water quality	0.46
Global stewardship: global-scale funding/participation	0.44
Eco-efficiency	0.18

All correlations are significant at .05 level or greater.

Anomalies

Some of the results appear anomalous and provide an opportunity for probing the strengths and weaknesses of the ESI, as illustrated in the following example.

Russia's ESI score of 56.2 gave it a rank of 33 out of 122. This is far higher than most observers would have expected. As Feshbach (1995) and others have documented, Russian environmental conditions are catastrophic; therefore, faulty and missing data are most likely driving this anomaly. Consider the environmental health measures: Russia is one of the only industrial countries in history ever to experience a decline in life expectancy, and environmental health problems are rampant. Yet Russia reports to the World Health Organization a set of deaths from intestinal infections about equal to the world median (in between the United Kingdom and Norway) and does not report any information on deaths from acute respiratory diseases. Russia's self-reported water quality data are similarly out of sync with its well-documented water quality problems.

We could have adjusted these scores based on the individual studies that have been done on Russia's environmental conditions, but we deliberately chose not to because we thought it would dilute the ability of the ESI to measure conditions in a comprehensive, global, and consistent manner. For each country-specific change we might implement based on particular knowledge of that country, there would be an unknown number of equally compelling changes that we didn't know enough to make. Over time, we are committed to strategies that will reduce anomalies across the board by improving the data and methods for all countries. In the meantime, the ESI values will be vulnerable to these data quality problems.

Singapore, by contrast, had a low score that surprised some people. It is considered a well-managed, prosperous country with a number of effective environment and natural resource policies. Its ESI value of 46.8 made it a stark outlier in the general relationship between the ESI and per capita income. Some commentators suggested that Singapore's unexpectedly low score reflected a flaw in the Index's methodology. They suggested that, had we adequately taken into account Singapore's high population density, its existence as a city-state with virtually complete urbanization, and the fact that it occupies a small island, we would have arrived at a higher, more accurate, score for Singapore (Ho, 2001).

We do not agree with these suggestions. There are compelling analytical reasons to believe that small islands with large populations and considerable economic activity will approach, if not exceed, the limits of environmental sustainability. Singapore, for example, has an inadequate water supply given its population, a distinct lack of open space, and faces a number of other sustainability thresholds at close range. We do not wish to "control" for such factors; in fact, we wish to do precisely the opposite—to illuminate cases where such limits are being approached.

This does not mean that we are critical, either implicitly or explicitly, of the choices Singapore has made or of its environmental management. To the contrary, given its limited environmental endowments and many natural resource challenges, Singapore performs remarkably well. In a number of critical areas, especially ones that go to performance such as the eco-efficiency of the economy, Singapore's results are top-notch. In fact, if one estimates ESI as a function of income and population density, for example, Singapore's observed ESI is higher than its estimated score. Singapore's data demonstrate that wise management can dramatically reduce a nation's exposure to environmental threats even where critical sustainability thresholds are near. The fact remains, however, that any country experiencing the extreme levels of environmental stress (especially in terms of water) that one observes in Singapore is in danger of exceeding fundamental environmental limits. We believe the ESI as constructed accurately flags such danger points.

Belgium's very low score also generated controversy. Its ESI score of 44.4 gave it a rank of 78, the lowest of any industrial country by far and lower than many developing countries. These results generated intense controversy within Belgium (de Muelenaere, 2001), and sparked a detailed examination of Belgian environmental conditions as experts and commentators sought to understand them. Because there is no independent, verifiable measure of environmental sustainability with which to judge whether the low Belgian score is appropriate, a few observations can be made. We acknowledge that it would be wrong to interpret a country's ESI score or rank as representing knife-edge precision, especially when comparing very different countries. The ESI places Belgium in between Cameroon and Romania, for example, yet these three countries clearly have very different environmental dynamics, making pinpoint precision extremely difficult. But the more important point is whether Belgium's rank with respect to other industrial countries is defensible, and here the controversy was instructive. Although there were minor disagreements about the choice of variables, on balance, scrutiny of the Belgian data showed that, on many critical measures, the values were far worse than those in similar countries, and that the number of sub-par values was high. Values for reducing stress on air, water, and ecosystems were especially low, as was that for water quality. The fact that the ESI helped draw attention to these trends by showing how alarming they were in the aggregate, reveals one of the key benefits from engaging in the exercise.

Critiques and Responses

In addition to the attention given to the values that appeared anomalous, a number of more fundamental criticisms have been leveled at the ESI. One set of critics charges that the ESI includes too many separate issues. Others argue that the ESI tilts too strongly toward the present at the expense of the future. *The Ecologist* and Friends of the Earth published a critique that was representative of this line of thought. They argued that the ESI overly weighted socioeconomic factors such as infant mortality and access to clean drinking water, and gave insufficient weight to global change indicators such as greenhouse gas emissions ("Keeping Score," 2001).[5]

There is, of course, no easy way to resolve the debate over how to define "sustainability" and what should be included in an index meant to measure performance in this regard. These are instrumental and value-based decisions, not scientific ones. Where cross-sectoral interests are minor (if one does not care about capturing a broad range of environmental phenomena), and transgenerational concerns are paramount (if one is narrowly concerned with long-term carrying capacity), a definition of environmental sustainability that emphasizes resource consumption and greenhouse gas emissions is not just adequate but essential. Some indicators that adopt definitional strategies along these lines are quite sophisticated and capable of generating useful insights. Indicators that reduce sustainability metrics to total materials flow, for example, or that calculate "ecological footprints" based on consumption patterns relative to territorial size, or that reduce sustainability to energy efficiency, have been effective at shedding light on important patterns and dynamics (United Nations, 2001).

Competing definitions help to highlight the competing goals that are at issue. Precisely because environmental sustainability, as formulated in policy objectives by international and national actors, combines multiple dimensions, there are competing purposes at play and, therefore, competing definitions. Global society is in the middle of an intense debate over climate change, for example, so greenhouse gas–based indicators have a vital role to play. But it is important not to equate any specific instrumental goal with a broader philosophical truth. It is worthwhile to fight greenhouse gas emissions, but that doesn't mean that reducing greenhouse gas emissions becomes equivalent to achieving sustainability. To have a productive debate, it is important to evaluate definitions according to their purposes, and to place such evaluations in the context of the broad efforts underway.

In the case of the ESI, we deliberately avoided such an approach as a result of our instrumental calculus and value judgment. The instrumental calculus was based on a judgment that there was a genuine interest in the broad set of environmental issues that dominated the negotiation of Agenda 21, the ambitious global action plan on sustainable development prepared for the 1992 Earth Summit. Agenda 21 covers many environmental issues, from local to global, from ecologically focused to human-focused, from present-oriented to future-oriented. The Commission on Sustainable Development (CSD), charged with steering the implementation of Agenda 21, has taken this breadth seriously; the CSD's efforts to formulate indicators of sustainable development have always covered a large number of environmental issues and given strong weight to local as well as global concerns.[6]

The criticism that the ESI combines measures of both processes and outcomes (by mixing factors such as capacity and environmental systems) is a fair one, but only in a limited context. If one had an unambiguous definition of environmental sustainability that could be tied to an equally unambiguous observable outcome, then adding process measures to that outcome would be diversionary at best and misleading at worst. We think it is unlikely that such a state of affairs will ever exist, though. That does not mean that there is not a useful role for strictly outcome-based measurements; it just means that, for the broad set of issues contained within the environmental sustainability agenda, they are unlikely to be effective. Environmental sustainability, as used within the Agenda 21 framework and within the broader environmental movement, is not a one-dimensional construct analogous to wealth. In this context, it is better understood as a multidimensional concept with a poorly understood goal and a weak understanding of the interactions among its multiple dimensions. By and large, countries that seek environmental sustainability as an overall goal cannot specify where they want to end up along these multiple dimensions because they do not understand the trade-offs that may affect how they interact, nor do they understand the constraints that may limit their ability to achieve such a goal. A better analogy than wealth is health. When physicians seek to develop quantitative metrics of a patient's overall health trajectory, they combine measures of current processes, such as electrocardiac activity; they measure current outcomes such as body weight and cholesterol level; and they learn about personal responses, such as smoking, diet, and exercise, that will affect future outcomes. Measuring national trajectories of environmental sustainability requires similar breadth.

It is worth emphasizing this point: the ESI is best thought of as an empirically based assessment of the long-term prognosis for environmental conditions within a country, given current available information and measured in comparison to other countries. It is *not* meant to calculate planetary sustainability or the severity of the planetary predicament. It also is incapable of signaling whether a country at any location in the spectrum is environmentally sustainable in an absolute sense.[7]

Another set of criticisms has focused on the weighting scheme used by the ESI. Nominally, all the inputs into the Environmental Sustainability Index receive equal weight. The 22 indicators are calculated by averaging the values of the appropriate variables, and the Index score is calculated by averaging the 22 indicators. No variable or indicator gets more weight than any other.

In fact, however, there are implicit weights that derive from the structure we impose on the Environmental Sustainability Index, just as there are implicit weights in any index that aggregates multiple measures. We identify seven separate indicators of social and institutional capacity, for example, but only two indicators of human vulnerability. Implicitly, we are giving capacity measures more than three times the weight of vulnerability measures.

Not everyone will agree with the implicit weights reflected in the ESI. In particular, many critics have argued that the ESI assigns inappropriately low weights to greenhouse gas emissions and resource consumption measures. Unfortunately, disputes over the appropriateness of our weights cannot be resolved easily. There is no agreed prioritization among competing environmental issues, and there are no independent measures of environmental sustainability to use as an empirical check. While arbitrary in some sense, our weighting scheme derives from careful analysis, systematic attention to the environmental literature, and significant expert judgment. We took three separate steps to address the issue of weights.

First, we explored two techniques for assigning weights based on empirical relationships among the variables. Both of these techniques proved to be unfruitful.

In the first instance, we sought to construct a time series on a subset of the data that spanned the five components, in an effort to identify causal relationships that could be used as the basis for assigning differential weights. Variables with stronger causal impacts would receive stronger weights. In the end, this effort failed for the simple reason that robust time series data across a relevant range of indicators was impossible to construct.

The other such technique we used was the principal components analysis referred to above: we sought to identify statistically patterns of variation within the data that would assign differential weights to variables based on their ability to discriminate efficiently among the observations. If the resulting principal components had been compatible with our theoretical understanding of environmental sustainability, the resulting variable weights could have been used to assign differential weights. However, for the reasons discussed above, the results were not considered useful.

In addition to seeking a rationale for differential weights within the data, we conducted a survey of environmental experts and members of the business community to determine their views on the relative importance of the indicators used in the Index. The survey asked respondents to rank the relative importance of the indicators that comprise the Index. As the discussion in Annex 1 shows, for the most part, the results revealed very small differences among the indicators. The results were influential in the decision to drop one indicator that was judged to be of very low importance (exposure to natural disasters), but otherwise the survey results did not provide justification for differential weights.

We thus arrived at five broad categories of indicators based on strong theoretical foundations and carefully considered analytical judgments. Within these broad categories, we have identified a set of 22 indicators based on a systematic review of the factors that play a causal role in environmental sustainability dynamics; on their overall substantive importance to environmental sustainability; a rigorous assessment of the parameters with substantive importance to environmental sustainability; and on the quality and viability of available data. We assigned these 22 indicators equal weight.

We acknowledge, however, that many people will wish to apply weights in a different way. This is a major motivation behind our commitment to complete transparency and our decision to release the ESI data through the Web. We plan to expand our efforts to make the data available in a readily usable form so that those who wish to do analysis using other variable weights can do so.[8]

Conclusions

Societies are setting ambitious goals concerning sustainability. The ESI is intended to contribute to the success of these efforts by:

- providing tangible measures of environmental sustainability, filling a major gap in the environmental policy arena;

- making it more feasible to quantify goals, measure progress, and benchmark performance;

- facilitating more refined investigation into the drivers of environmental sustainability, helping to draw special attention to "best practices" and areas of success as well as lagging performance and potential disasters;

- helping to build a foundation for shifting environmental decision-making onto a more analytically rigorous foundation;

- combining a single measure of environmental sustainability with three additional levels of aggregation to meet a wide range of policy and research needs;

- striking a useful balance between the need for broad country coverage and the need to rely on high-quality data that are often of more limited country coverage; and

- building on an easily understood database using a methodology that is transparent, reproducible, and capable of refinement over time.

The Index is not without its weaknesses, however. In particular, the ESI:

- assumes a particular set of weights for the Index constituents that imply a set of priorities and values that may not be shared universally;

- relies in some instances on data sources of less than desirable quality and limited country coverage;

- suffers from substantive gaps attributable to a lack of comparable data on a number of high-priority issues; and

- lacks time series data, preventing any serious exercise in validation and limiting its value as a tool for identifying empirically the determinants of good environmental performance.

The ESI remains a "work in progress." A number of refinements of the analysis need to be undertaken to deepen our understanding of environmental sustainability and how to measure it. Specifically, we see a need for a number of actions:

1. The world needs a major new commitment to data gathering and data creation. We recommend a pluralistic approach to filling critical data gaps, making use of existing international organizations where they are capable, but filling in where they are not with strategies that draw on networks of scientists, local and regional officials, industries, and nongovernmental organizations. The world is sufficiently better connected, better skilled and better equipped that we need not rely on conventional bureaucracies to perform these tasks alone.

2. Because there are a variety of value judgments and significant scientific uncertainties about causality, it is necessary to augment the Environmental Sustainability Index with a flexible information system that permits users to apply their own value judgments or to experiment with alternative causal hypotheses. We have tried to advance this objective by creating an interactive version of the Index that operates on a desktop computer and by making our data and methods as transparent as possible. More could be done along these lines, including producing tools to facilitate more powerful integration of environmental sustainability data from different sources.

3. We need more sophisticated methods for measuring and analyzing information that comes from different spatial scales. Environmental sustainability is a function of the interaction of mechanisms that operate at the levels of ecosystems, watersheds, firms, households, economic sectors, and other phenomena that we are not well equipped to understand as parts of a whole. The modest efforts to integrate information from different spatial scales used in this Index need to be evaluated, improved on, and supplemented.

4. Consistent measurements over time are vital to create the ability to carry out robust investigations into cause-effect relationships. These measurements should evolve as data availability and aggregation techniques improve, but they must remain fully transparent and adequately archived for meaningful scientific investigation to be conducted. In addition to continuing measurements into the future, it is possible that retrospective measurements of certain variables could permit more rigorous causal analysis.

References

Cohen, Joel E., *How Many People Can the Earth Support?* New York: Norton, 1995.

de Muelenaere, Michel, "Bonnet d'âne pour la Belgique," *Belgique* (February 1, 2001), p. 2.

Esty, Daniel C., "A Term's Limits," *Foreign Policy* (September-October 2001).

Esty, Daniel C., and Michael E. Porter, "Measuring National Environmental Performance and Its Determinants," in Michael Porter and Jeffrey Sachs (eds.), *The Global Competitiveness Report 2000*, New York: Oxford University Press, 2000.

Feshbach, Murray, *Ecological Disaster: Cleaning up the Hidden Legacy of the Soviet Regime*, New York: Twentieth Century Fund, 1995.

Ho, Andy, "Just How Clean and Green Is Singapore?" *The Straits Times* (Singapore), February 15, 2001, p. 17.

Jackson, Ira, *Capitalism with a Conscience: A New Manifesto for Making a Profit While Making a Difference*, New York: Currency, forthcoming.

Kaufmann, Daniel, Aart Kraay, and Pablo Zoido-Lobaton, "Aggregating Governance Indicators," Washington: World Bank, 2000.

"Keeping Score," *The Ecologist* (April 2001).

Lomborg, Bjørn, *The Skeptical Environmentalist: Measuring the Real State of the World*, Cambridge: Cambridge University Press, 2001.

National Research Council, *Our Common Journey: A Transition Toward Sustainability*, Washington: National Academy Press, 1999.

National Research Council, *Ecological Indicators for the Nation*, Washington: National Academy Press, 2000.

Organization for Economic Cooperation and Development, "Shaping the 21st Century: The Contribution of Development Co-operation," Paris, OECD, 1996.

Porter, Michael, and Jeffrey Sachs (eds.), *The Global Competitiveness Report*, New York: Oxford University Press, 2000.

Porter, Michael, and Jeffrey Sachs (eds.), *The Global Competitiveness Report 2001–2002*, New York: Oxford University Press, 2001.

Porter, Michael, and C. van der Linde, "Green and Competitive: Ending the Stalemate," *Harvard Business Review* (September-October 1995), pp. 120–134.

"Achieving Sustainability: Poverty Elimination and the Environment,"London, United Kingdom Department for International Development, 2000.

United Nations, "Report On Aggregation Indicators for Sustainable Development," New York: UN Division on Sustainable Development, 2001.

World Commission on Environment and Development ("Brundtland Commission"), *Our Common Future*, Oxford: Oxford University Press, 1987.

World Economic Forum, *Environmental Sustainability Index*, January, 2001 (available at http://www.ciesin.columbia.edu/indicators/ESI/)

World Economic Forum, *Pilot Environmental Sustainability Index*, January 31, 2000 (available at http://www.ciesin.columbia.edu/indicators/ESI/archive.html)

Endnotes

1   I am grateful to Dan Esty for helpful suggestions on earlier drafts of this chapter. The ESI was a joint initiative involving the World Economic Forum's Global Leaders for Tomorrow (GLT) Environment Task Force, the Yale Center for Environmental Law and Policy, and the Columbia University Center for International Earth Science Information Network (CIESIN). Kim Samuel-Johnson chaired the GLT Environment Task Force; Dan Esty led the Yale team and served as overall project director; and other members of the Yale team included Brian Fletcher, Maria Ivanova, Lisa Max, Kanet McGarry, Anita Padmenhaben, and Colleen Ryan. Marc Levy led the CIESIN team; other team members included Kobi Abayomi, Deborah Balk, Bob Chen, Alex de Sherbinin, Francesca Pozzi, Maarten Tromp, and Antoinette Wannebo. We are especially grateful to the many experts who reviewed the ESI effort during its development, including Alan AtKisson, Christian Avérous, Steve Charnovitz, Peter Cornelius, Frank Dixon, André Dua, Raimundo Florin, Tom Graedel, Kirk Hamilton, Allen Hammond, Peter Hardi, Theodore Heintz, Jochen Jesinghaus, Kai Lee, Maria Leicher, Victor Lichtinger, Bedrich Moldan, William Nordhaus, Theo Panayoutou, Tom Parris, José Luis Samaniego, Md. Rumi Shammin, Gus Speth, Adreas Sturm, Carla Tabossi, Simon Tay, Phillip Toyne, and Claas van der Linde. Peter Cornelius played a critical role at the World Economic Forum headquarters.

2   From the beginning, the ESI team was aware of the challenges that lay in the way of measuring environmental sustainability and realized that the initial efforts had to be formulated in a way that permitted learning and debate. Following consultation with a range of international experts and a review of the literature, the first effort was a Pilot Environmental Sustainability Index, released in January 2000. An accompanying report detailed the analytical structure adopted, methodology used, and data sources; this report was circulated in paper and through the World Wide Web (World Economic Forum, 2000). This report generated useful debate on the choices made in the Pilot ESI, which facilitated creation of the first full Environmental Sustainability Index in January 2001. Based on a wide range of informal feedback, further methodological refinement, and a formal peer review, the 2001 ESI adopted a number of innovations over the Pilot ESI. As with the Pilot ESI, the report and data have been made available through the Web (World Economic Forum, 2001).

3   Esty and Porter provide some preliminary analysis in this regard in Chapter 3 of this volume.

4   Because the corruption measure was used to estimate missing values in the ESI data set, the correlations reported here were calculated on a version of the data set in which missing values were not estimated to avoid drawing spurious conclusions.

5   They also argued that including several capacity measures in the ESI was misguided because they diverted attention from concrete environmental results (they were a process measure, not an outcome measure), and because they failed to account for the fact that capacity can be used for bad as well as good. Finally, they strongly criticized the low emphasis given to global change indicators, such as greenhouse gas emissions, concluding that the definition of environmental sustainability adopted in the ESI was "designed to make dirty nations look clean." An alternative index was proposed that removed all the socioeconomic and capacity measures as well as many other variables thought to be problematic, with greater emphasis placed on global change related measures. Rich countries were on the bottom of this 15-variable, seven-indicator index (the United States fell to 112, for example), while the Central African Republic, Bolivia, and Mongolia moved to the top. We find this approach interesting but, at best, too narrow and theoretically unsatisfactory.

6   It is somewhat ironic for *The Ecologist* to accuse the ESI of being designed to benefit the rich, when the indicator set used by the ESI comes closer to the CSD set than to any other indicator effort, and the CSD is a UN body in which the poor countries of the world have a dominant voice.

7   In this regard, the misuse of the ESI data by Bjørn Lomborg (2001) is particularly distressing. Lomborg cites the ESI data and our analysis showing a positive relationship between ESI score and GDP per capita to support his contention that "only when we get sufficiently rich can we afford the relative luxury of caring about the environment" (p. 33), which is embedded in his overall argument that "things are getting better." As we argue below, this is most definitely not a proposition that we think the ESI data support—knowing that Finland is more likely than Haiti to have hospitable environmental conditions several generations into the future does not mean that conditions globally might not be worse in the future than they are now, or that poor countries are incapable of improving their conditions until they become richer.

8   Although we continue to disagree with some of the critics who have proposed alternative weighting schemes that discount local issues such as water quality and increase the impact of global issues such as climate change, we are gratified that they have been able to take our indicators and combine them in different ways.

Chapter 3
National Environmental Performance Measurement and Determinants

Daniel C. Esty and Michael E. Porter

Environmental performance, encompassing the control of pollution and stewardship of natural resources, is of growing concern in both advanced and developing economies. Environmental quality plays a major role in quality of life, with a direct impact on the health and safety of a nation's citizens as well as its attractiveness as a place to live. Many governments work hard to address environmental issues, but public expectations often outstrip results. As a result, pollution control and natural resource management have increasingly become core governmental challenges.

Despite high interest across almost all countries, setting environmental policymaking remains more an art than a science. Statistical analyses of the determinants of environmental performance across nations have been rare—indeed, almost non-existent. Data on which policies and programs are successful—and which are not—are hard to come by. Research in the environmental realm has relied heavily on anecdotal evidence and case studies. Thus, there is precious little information on which to base environmental judgments at either the public policy or corporate level. This may explain why the environmental field remains mired in deep controversies over the best path forward, with debate often dominated by emotional claims and heated rhetoric. We believe that more sophisticated use of environmental indicators and statistical tools to develop systematic and objective ways to gauge results offers a constructive way out of the current stasis.

This essay builds on our previous effort to investigate statistically the underpinnings of environmental performance, and to use the findings to rank countries in terms of environmental outcomes and policies.[1] In particular, we seek to explain differences in national environmental results—as measured by levels of air pollution (particulates and SO_2) and energy use—based on national policy choices in environmental regulation as well as in broader economic, political, and legal structures. We also explore empirically the question of whether strong environmental performance must come at the expense of competitiveness and economic development, as traditional economic theory has suggested (Jaffe et al., 1995). More broadly, we aim to put environmental decision-making on a firmer analytic footing and to encourage further efforts to generate better data and improve statistical methods.

Although hampered by imperfect data, a lack of time series data that would permit more definitive tests of causality, and the need to use relatively crude methods, we find substantial evidence that environmental performance varies systematically with both the quality of a country's environmental regulatory regime and its broader economic and legal context. We use our model to create a framework for measuring the quality of national environmental regulation and to rank countries on both the quality of regulation and environmental performance (see Table 8). We find a significant correlation between income and environmental performance, suggesting that alleviating poverty should be seen as a priority for environmental policymakers. However, dramatic differences in environmental performance occur even among countries with similar economic levels. This finding implies that environmental improvement is not merely a function of economic development but benefits from carefully constructed policy choices. Our analysis also suggests that a country's broader economic, legal, and other institutional underpinnings are also important determinants of environmental performance. On the trade-off between being green and being competitive, we find no evidence that improving environmental quality compromises economic strength. In fact, higher levels of environmental performance appear to be correlated positively with competitiveness.

Modeling Environmental Performance and Its Causes

We employ three measures of environmental performance (environmental "output") that are available with broad country coverage: the level of urban particulates, urban SO_2 concentrations, and energy use per unit of GDP.[2] These measures constitute the dependent variables for the analysis.

Building on theoretical work in the economic, legal, regulatory, and environmental domains, we then assemble data on policy variables that potentially determine environmental outcomes.[3] The framework for the analysis is shown in Figure 1. Environmental performance is hypothesized to result from two broad sets of independent variables. One set, which we call the *environmental regulatory regime*, comprises measures of various aspects of a country's environmental regulatory system, including standards, implementation and enforcement mechanisms, and associated institutions. These variables capture regulatory elements that directly affect pollution control and natural resource management.

The second set of independent variables, which we term *economic and legal context*, are indicators of a country's more general administrative, scientific, and technical institutions and capabilities. These include measures of the extent of the rule of law, protection of property rights, and technological strength. Our hypothesis is that a nation's environmental reg-

ulatory regime will be more effective in producing the desired outcomes if the economic and legal context is sound. Hence, context indirectly (but perhaps importantly) determines environmental performance.

The dotted arrows in Figure 1 represent the final stage of the analysis, in which we examine the connection between environmental performance and economic success. We explore, in particular, the relationship between our environmental quality measures and GDP per capita, as well as the relationship between an index measuring the overall environmental regulatory regime (environmental regulatory regime index [ERRI]) and GDP per capita. We also examine the relationship between the ERRI and competitiveness as measured by the *Global Competitiveness Report 2000* (Porter et al., 2001). These relationships shed light on the longstanding debate over the extent of the trade-off between environmental progress and economic success—a question of particular interest in the developing world.

Environmental Outcomes. Environmental output data are notoriously spotty, unreliable, and uneven, as are data on the

Figure 1

Determinants of Environmental Performance

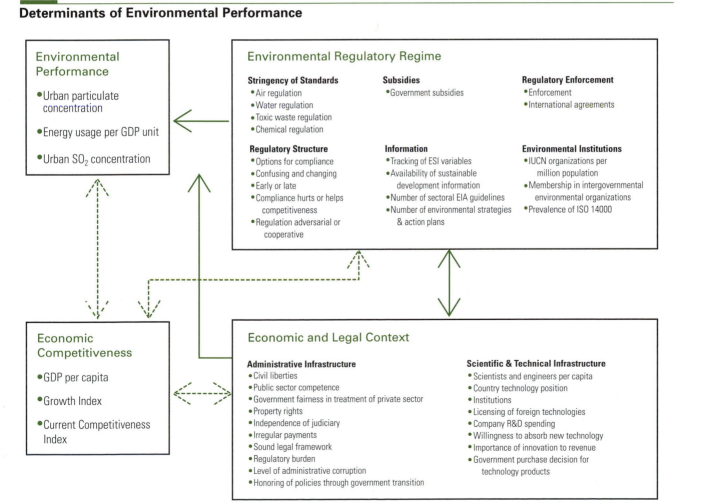

Table 1

Absolute Environmental Performance by Country

Urban Particulate Concentration* (Per City Population)			Urban SO₂ Concentration* (Per City Population)			Energy Usage (Per Million $ GDP)		
Rank	**Country**	**Annual Mean**	**Rank**	**Country**	**Annual Mean**	**Rank**	**Country**	**Bil. Btu**
1	Sweden	9.0	1	Argentina	1.02	1	Denmark	4.84
2	Norway	10.3	2	Lithuania	2.10	2	Switzerland	5.19
3	France	14.2	3	New Zealand	3.49	3	Japan	6.55
4	Iceland	24.0	4	Finland	4.38	4	Italy	6.66
5	New Zealand	27.3	5	Iceland	5.00	5	Ireland	6.85
6	Switzerland	30.7	6	Sweden	5.23	6	Austria	7.09
7	Canada	31.3	7	Latvia	5.36	7	Germany	7.28
8	Netherlands	40.0	8	Norway	5.47	8	France	7.39
9	Australia	43.2	9	Denmark	7.00	9	Finland	8.37
10	Germany	43.3	10	Portugal	9.22	10	United Kingdom	8.59
11	Japan	43.6	11	Netherlands	10.00	11	Spain	8.73
12	Austria	45.7	12	Romania	10.00	12	Honduras	8.97
13	Finland	49.9	13	Spain	11.00	13	Mauritius	9.11
14	Argentina	50.0	14	Thailand	11.00	14	Sweden	9.14
15	Portugal	50.4	15	Switzerland	11.34	15	Israel	9.96
16	Venezuela	53.0	16	Germany	12.80	16	Peru	10.81
17	Czech Republic	58.4	17	Canada	12.87	17	Netherlands	11.01
18	Denmark	61.0	18	Australia	13.17	18	Slovenia	11.26
19	Hungary	63.7	19	Austria	13.21	19	Australia	11.46
20	Slovak Republic	64.5	20	France	13.89	20	Guatemala	11.52
21	Spain	72.7	21	United States	15.43	21	Portugal	11.77
22	Romania	82.0	22	Italy	15.55	22	Belgium	11.83
23	Korea	83.8	23	Ireland	18.89	23	Norway	12.17
24	Italy	86.9	24	Singapore	20.00	24	Argentina	12.22
25	Malaysia	91.6	25	Malaysia	20.49	25	Uruguay	12.86
26	Latvia	100.0	26	Belgium	21.02	26	Greece	12.95
27	Russia	100.0	27	Ecuador	21.52	27	Bangladesh	13.15
28	Brazil	106.2	28	United Kingdom	21.96	28	United States	13.41
29	Lithuania	114.3	29	South Africa	22.37	29	Sri Lanka	13.70
30	Colombia	120.0	30	Slovak Republic	22.66	30	El Salvador	13.75
31	Ecuador	125.7	31	Japan	24.33	31	Brazil	14.01
32	Greece	178.0	32	Czech Republic	27.34	32	Iceland	14.49
33	Bulgaria	199.2	33	India	27.55	33	New Zealand	15.09
34	Philippines	200.0	34	Chile	29.00	34	Paraguay	15.32
35	Thailand	223.0	35	Philippines	33.00	35	Estonia	16.09
36	Costa Rica	244.5	36	Venezuela	33.00	36	Costa Rica	16.13
37	Indonesia	271.0	37	Greece	34.00	37	Chile	16.63
38	Guatemala	272.3	38	Hungary	37.33	38	Canada	17.54
39	India	277.5	39	Costa Rica	38.84	39	Mexico	17.72
40	Mexico	279.0	40	Korea	52.41	40	Korea	17.91
41	China	310.8	41	Bulgaria	52.45	41	Bolivia	18.41
42	Honduras	320.0	42	Poland	54.72	42	Dominican Republic	18.68
			43	Egypt	69.00	43	Panama	18.70
			44	Mexico	74.00	44	Thailand	19.29
			45	Brazil	75.78	45	Philippines	19.74
			46	China	97.07	46	Singapore	20.41
			47	Russia	97.55	47	Zimbabwe	22.34
						48	Malaysia	22.88
						49	Indonesia	22.96
						50	Nigeria	23.66
						51	Colombia	23.98
						52	Latvia	25.01
						53	Ecuador	27.57
						54	India	28.13
						55	Egypt	31.03
						56	Hungary	32.29
						57	Jordan	34.52
						58	Jamaica	35.58
						59	Nicaragua	36.46
						60	South Africa	37.92
						61	China	39.10
						62	Venezuela	44.11
						63	Poland	45.05
						64	Lithuania	54.92
						65	Czech Republic	56.22
						66	Romania	58.39
						67	Bulgaria	60.71
						68	Slovak Republic	63.95
						69	Vietnam	64.57
						70	Russia	74.19
						71	Ukraine	96.53

*Not all data were available for all countries.

characteristics of national regulatory regimes. Hence, establishing a sufficient database for a broad empirical analysis is no small undertaking. The performance measures used in this study are drawn from data assembled for the World Economic Forum's Environmental Sustainability Index (ESI) project.[4]

Three measures of environmental performance emerge as reliable enough and available in a large enough number of countries to use in our analysis. The first is urban particulate concentration, derived from World Bank and World Health Organization (WHO) data sources. This measure provides the mean total suspended particulate concentrations in the air (airborne dust) normalized by a country's urban population. A higher concentration indicates more pollution and, thus, worse air quality.

The second performance measure is mean SO_2 concentration normalized by urban population. This measure is also drawn from World Bank and WHO data. Again, higher figures represent worse air pollution.

The third environmental performance measure gauges energy efficiency. Using U.S. Department of Energy data, we measure total energy consumption per unit of a country's GDP. Higher figures represent more energy consumed per unit of economic output and, thus, greater energy inefficiency. In comparing this measure across countries, we need to account for the fact that Russia and the countries of the former Soviet bloc operated for decades under an energy regime with prices set well below market levels. This history has left a legacy of energy inefficiency in these countries that is only slowly being corrected. We therefore include a dummy variable in our model to control for this history, which proves to be highly significant statistically.

Table 1 provides absolute rankings by country for each of the three environmental performance measures. Urban particulate data are available for just 42 of the 75 countries covered by the *Global Competitiveness Report 2001–2002* (*GCR*). The United States and the United Kingdom track particulates, but on a more refined basis than the rest of the world, which prevents them from being comparable. Therefore, these countries are excluded from the urban particulate analysis.[5] Sweden and Norway are at the top of the particulate ranking, with China and Honduras at the bottom.

The SO_2 rankings cover 47 countries. Argentina and Lithuania rank at the top on this measure; China and Russia face the most severe SO_2 problems.

Energy use data are available for 72 countries. Denmark and Switzerland rank highest in energy efficiency, and Russia and Ukraine emerge as the most energy-inefficient countries.

Figure 2

Relationship Between Urban Particulate Concentration and GDP per Capita

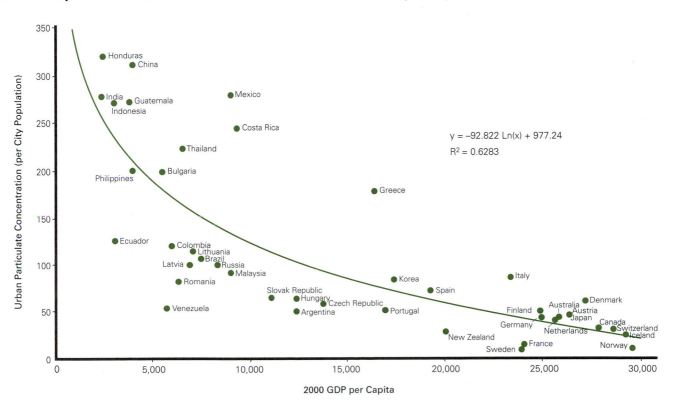

$y = -92.822 \, Ln(x) + 977.24$

$R^2 = 0.6283$

Figure 3

Relationship between Urban SO₂ Concentration and GDP per Capita

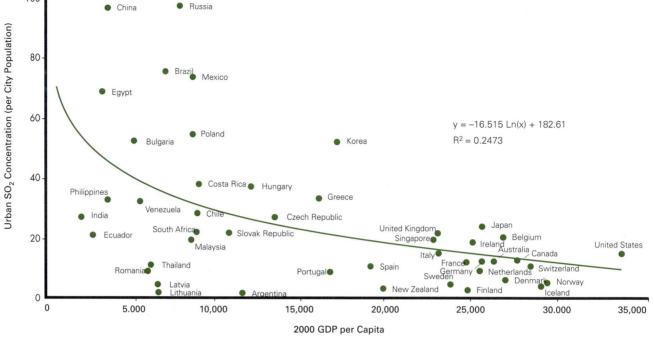

$$y = -16.515 \, \text{Ln}(x) + 182.61$$
$$R^2 = 0.2473$$

Figure 4

Relationship between Energy Usage and GDP per Capita (log model)

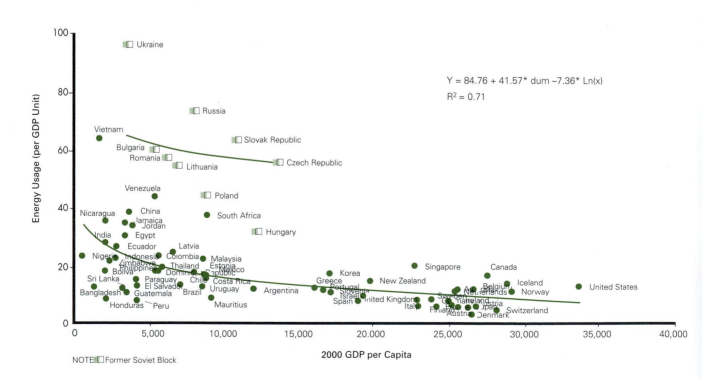

$$Y = 84.76 + 41.57^* \, \text{dum} - 7.36^* \, \text{Ln}(x)$$
$$R^2 = 0.71$$

NOTE: ▦ Former Soviet Block

Table 2

Energy Usage Relative to Expected Given GDP per Capita, Listed by Income Groups

Low Income Countries (≤ $6,500)			Middle Income Countries ($6,500–23,000)			High Income Countries (≥ $23,000)		
Rank	Country	Residual	Rank	Country	Residual	Rank	Country	Residual
1	Honduras	-18.29	1	Hungary	-24.70	1	Denmark	-4.78
2	Bangladesh	-17.48	2	Poland	-14.29	2	Italy	-4.08
3	Guatemala	-12.60	3	Mauritius	-8.22	3	Switzerland	-4.06
4	Peru	-11.57	4	Lithuania	-6.24	4	Japan	-3.44
5	Nigeria	-11.28	5	Brazil	-5.19	5	Ireland	-3.31
6	Sri Lanka	-10.96	6	Uruguay	-4.96	6	France	-3.12
7	El Salvador	-9.13	7	Spain	-3.44	7	Germany	-2.96
8	Bolivia	-9.04	8	Argentina	-3.21	8	Austria	-2.75
9	Paraguay	-7.69	9	Israel	-2.06	9	United Kingdom	-2.18
10	Zimbabwe	-4.27	10	Slovenia	-1.74	10	Finland	-1.89
11	Philippines	-4.05	11	Estonia	-1.51	11	Sweden	-1.42
12	Romania	-3.53	12	Costa Rica	-1.42	12	Netherlands	0.96
13	Indonesia	-2.84	13	Portugal	-1.34	13	Australia	1.46
14	Bulgaria	-2.26	14	Chile	-0.96	14	Belgium	2.16
15	Dominican Republic	-2.09	15	Greece	-0.41	15	Norway	3.17
16	Panama	-1.82	16	Mexico	-0.09	16	Iceland	5.41
17	Thailand	-0.88	17	Czech Republic	0.02	17	United States	5.43
18	India	0.67	18	New Zealand	3.24	18	Canada	8.10
19	Ecuador	1.90	19	Korea	4.98			
20	Colombia	3.17	20	Malaysia	5.08			
21	Egypt	6.55	21	Latvia	5.25			
22	Nicaragua	8.98	22	Slovak Republic	6.14			
23	Jordan	10.96	23	Singapore	9.58			
24	Jamaica	11.21	24	Russia	14.20			
25	China	15.30	25	South Africa	20.33			
26	Venezuela	22.98						
27	Ukraine	30.66						
28	Vietnam	35.66						

Figures 2, 3, and 4 plot the relationships between each measure of environmental performance and GDP per capita. One pattern, immediately discernable across all three measures, is that richer countries achieve better results than poorer ones. The improvement of environmental performance as income rises is most pronounced with regard to urban particulates and energy efficiency, and least pronounced for SO_2 emissions. Among lower-income countries, variance on all three measures is higher than among more prosperous countries. This suggests that environmental performance can be *improved substantially* in many low-income countries, independent of the gains that come with economic development.

The regression relationship between environmental performance and GDP per capita provides an interesting perspective on how each country performs *relative* to its wealth. Countries above the regression line in Figures 2, 3, and 4 exhibit weaker environmental results on the particular performance measure than would be expected, given their level of GDP; those countries below the regression line demonstrate better performance. These results are shown in Tables 2, 3, and 4.

With regard to particulate levels, Italy, Greece, Mexico, Costa Rica, China, and Denmark are notable laggards relative to income. Sweden, Norway, Argentina, Latvia, Ecuador, and Venezuela show relatively strong performance.

In terms of SO_2 performance, Russia, Brazil, Mexico, Korea, China, Egypt, Japan, and Belgium lag relative to income. The United States is also a weak performer. Iceland, Finland, Sweden, Argentina, Latvia, Lithuania, Thailand, Romania, and Ecuador show relatively strong results.

In energy efficiency, Denmark, Switzerland, Japan, Italy, Hungary, Poland, Honduras, and Bangladesh, appear to be more energy efficient than would be expected, given their level of income. The United States, Canada, Singapore, Russia, South Africa, Venezuela, Ukraine, and Vietnam emerge as poor performers relative to income. As seen in Figure 4, the dummy variable for former Soviet bloc countries is highly significant, suggesting that the countries that faced artificially low energy prices suffered a common fate of huge energy inefficiency.

Taken together, these findings are consistent with established theory that suggests that pollution control improves with economic development (World Commission on Environment and Development, 1987). Our data, however, do not reveal an inverted U-shaped environmental "Kuznets curve." A number of other studies have found such a pattern, characterized by rising emissions in the early stages of development and improving environmental performance after middle-income levels have been reached (Grossman and Krueger, 1995; Harbaugh et al., 2000). The results here

Table 3

Urban Particulate Concentration Relative to Expected Given GDP per Capita, Listed by Income Groups*

Low Income Countries (≤ $6,500)			Middle Income Countries ($6,500–23,000)			High Income Countries (≥ $23,000)		
Rank	**Country**	**Residual**	**Rank**	**Country**	**Residual**	**Rank**	**Country**	**Residual**
1	Venezuela	-121.87	1	Latvia	-57.59	1	Sweden	-32.50
2	Ecuador	-106.27	2	Argentina	-52.99	2	France	-26.77
3	Romania	-83.06	3	Slovak Republic	-48.69	3	Norway	-11.65
4	Colombia	-50.93	4	Brazil	-44.20	4	Iceland	1.05
5	Philippines	-8.40	5	Malaysia	-41.31	5	Canada	3.80
6	Bulgaria	20.92	6	Lithuania	-41.17	6	Netherlands	4.93
7	India	22.78	7	Russia	-40.59	7	Switzerland	5.62
8	Indonesia	37.35	8	Hungary	-39.10	8	Germany	5.75
9	Guatemala	59.80	9	Czech Republic	-34.57	9	Australia	8.73
10	Thailand	60.26	10	New Zealand	-30.61	10	Japan	9.28
11	Honduras	67.85	11	Portugal	-23.31	11	Finland	12.13
12	China	102.36	12	Spain	10.93	12	Austria	13.20
			13	Korea	12.41	13	Denmark	31.29
			14	Greece	101.19	14	Italy	43.13
			15	Costa Rica	114.79			
			16	Mexico	146.02			

*Not all data were available for all countries.

Table 4

Urban SO₂ Concentration Relative to Expected Given GDP per Capita, Listed by Income Groups*

Low Income Countries (≤ $6,500)			Middle Income Countries ($6,500–23,000)			High Income Countries (≥ $23,000)		
Rank	**Country**	**Residual**	**Rank**	**Country**	**Residual**	**Rank**	**Country**	**Residual**
1	Ecuador	-28.49	1	Lithuania	-34.29	1	Finland	-11.07
2	Romania	-28.10	2	Latvia	-31.41	2	Sweden	-10.89
3	Thailand	-26.69	3	Argentina	-26.04	3	Iceland	-7.81
4	India	-26.49	4	New Zealand	-15.54	4	Norway	-7.16
5	Philippines	-12.81	5	Portugal	-12.63	5	Denmark	-7.02
6	Venezuela	-6.84	6	Malaysia	-11.88	6	Netherlands	-4.97
7	Bulgaria	11.99	7	South Africa	-9.52	7	Germany	-2.60
8	Egypt	21.64	8	Spain	-8.72	8	France	-2.12
9	China	51.25	9	Slovak Republic	-6.21	9	Switzerland	-1.85
			10	Chile	-2.89	10	Australia	-1.70
			11	Czech Republic	2.07	11	Austria	-1.31
			12	Singapore	3.26	12	Italy	-0.97
			13	Costa Rica	7.03	13	Canada	-0.75
			14	Hungary	10.30	14	Ireland	3.67
			15	Greece	11.60	15	United States	5.09
			16	Poland	22.43	16	United Kingdom	5.37
			17	Korea	30.98	17	Belgium	6.91
			18	Brazil	40.29	18	Japan	9.49
			19	Mexico	41.61			
			20	Russia	63.80			

*Not all data were available for all countries.

may be explained by the fact that our sample of countries contains relatively few countries in the "early industrialization" stage of development during which emissions and energy use would be low and rising, especially for the air pollution measures.

The relationship between environmental performance and level of development supports several preliminary but important policy conclusions. First, the evidence that poorer countries uniformly perform less well on all three environmental quality measures supports an emphasis on alleviating poverty as a core policy goal from the perspective of environmental progress.

Second, the wide variations in environmental performance among countries at similar levels of economic development suggest that income or development stage affects, but does not alone determine, environmental outcomes. Some rich countries seem to have learned how to advance environmental quality ahead of their economic progress; others have not. Similarly, some developing countries appear to have achieved far better environmental quality relative to their level of development, while other countries seem to be sacrificing environmental goals in pursuit of economic growth. We explore whether this approach is effective later in this essay.

Third, it is notable that environmental performance improves with rising income most quickly for the most localized problem (particulates), and least rapidly with regard to environmental impact (energy use) that generates the harms (CO_2 emissions from fossil fuel burning) spread most widely over space and time.[6] Intermediate results occur for the variable (SO_2) that arises on an intermediate spatial and temporal scale. This pattern comports with the theoretical prediction that the geographic and temporal spread of an environmental issue represents critical policy variables. Where harms have a transboundary or intertemporal dimension, they constitute "super externalities," which raise special collective action problems and often prove especially difficult to address (Dua and Esty, 1997).

Determinants of Environmental Performance

Data on the environmental regulatory regime and the broader economic and legal context are drawn from both the ESI project and the *Global Competitiveness Report 2001–2002* annual survey of business and government leaders.[7] We divide the determinants of environmental performance into two broad groups: measures related to a country's environmental regulatory regime, and measures of its economic and legal context. The full list of variables, along with their definitions and sources, can be found in Annex 1.

For purposes of analysis, we separate the environmental regulatory regime variables into a number of categories representing different aspects of a country's regulatory approach:

- stringency of environmental pollution standards

- sophistication of regulatory structure

- quality of available environmental information

- extent of subsidization of natural resources

- strictness of enforcement

- quality of environmental institutions

The stringency of standards category includes measures of the perceived rigor of a nation's air pollution, water pollution, toxic waste, and chemical regulations. This information is drawn from the *GCR* Survey. We expect a negative relationship between each of the measures of regulatory stringency and our dependent variables because more rigorous standards should lead to lower levels of urban particulates, lower SO_2 concentrations, and lower energy use per unit of GDP.

The regulatory structure category measures the degree to which a nation's environmental regulations are flexible, clear, consistent, progressive, structured to help competitiveness, and designed to promote cooperative rather than adversarial business-government relations. In each case, we anticipate a negative relationship between these variables and our measures of environmental performance because a more refined and sophisticated regulatory structure is expected to produce less pollution and energy use.

The information category attempts to measure the degree to which a nation has a sufficient data foundation to determine policy and to support enforcement of environmental regulations. There are no direct measures of the quality of the information underlying each country's environmental regime, and we rely on four proxy variables drawn from the ESI data set: (1) the degree to which a country collects data in the 65 categories tracked by the ESI analysis; (2) the extent of sustainable development information and the existence of plans to support national environmental decision-making (as called for in the Rio Earth Summit's Agenda 21 process); (3) the prevalence of guidelines for sectoral environmental impact assessments; and (4) the breadth of environmental action plans. All of these information indicators are relatively crude, but they should provide some basis for gauging whether a nation seeks to make environmental judgments on an analytically rigorous basis. We expect a negative relationship between these information variables and our environmental performance measures.

The subsidies measure is derived from the *GCR* Survey data on the extent of a country's subsidization of energy and other materials. Where price signals are distorted, we expect to see greater inefficiency and higher levels of pollution. Thus, we anticipate a positive relationship among the level of subsidies and particulate levels, SO_2 concentrations, and energy use.

Strictness of enforcement measures is drawn from the *GCR* Survey. The first measure gauges how aggressively a nation's environmental regulations are enforced, and the second gauges the depth of a country's commitment to treaty requirements and other international environmental obligations. We expect a negative relationship between these measures of enforcement rigor and our dependent variables because those countries that take environmental regulations (whether domestic or international) seriously should experience better pollution control and more efficient energy use.

The final regulatory regime category, institutional quality, seeks to measure the degree to which intergovernmental (international) organizations and nongovernmental entities (environmental groups, community organizations, business associations, and other elements of civil society) reinforce government's environmental efforts. The mechanisms for such reinforcement are diverse (Esty, 1998). In some cases, these entities directly undertake environmental activities and, thus, substitute for government action. Environmental groups, for instance, may identify harms, highlight issues that demand attention, undertake data gathering and analyses, or spotlight causes of environmental harm. NGOs also may strengthen a society's capacity for pollution control by providing environmental education to the public or technical assistance to polluters. Of course, such entities may play counterproductive roles as well, especially if they pursue extreme positions and use solely adversarial approaches, unnecessarily increasing costs.

Our capacity to measure the degree of institutional quality is limited, and the variables in this category are, of necessity, somewhat crude proxies. We use data from the ESI database on the number of entities (scaled by population) that participate in the World Conservation Union (IUCN), an umbrella organization of environmental NGOs and research centers. We also draw on ESI data that provide a measure of the breadth of a country's engagement with intergovernmental environmental bodies. A third institutional quality variable comes from the *GCR* Survey and gauges the extent to which a nation's companies utilize the ISO 14000 certification process for environmental management. We expect a negative relationship between these measures and our dependent variables gauging environmental outcomes.

The second broad group of independent variables tracks potentially significant dimensions of a country's economic and legal context. We analyze this broader set of societal variables based on a growing theoretical literature that suggests that a country's underlying political, legal, and economic structures may contribute as much to environmental protection as the details of its regulatory regime (Esty, 1997; Sachs, 1998; Esty and Porter, 2000).

Under the economic and legal context, there are two categories of variables. First, we analyze what we call administrative infrastructure. In this category, we assemble data on civil and political liberties drawn from the ESI and measures (from the *GCR* Survey) of public-sector competence, degree of governmental favoritism, how vigorously private property is protected, independence of the judiciary, demands for irregular payments as a price for doing business, extent of the rule of law, burdensome regulations, corruption, and the degree to which new governments honor the obligations of earlier administrations. For each of these variables, we anticipate a negative relationship vis-à-vis our particulates, SO_2, and energy use measures.

The second group of variables under legal and economic context addresses various aspects of a country's technical capacity. Again, it is hard to measure scientific and technological sophistication directly, so we rely on a series of proxies, including ESI data on the number of scientists and engineers (scaled by population) in each country and *GCR* Survey data that gauge a country's technology position, strength of its scientific community, degree to which foreign technology is commonly licensed, intellectual property protection, research and development spending, willingness to absorb new technologies, business commitment to innovation, and government commitment to technology development and innovation. We expect each of these technical

capacity measures to correlate negatively with environmental impacts, because greater technical strength should lead to better environmental performance.

As noted above, the independent variables are far from perfect measures of the potential determinants of national environmental outcomes. These variables are the best ones currently available, however, and represent, in some cases, a significant improvement over earlier efforts to model the policy levers and other drivers of environmental performance. Despite their limitations, the data allow us to begin to identify empirically the variables that determine a nation's success in controlling pollution and improving energy efficiency.

Statistical Methodology

Our analytic approach unfolds in several stages. First, we use bilateral regressions (Tables 5, 6, and 7) to explore whether there is a statistically significant relationship between each independent variable and energy use, urban particulate levels, and SO_2 concentrations. Because many of the independent variables are collinear and the degrees of freedom are limited, multiple regression techniques cannot easily be used to examine the joint influence of all the variables. Instead, as a second stage of analysis, we "roll up" the significant independent variable in each category into a subindex using common factor analysis. Then, we regress these subindexes against the dependent variables.[8] (The percentage of covariance explained by the first factor and the first factor coefficient for each index variable is reported in Esty and Porter (2001)). Finally, the statistically significant category subindexes are rolled up into an overall environmental regulatory regime index (ERRI) and an overall economic and legal context index (ELCI).

In light of the significant association between per capita GDP and environmental performance, we also analyze performance relative to a peer group of countries defined by income level. We regress ERRI against GDP per capita (graphed in Figure 5) and calculate the residuals (distance above or below the regression line) for each country (Table 9). This allows us to analyze each country's performance against expectations established by its income level. We also examine the relationship between the ELCI and ERRI, and the relationship between ELCI and GDP per capita.

Results for Individual Measures and Indexes

The bilateral regression results are shown in Tables 5, 6, and 7, and the energy efficiency regressions are shown in Table 5. Many of the independent variables show a statistically significant relationship with energy use with the expected negative sign and a reasonable degree of explained variance. All of the elements of the regulatory stringency category show particular significance, as do the enforcement variables. Most of the regulatory structure measures also prove to be

Table 5

Bilateral Regressions: Energy Usage*

	2001 Dependent Variable: Energy Usage (per Unit GDP)			
	(b)	R²	Sig.	df
Environmental Regulatory Regime Index	**-5.281**	**0.67**	**0.000**	**68**
Stringency Subindex	-5.632	0.68	0.000	68
Air regulation	-4.044	0.69	0.000	68
Water regulation	-3.859	0.68	0.000	68
Toxic waste regulation	-3.576	0.67	0.000	68
Chemical regulation	-3.902	0.68	0.000	68
Overall regulation	-3.917	0.67	0.000	68
Regulatory Structure Subindex	-4.480	0.64	0.002	68
Options for compliance	-4.005	0.60	0.102	68
Confusing and changing	-4.982	0.65	0.001	68
Early or late	-4.058	0.67	0.000	68
Compliance Hurts or Helps competitiveness	-6.094	0.62	0.016	68
Regulation Adversarial or cooperative	-6.355	0.63	0.007	68
Information Subindex	-2.507	0.61	0.081	68
ESI-Variables %-available	-0.271	0.62	0.020	68
Sustainable development info	-1.009	0.58	0.764	41
Number of sectoral EIA guidelines	0.041	0.59	0.923	68
Number of environmental strategies & action plans	-0.197	0.59	0.815	68
Subsidies Subindex				
Government subsidies	7.065	0.66	0.000	68
Regulatory Enforcement Subindex	-4.466	0.65	0.001	68
Enforcement	-3.890	0.65	0.001	68
International agreements	-3.976	0.64	0.002	68
Environmental Institutions Subindex	-4.740	0.65	0.001	68
IUCN	-1.392	0.60	0.300	68
Memberships	-0.699	0.65	0.001	67
Prevalence of ISO 14000	-3.994	0.63	0.011	68
Economic and Legal Context Index	**-4.836**	**0.65**	**0.001**	**68**
Administrative Infrastructure Quality Subindex	-5.647	0.68	0.000	68
Civil liberties	-5.190	0.75	0.000	68
Public sector competence	-2.383	0.59	0.333	68
Government favor private sector firms	-4.200	0.64	0.003	68
Property rights	-4.756	0.71	0.000	68
Independent judiciary	-3.426	0.66	0.000	68
Irregular payments	-4.973	0.68	0.000	68
Legal framework	-3.880	0.66	0.000	68
Regulatory burden	-5.144	0.63	0.006	68
Level of administrative corruption	-5.695	0.69	0.000	68
Honoring of policies through government transition	-4.558	0.65	0.001	68
Scientific and Research Infrastructure Subindex	-3.788	0.63	0.008	68
Scientists and engineers	-0.003	0.64	0.004	64
Technology position	-3.636	0.66	0.000	68
Institutions	-3.341	0.62	0.018	68
Licensing of foreign technology	-3.692	0.61	0.055	68
Company R&D spending	-4.207	0.64	0.002	68
Willingness to absorb new technology	-3.803	0.62	0.033	68
Importance of innovation to revenue	-6.158	0.62	0.020	68
Government purchase decisions for technical products	-2.962	0.60	0.160	68

*Refer to Annex 1 for definitions of variables.

Table 6

Bilateral Regressions: Urban Particulates*

| | 2001 Dependent Variable: Urban Particulates (per City Pop) | | | |
	(b)	R²	Sig.	df
Environmental Regulatory Regime Index	**-58.19**	**0.44**	**0.000**	**40**
Stringency Subindex	-67.58	0.52	0.000	40
Air regulation	-46.86	0.52	0.000	40
Water regulation	-46.44	0.53	0.000	40
Toxic waste regulation	-45.10	0.52	0.000	40
Chemical regulation	-46.24	0.51	0.000	40
Overall regulation	-47.54	0.51	0.000	40
Regulatory Structure Subindex	-52.54	0.35	0.000	40
Options for compliance	-89.06	0.33	0.000	40
Confusing and changing	-60.31	0.42	0.000	40
Early or late	-45.23	0.47	0.000	40
Compliance hurts or helps competitiveness	-61.14	0.17	0.007	40
Regulation adversarial or cooperative	-46.15	0.12	0.028	40
Information Subindex	-56.07	0.22	0.002	40
ESI-variables %-available	-3.86	0.15	0.011	40
Sustainable development Info	-58.76	0.18	0.028	25
Number of sectoral EIA guidelines	-0.99	0.00	0.825	40
Number of environmental strategies & action plans	4.94	0.01	0.525	40
Subsidies Subindex				
Government subsidies	65.95	0.31	0.000	40
Regulatory Enforcement Subindex	-58.31	0.43	0.000	40
Enforcement	-52.79	0.45	0.000	40
International agreements	-49.93	0.38	0.000	40
Environmental Institutions Subindex	-47.86	0.29	0.000	40
IUCN	-16.40	0.05	0.150	40
Memberships	-6.40	0.22	0.002	39
Prevalence of ISO 14000	-47.01	0.25	0.001	40
Economic and Legal Context Index	**-58.94**	**0.40**	**0.000**	**40**
Administrative Infrastructure Quality Subindex	-57.48	0.39	0.000	40
Civil liberties	-42.67	0.37	0.000	40
Public sector competence	-42.49	0.07	0.095	40
Government favor private sector firms	-53.99	0.36	0.000	40
Property rights	-45.62	0.48	0.000	40
Independent judiciary	-32.47	0.30	0.000	40
Irregular payments	-59.91	0.46	0.000	40
Legal framework	-40.45	0.35	0.000	40
Regulatory burden	-47.93	0.15	0.013	40
Level of administrative corruption	-54.64	0.38	0.000	40
Honoring of policies through government transition	-43.16	0.24	0.001	40
Scientific and Research Infrastructure Subindex	-58.15	0.38	0.000	40
Scientists and engineers	-0.04	0.42	0.000	39
Technology position	-42.94	0.40	0.000	40
Institutions	-57.57	0.36	0.000	40
Licensing of foreign technology	-56.20	0.15	0.010	40
Company R&D spending	-49.65	0.32	0.000	40
Willingness to absorb new technology	-75.25	0.41	0.000	40
Importance of innovation to revenue	-63.51	0.15	0.012	40
Government purchase decisions for technical products	-68.82	0.26	0.001	40

*Refer to Annex 1 for definitions of variables.

Table 7

Bilateral Regressions: Urban SO$_2$ Concentration*

	2001 Dependent Variable: Urban SO$_2$ (per City Pop)			
	(b)	R^2	Sig.	df
Environmental Regulatory Regime Index	**-11.351**	**0.21**	**0.001**	**45**
Stringency Subindex	-13.857	0.28	0.000	45
Air regulation	-9.407	0.27	0.000	45
Water regulation	-9.592	0.28	0.000	45
Toxic waste regulation	-9.283	0.27	0.000	45
Chemical regulation	-9.538	0.27	0.000	45
Overall regulation	-9.839	0.27	0.000	45
Regulatory Structure Subindex	-9.686	0.16	0.005	45
Options for compliance	-9.312	0.05	0.130	45
Confusing and changing	-11.905	0.20	0.002	45
Early or late	-10.105	0.28	0.000	45
Compliance hurts or helps competitiveness	-11.584	0.09	0.038	45
Regulation adversarial or cooperative	-11.128	0.11	0.022	45
Information Subindex	-10.206	0.10	0.029	45
ESI-variables %-available	0.207	0.00	0.662	45
Sustainable development Info	-21.624	0.25	0.004	29
Number of sectoral EIA guidelines	-0.708	0.01	0.464	45
Number of environmental strategies & action plans	0.722	0.00	0.732	45
Subsidies Subindex				
Government subsidies	12.301	0.15	0.008	45
Regulatory Enforcement Subindex	-10.989	0.18	0.003	45
Enforcement	-8.960	0.17	0.004	45
International agreements	-10.221	0.19	0.003	45
Environmental Institutions Subindex	-6.921	0.08	0.053	45
IUCN	-6.270	0.10	0.030	45
Memberships	-0.684	0.04	0.194	44
Prevalence of ISO 14000	-8.027	0.10	0.034	45
Economic and Legal Context Index	**-11.738**	**0.19**	**0.002**	**45**
Administrative Infrastructure Quality Subindex	-12.815	0.23	0.001	45
Civil liberties	-12.206	0.47	0.000	45
Public sector competence	-3.364	0.01	0.553	45
Government favor private sector firms	-10.056	0.15	0.008	45
Property rights	-9.644	0.27	0.000	45
Independent judiciary	-7.166	0.18	0.003	45
Irregular payments	-12.413	0.26	0.000	45
Legal framework	-9.343	0.23	0.001	45
Regulatory burden	-9.259	0.10	0.032	45
Level of administrative corruption	-12.877	0.27	0.000	45
Honoring of policies through government transition	-8.685	0.11	0.021	45
Scientific and Research Infrastructure Subindex	-10.010	0.14	0.009	45
Scientists and engineers	-0.006	0.09	0.038	45
Technology position	-7.931	0.18	0.003	45
Institutions	-8.883	0.11	0.025	45
Licensing of foreign technology	-11.980	0.08	0.049	45
Company R&D spending	-7.802	0.12	0.020	45
Willingness to absorb new technology	-15.067	0.20	0.002	45
Importance of innovation to revenue	-15.770	0.13	0.011	45
Government purchase decisions for technical products	-9.316	0.06	0.109	45

*Refer to Annex 1 for definitions of variables.

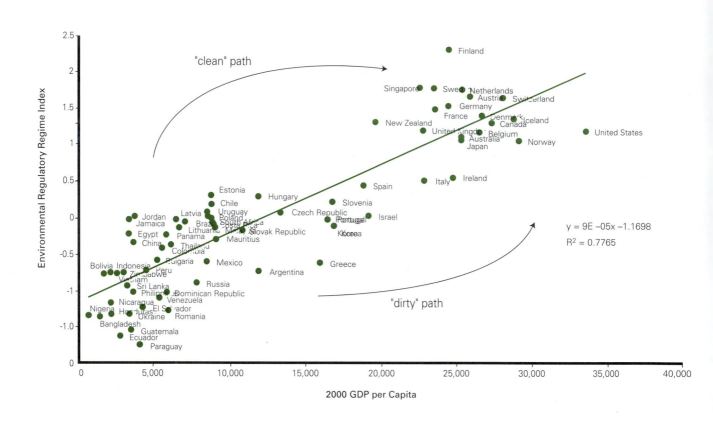

Figure 5

Relationship between the Environmental Regulatory Regime Index and GDP per Capita

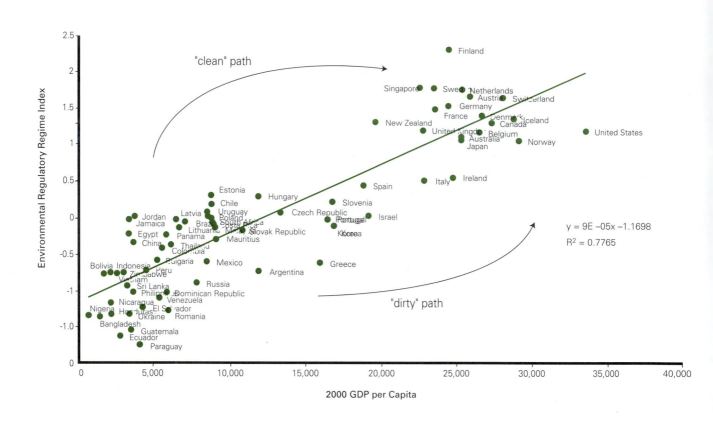

highly significant. These categories of variables account for the highest amount of explained variance. The subsidies variable is highly significant and has the expected positive sign. Consistent with economic theory, this result suggests that mispriced resources will be used inefficiently, and that subsidies represent a major policy error.

The information and institutions measures do not perform as well. In the information category, one variable (percentage of ESI variables available) emerges as significant, while the other three measures do not. In the institutional category, IUCN membership fails to show significance, while the other two measures of institutional capacity are significant.

Among the economic and legal context variables, all but one (public-sector competence) emerge as highly significant with the expected negative sign. The variables measuring corruption and whether new governments honor the commitments of earlier administrations prove to be statistically significant. In the scientific and technical capacity category, all of the variables except one (government commitment to technology development and innovation) show a reasonable degree of significance and the expected negative sign.

To build the indexes and subindexes, we use only statistically significant variables. All of the subindexes are highly significant in explaining energy use, have the expected negative sign, and account for substantial explained variance. The ERRI and ELCI register similarly high levels of significance with the expected negative signs, and a substantial degree of explained variance.

Although preliminary, the latter results provide some empirical support for the hypothesis that a nation's underlying economic and legal structure may be as important to environmental success as the specific details of its environmental regulatory regime. This conclusion argues for more attention to "fundamentals," such as eliminating corruption and building functioning market economies, and to "governance," such as strengthening the rule of law and developing mechanisms to protect property rights, in setting development priorities and targeting development assistance. It is interesting to note that this is the direction recent policies of the United Nations Development Programme are taking.

The ERRI and the ELCI prove to be highly correlated and show similar levels of significance and explained variance. Hence, it appears that environmental regulation and overall economic

Table 8

Environmental Regulatory Regime Index by Country, Absolute Ranking

Environmental Regulatory Regime Index		
Rank	Country	Score
1	Finland	2.303
2	Sweden	1.772
3	Singapore	1.771
4	Netherlands	1.747
5	Austria	1.641
6	Switzerland	1.631
7	Germany	1.522
8	France	1.464
9	Denmark	1.384
10	Iceland	1.354
11	New Zealand	1.299
12	Canada	1.297
13	United Kingdom	1.185
14	United States	1.184
15	Belgium	1.159
16	Australia	1.083
17	Japan	1.057
18	Norway	1.045
19	Ireland	0.546
20	Italy	0.498
21	Spain	0.437
22	Estonia	0.296
23	Hungary	0.283
24	Slovenia	0.209
25	Chile	0.177
26	Czech Republic	0.073
27	Uruguay	0.059
28	Israel	0.021
29	Poland	0.005
30	Jordan	0.002
31	Portugal	-0.028
32	South Africa	-0.029
33	Latvia	-0.036
34	Jamaica	-0.037
35	Brazil	-0.077
36	Costa Rica	-0.078
37	Korea	-0.121
38	Malaysia	-0.127
39	Lithuania	-0.146
40	Slovak Republic	-0.177
41	Egypt	-0.224
42	Panama	-0.242
43	Mauritius	-0.290
44	China	-0.348
45	Thailand	-0.389
46	Colombia	-0.416
47	Bulgaria	-0.584
48	Mexico	-0.602
49	Greece	-0.619
50	Peru	-0.722
51	Argentina	-0.732
52	Zimbabwe	-0.732
53	Bolivia	-0.743
54	Indonesia	-0.758
55	India	-0.759
56	Vietnam	-0.770
57	Russia	-0.895
58	Sri Lanka	-0.936
59	Philippines	-1.014
60	Dominican Republic	-1.014
61	Venezuela	-1.079
62	Nicaragua	-1.164
63	El Salvador	-1.215
64	Romania	-1.268
65	Ukraine	-1.297
66	Honduras	-1.300
67	Nigeria	-1.314
68	Bangladesh	-1.331
69	Guatemala	-1.532
70	Ecuador	-1.616
71	Paraguay	-1.743

and legal context generally improve in parallel. We explored the joint influence of ERRI and ELCI on environmental performance. In practice, the high correlation between the two indexes (as shown in Figure 7) means that their effects on energy use could not be distinguished statistically.

Table 6 presents the second set of bilateral regressions for urban particulate concentrations. Again, the vast majority of variables are significant with the expected sign and account for a reasonable degree of explained variance. All of the measures of regulatory stringency and structure are highly significant, with the stringency variables accounting for the greatest level of explained variance. The subsidies measure is highly significant, has the anticipated positive sign, and accounts for a reasonable degree of explained variance.

In the information category, two variables emerge as significant with the expected negative sign, but do not account for as high a degree of explained variance. In the institutional reinforcement category, the number of IUCN memberships is, again, not significant, while the other two variables (participation in intergovernmental environmental bodies and corporate participation in environmental management systems) emerge as highly significant.

The regulatory stringency, regulatory structure, information, enforcement, and institutional subindexes all emerge as highly significant with the expected negative sign, as does the cumulative ERRI. Across all of these subindexes, however, the degree of explained variance is somewhat lower in the urban particulate regressions than in the energy use ones. Two of the subindexes—information foundations and institutional reinforcement—perform notably less well than the others. This may reflect the fact that these variables are imperfect proxies or that information and institutions play more mixed roles.

All of the variables in the economic and social context regression emerge as significant in the urban particulates regulations. All have the expected negative sign, with many accounting for a substantial degree of explained variance. The administrative infrastructure and technical capacity subindexes both show very high levels of significance, the expected negative sign, and a high degree of explained variance. The ELCI similarly emerges as highly significant and accounts for almost as much explained variance as the ERRI. However, both the ERRI and the ELCI explain a somewhat smaller proportion of variations in urban particulate concentrations than energy use. Again, the independent effects of ERRI and ELCI could not be distinguished statistically.

The SO_2 regression results are presented in Table 7. Most of the independent variables are once again significant with the expected negative sign. The degree of explained variance is

Table 9

Environmental Regulatory Regime Index Relative to Expected Results Given GDP per Capita, Listed by Income Groups

Low Income Countries (≤ $6,500)			Middle Income Countries ($6,500–23,000)			High Income Countries (≥ $23,000)		
Rank	Country	Residual	Rank	Country	Residual	Rank	Country	Residual
1	Jordan	0.794	1	Singapore	0.806	1	Finland	1.165
2	Jamaica	0.793	2	Estonia	0.614	2	Sweden	0.725
3	Egypt	0.612	3	New Zealand	0.612	3	Netherlands	0.541
4	China	0.455	4	Latvia	0.499	4	France	0.404
5	Panama	0.355	5	Chile	0.494	5	Germany	0.377
6	Vietnam	0.216	6	Brazil	0.407	6	Austria	0.368
7	Colombia	0.204	7	Uruguay	0.402	7	United Kingdom	0.202
8	Bolivia	0.204	8	Lithuania	0.374	8	Switzerland	0.154
9	India	0.188	9	Poland	0.343	9	Denmark	0.037
10	Zimbabwe	0.187	10	Hungary	0.308	10	Canada	-0.112
11	Thailand	0.180	11	South Africa	0.288	11	Australia	-0.138
12	Indonesia	0.132	12	Costa Rica	0.235	12	Japan	-0.168
13	Bulgaria	0.078	13	Malaysia	0.214	13	Belgium	-0.173
14	Peru	0.002	14	Mauritius	-0.003	14	Iceland	-0.184
15	Sri Lanka	-0.092	15	Czech Republic	-0.031	15	Italy	-0.495
16	Philippines	-0.211	16	Slovak Republic	-0.032	16	Norway	-0.523
17	Nicaragua	-0.217	17	Spain	-0.175	17	Ireland	-0.623
18	Nigeria	-0.225	18	Slovenia	-0.211	18	United States	-0.792
19	Bangladesh	-0.307	19	Mexico	-0.259			
20	Honduras	-0.359	20	Portugal	-0.426			
21	Dominican Republic	-0.397	21	Russia	-0.487			
22	Venezuela	-0.436	22	Korea	-0.558			
23	El Salvador	-0.461	23	Israel	-0.626			
24	Ukraine	-0.470	24	Argentina	-0.705			
25	Romania	-0.684	25	Greece	-0.964			
26	Guatemala	-0.714						
27	Ecuador	-0.730						
28	Paraguay	-0.981						

generally much lower for SO_2, however, than for either energy use or particulate concentrations. This finding may reflect the fact that the benefits of SO_2 control (reduced acid rain) accrue downwind—frequently beyond the territorial boundaries of the jurisdiction undertaking regulatory action. Thus, from a cost-benefit perspective, the regulating entity has less to gain there than in controlling particulates or investing in energy efficiency, both of which provide more localized benefits.

The subsidies measure again shows a high level of significance and the expected positive sign, but accounts for less variance than with the other measures of pollution. In the information category, three of the four measures are not statistically significant. Again, the looser fit may suggest that even a well-informed government that is serious about environmental protection has less of an incentive to address SO_2, given its geographic dispersion, than other, more localized issues.

All of the environmental regulatory regime subindexes are significant and have the expected negative sign in the SO_2 regressions. Only the regulatory stringency subindex, however, accounts for a reasonable degree of explained variance. ERRI once again proves to be highly significant, although the degree of explained variance is not high. As a general matter, the regression fit for SO_2 appears weaker than for particulates or energy use, perhaps reflecting the more limited regulatory payoffs noted above.

Among the variables in the economic and legal context grouping, all but one (public sector competence) emerge with high statistical significance and the expected negative sign in the SO_2 regressions. Some of the measures account for a reasonable degree of explained variance (for example, civil liberties, property rights, and irregular payments). In general, administrative infrastructure variables show greater significance and higher degrees of explained variance than technical capacity measures. The administrative infrastructure subindex is highly significant with a reasonable degree of explained variance. The technical capacity subindex shows a high degree of significance but does not account for an especially large amount of explained variance. The overall ELCI is significant and explains a reasonable amount of the variance in SO_2 concentrations.

Ranking Environmental Regulatory Quality

The bilateral, subindex, and index regressions establish a statistically significant relationship between the various policy measures and environmental performance. The next stage in the analysis is to use ERRI to explore the differences across countries in environmental regulatory quality.

Table 8 presents countries ranked by absolute ERRI scores. This index (combining the always significant regulatory

Figure 6

Relationship between the Environmental Regulatory Regime Index and Economic and Legal Context Index

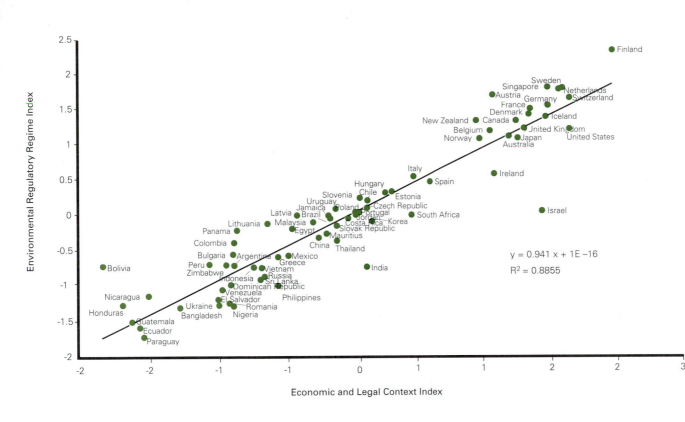

$y = 0.941 x + 1E{-16}$

$R^2 = 0.8855$

stringency, structure, subsidies, and enforcement subindexes) represents a summary performance measure of the quality of the environmental regulatory system in a country. Among the top-ranked countries are Finland, Sweden, and Singapore; countries at the bottom include Guatemala, Ecuador, and Paraguay.

Given the significant relationship between level of development and environmental performance, we would expect a similar relationship with environmental regulatory quality. What is most interesting in Table 8, then, is not so much the fact that Finland outranks Paraguay on the stringency of environmental regulation, but the reasons why countries with similar incomes perform so differently. For instance, why does Costa Rica (36th place) do better than Panama (42nd place) and Peru (50th place)? Similarly, why do Spain (21st) and Portugal (31st) outperform Greece (49th) so dramatically? Likewise, Chile (25th) distinctly outperforms Argentina (51st), and Poland (29th) comes in way ahead of Russia (57th). The last two pairings reveal a general pattern suggesting that more aggressively market-oriented economies (Chile and Poland) may outperform those (Argentina and Russia) where a more interventionist economic tradition persists.

To control for income differences and, hence, level of economic development, Table 9 ranks countries by their residuals from the regression of ERRI and GDP per capita (plotted in Figure 5). This relative ranking represents a measure of environmental regulatory quality relative to expectations established by income level. Among low-income countries, Jordan and Jamaica come out on top, while Ecuador and Paraguay trail. Among middle-income countries, Singapore, Estonia, and New Zealand rank high, and Israel, Argentina, and Greece lag. Among the wealthiest nations, Finland, Sweden, and the Netherlands lead, while Italy, Norway, and Ireland rank low. The United States occupies the bottom rung of the high-income group ladder.

As noted earlier, ERRI and ELCI are highly correlated, as shown in Figure 6. Nevertheless, it is evident that some countries have advanced their economic and legal context ahead of their environmental regulatory quality, while others have pushed environmental regulation faster than the overall context. In Israel, India, Ireland, the United States, South Africa, the Philippines, and Nigeria, environmental regulation lags overall economic/legal context, while in Finland, Austria, New Zealand, Panama, and Bolivia, environmental regulatory quality is ahead of improvements in the broader economic and legal context.

The Relationship between Environmental Performance and Competitiveness

Finally, we turn to the question of whether environmental regulatory stringency detracts from or contributes to economic progress. Figure 7 shows that the quality of a nation's

Figure 7

Relationship between the Environmental Regulatory Regime Index and Current Competitiveness

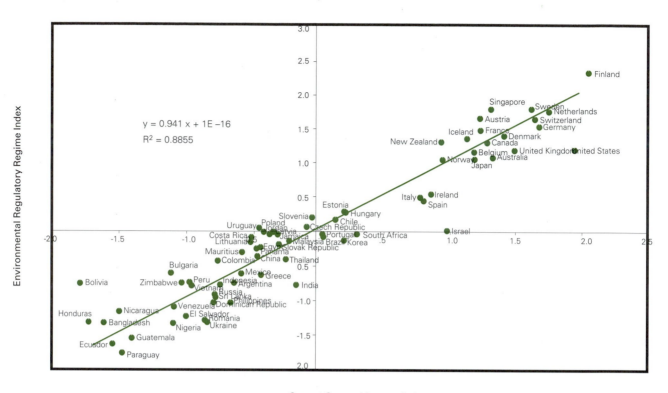

Current Competitiveness Index

environmental regulatory regime correlates strongly and positively with its competitiveness. Many of the nations with top-tier competitiveness rankings also have strong environmental performance scores. Finland, for example, ranks at the top of the ERRI and at the top of the current competitiveness index (CCI). The United States stands out as an exception, with a high competitiveness rank and a relatively low environmental regulation score. Figure 5 tells a similar story about how high levels of per capita income and economic development correlate with high environmental regulatory quality.

The correlations revealed in Figures 5 and 7 do not, of course, prove causation. But the finding that a strong environmental regulatory regime is not inconsistent with top-tier economic performance is itself interesting. Indeed, the fact that the top environmental performers do not appear to have suffered economically strongly supports the "soft" version of the "Porter hypothesis," which argues that environmental progress can be achieved without sacrificing competitiveness (Porter, 1991; Porter and van der Linde, 1995). Testing the "hard" version of this hypothesis—that countries with forward-leaning environmental policies and programs actually will enhance their competitiveness—requires time series data that are not yet available.

Figure 5 highlights the development policy choice that every nation faces. Countries would like to move from the lower-left corner of the chart (which represents low levels of environmental performance and low national income) to the upper-right quadrant (which represents high levels of environmental performance and high income). The question is what path to take. Or, to put it differently, must the environment be sacrificed to achieve economic progress? Those countries above the regression line can be seen as having chosen a "clean" development trajectory in which environmental regulatory quality advances ahead of economic advancement. Those below the line have chosen a relatively "dirty" path to growth, with relatively lax environmental regulation in the hope of growing faster.

In addressing this choice, we are able to provide a crude test using the available data. We regress a number of control variables on GDP per capita growth between 1995 and 2000, including the initial level of GDP per capita, gross fixed capital formation as a percent of GDP, and government spending as a percent of GDP. We then introduce a variable, which measures the residual from the regression of ERRI on GDP per capita (Table 10). Countries with positive residuals have higher ERRI scores than would be expected given their income, and vice versa. The residual has a positive sign with significance at virtually the 90 percent level. While tentative, this result suggests that the "clean" model may be more successful. Countries that pursue a stringent regulatory regime appear to achieve more rapid growth. While also tentative, this result suggests the possible superiority of the "clean" model. However, more years of data and better controls will be necessary to validate this finding.

Conclusion

The results presented here must be seen as preliminary. The data available suffer from many limitations that narrow the feasible statistical approaches. Precise causal linkages, moreover, remain unproven. A central conclusion of our research is that better environmental data are required at the global, national, local, and corporate levels if a more systematic approach to environmental improvement is to be implemented. A worldwide commitment to improved environmental data should be adopted as a priority initiative.

With these caveats, however, the relationships that do emerge as statistically significant are striking. The analysis provides considerable empirical evidence that cross-country differences in environmental performance are associated with the quality of the environmental regulatory regime in place. We find that the rigor and structure of environmental regulations have particular impact, as does the degree of emphasis on enforcement. The damaging effect of subsidies is also clear. While developing a strong and sophisticated regulatory regime that fully internalizes externalities presents real challenges, ending price-distorting, inefficiency-creating, and pollution-inducing subsidies is within the policy grasp of every nation. Environmental performance appears to improve with certain kinds of information and to the extent that a nation's environmental regime is reinforced by an environmentally oriented private sector and broad-based relationships with international environmental bodies. Information and institutions have somewhat less impact on environmental performance based on our analysis. This finding may be due in part to weaknesses in the available data.

Our results also suggest that environmental performance requires improvements in a country's institutional foundations. In practice, a nation's economic and legal context and its environmental regulatory regime go hand in hand. This association demands further exploration, but the preliminary evidence developed here suggests that countries would benefit *environmentally* from an emphasis on developing the rule of law, eliminating corruption, and strengthening their governance structures.

The strong association between income and environmental performance also carries important implications. Among other things, it provides powerful corroboration for a policy emphasis on poverty alleviation and promotion of economic growth as a key mechanism for improving environmental results.

The empirical evidence developed here suggests that the antiglobalization arguments of environmental protesters from Seattle to Genoa are off the mark. Limiting trade and the engagement of developing countries with the rest of the world is a recipe for environmental failure, not success. Rather, the more fully a country moves to modernize its economy, institutional structures, and regulatory system, the more quickly its environmental performance appears to improve—along with improvements in per capita income (Esty, 1997).

The country rankings that emerge from our analysis largely seem to square with observed reality. The variations in performance highlight the fact that countries vary widely in their environmental outcomes and policy choices, even after controlling for level of income. There are clearly better and worse ways to approach pollution control and natural resource management. The data provided here offer some important clues about where the search for "best practices" should begin. Moreover, our findings suggest that the environment need not be sacrificed along the road to economic progress. Quite to the contrary, the countries that have the most aggressive environmental policy regimes also seem to be the most competitive and economically successful. We also find preliminary evidence that a stringent environmental regime relative to income may speed up economic growth rather than detract from it.

This study highlights the promise of a more analytically rigorous approach to policymaking. More fundamentally, our analysis strongly supports the notion that the uncertainties that plague environmental decision-making can be narrowed, and that current levels of policy contention could be reduced as well. The environmental domain need not rely on guesswork.

Our preliminary efforts to use statistical methods to explain environmental successes and failures seem to confirm some aspects of the prevailing wisdom. For example, poverty emerges as a source of serious environmental degradation and, thus, deserves ongoing policy attention. Subsidies not only skew prices and distort trade, they also lead to inefficient production and unnecessary pollution. But some new priorities also emerge from this research. Notably, there appear to be significant gains to be had by moving environmental laggards toward the best practices of those jurisdictions whose performance is top-tier. This argues for much greater strategic emphasis on information development and dissemination. Likewise, the significance of economic and legal context to environmental results argues for a new focus on governance as the foundation for both environmental and economic progress. The results here suggest that there are ways to move beyond the ideological and emotional obstacles that stand in the way of faster environmental progress.

References

Dua, André, and Daniel C. Esty, *Sustaining the Asia-Pacific Miracle: Environmental Protection and Economic Integration,* Washington: Institute for International Economics, 1997.

Esty, Daniel C., "Environmental Protection during the Transition to a Market Economy," in Wing Woo, Stephen Parker, and Jeffrey Sachs, eds., *Economies in Transition: Asia and Europe*, Cambridge: MIT Press, 1997.

Esty, Daniel C., "NGOs at the World Trade Organization: Cooperation, Competition, or Exclusion?" *Journal of International Economic Law* 1 (March 1998).

Esty, Daniel C., and Michael E. Porter, "Measuring National Environmental Performance and Its Determinants," in Michael E. Porter, Jeffrey D. Sachs et al., eds., *The Global Competitiveness Report 2000,* New York: Oxford University Press, 2000.

Grossman, Gene M., and Alan B. Krueger, "Economic Growth and the Environment," *Quarterly Journal of Economics* 110 (May 1995), pp. 353–375.

Harbaugh, William, Arik Levinson, and David Wilson, "Re-examining the Empirical Evidence for an Environmental Kuznets Curve," NBER Working Paper No. 7711 (May 2000).

Jaffe, Adam B., Steven R. Peterson, Paul R. Portney, and Robert N. Stavins, "Environmental Regulation and the Competitiveness of U.S. Manufacturing: What Does the Evidence Tell Us?" *Journal of Economic Literature* 33 (March 1995), pp. 132–161.

Porter, Michael E., America's Green Strategy," *Scientific American* 264 (April 1991), p. 168.

Porter, Michael E., and Claas van der Linde, "Green and Competitive: Ending the Stalemate," *Harvard Business Review* 73 (September/October 1995), pp. 120–155.

Porter, Michael E., et al., "The Current Competitiveness Index," in Michael E. Porter, Jeffrey D. Sachs et al., eds., *The Global Competitiveness Report 2000,* New York: Oxford University Press, 2000.

Sachs, Jeffrey, "Globalization and the Rule of Law," *Yale Law School Occasional Papers* 2d series, no. 4, 1998.

World Commission on Environment and Development, *Our Common Future*, Oxford: Oxford University Press, 1987.

Endnotes

1 An earlier version of this study appears in Daniel C. Esty and Michael E. Porter, "Ranking National Environmental Regulation and Performance: A Leading Indicator of Future Competitiveness?" in Michael E. Porter, Jeffrey Sachs, et al., *Global Competitiveness Report 2001–2002* (New York: Oxford University Press, 2001). The analysis reported on here was first developed in Daniel C. Esty and Michael E. Porter, "Measuring National Environmental Performance and Its Determinants," in Michael E. Porter, Jeffrey Sachs, et al., *Global Competitiveness Report 2000*. As is explained in the pages that follow, the present study incorporates data and variables from the *Global Competitiveness Report 2001–2002* Annual Survey and World Economic Forum Global Leaders for Tomorrow Environmental Task Force, 2001. *Environmental Sustainability Index 2001* (Geneva: World Economic Forum, 2001) (available at www.yale.edu/envirocenter/esi) (hereinafter *ESI 2001*).

2 For further discussion of the data gaps that plague the environmental domain, see *ESI 2001*.

3 Again, the lack of systematic environmental data gathering in many countries and the limited information available with regard to a number of key issues constrains our model. Filling these data gaps—in terms of both depth and breadth—should be a policy priority. Better data remain a prerequisite for a more analytically rigorous approach to environmental decision-making.

4 This project, undertaken by the World Economic Forum's Global Leaders for Tomorrow Environmental Task Force, with the support of the Yale University Center for Environmental Law and Policy and the Center for International Earth Science Information Network (CIESIN) at Columbia University, ranks 122 countries on their "environmental sustainability" based on performance in 22 categories building on a dataset of 65 underlying variables (ESI 2001).

5 Both the United States and the United Kingdom track smaller particulates than the rest of the world. The U.S. and U.K. emphasis follows the most recent medical evidence, which suggests that smaller particles penetrate deep into the lungs and present a greater health threat.

6 Energy use also has highly localized effects insofar as efficiency directly affects competitiveness, and some of the harms (particulates and other local air pollutants) do not spread geographically.

7 The 2001 Survey, undertaken jointly by the World Economic Forum and Harvard University's Center for International Development and Institute for Strategy and Competitiveness, builds on questionnaire responses from more than 4,000 business, government, and nongovernmental organization leaders in 70 countries.

8 In developing the category subindexes, we use only those variables that appear appropriately grouped based on Eigen Value analysis. Thus, in developing the regulatory stringency subindex, we drop the overall regulation measure. The sectoral Environmental Impact Assessment (EIA) guidelines measure and the environmental strategies and action plans measure drop out of the information subindex. The measures of civil liberties, public sector competence, irregular payments, and regulatory burden are all dropped from the administrative infrastructure subindex, and the scientists/engineers, licensing of foreign technology, and business innovation measures fall out of the technical capacity subindex.

Chapter 4

Bridging the Gap: How Improved Information Can Help Companies Integrate Shareholder Value and Environmental Quality

Forest Reinhardt

44

Environmental problems do not look the same from the perspective of a corporate executive as they do from the point of view of government officials. The most important differences arise because the corporate executive is ordinarily answerable to shareholders who expect him or her to deliver financial returns on the money they have invested in the enterprise. Hence any environmental investments that the firm undertakes must normally be justified in terms of shareholder value creation or reduction of financial risk to the firm.[1]

As a result, company executives and government officials often use the same words to mean very different things. To a government official, for example, "sustainability" is likely to call to mind to the Brundtland report (World Commission on Environment and Development, 1987), with its injunctions not to ignore the linkages between economics and the environment, and not to leave later generations worse off to improve the lot of the present. To a company executive, by contrast, "sustainability" is the ability to deliver superior returns to shareholders over multiyear periods in the face of market saturation, disruptive technological change, and other competitive threats.[2] Similarly, "environmental risk" to a

government official means the risk to human health and ecosystems posed by pollution, resource degradation, or other environmental insults. To a company official, the same phrase is far more likely to mean the financial risk borne by the firm's shareholders as a result of potential environmental regulation or litigation.[3]

Because of these differences in perspective, the informational needs of corporate executives in the environmental arena are not identical to those of environmental managers in the government. Most of the other chapters in this book discuss the information that would be useful to public policy-makers at the regional, national, and international levels in devising better environmental regulatory institutions. While information of this sort is likely to be valuable to corporate executives as well, the corporate executives need, in addition, information that enables them to draw the connections between their own behavior in the environmental arena and the creation of shareholder value that is necessarily their main concern.

This chapter discusses environmental information needs and opportunities at the level of the corporation. It starts with a review of the basic structure of environmental problems as seen from a business perspective. This structure allows us to identify several strategic possibilities that can, *under some circumstances,* permit firms to reconcile shareholder value maximization with provision of improved environmental quality. The chapter then discusses these strategic possibilities in turn; in examining each one, it emphasizes the information that managers need to determine the possibility's applicability to their own circumstances, and to implement it if this appears to be warranted. The chapter concludes with some more general observations about corporate environmental performance that follow from the analysis.

Public Goods and Shareholder Value

The debate on business and the environment has been framed in simplistic yes-or-no terms: Does it pay to be green? The inapplicability of such categorical, all-or-nothing thinking would be obvious in almost any other business context. Does it pay to build your next plant in Singapore? To increase spending on research and development? To sue your competitors for patent infringement? The answer, of course, is: it depends. And so it is with environmental questions: the right policy depends on the circumstances confronting the company and the strategy it has chosen.

To assert that it always pays to be green is to confuse what is desirable with what is realistic. It would be convenient if environmental problems could always be addressed profitably. In that case, only the myopia of corporate managers would prevent them from seizing an unlimited number of "win-win" opportunities for environmental improvement.

Unfortunately, environmental problems do not automatically create opportunities to make money. The reason is that pollution and some forms of natural resource depletion are classic examples of externalities: that is, social costs that are not automatically reflected in private market prices. Put another way, environmental quality is ordinarily a public good, like street lighting or national defense. The reason that companies can make money producing private goods and services like hamburgers and haircuts is that it is very easy to refuse to provide them to consumers who don't offer payment in exchange. But once a public good is produced, there is no practical way to exclude those who have not contributed to its provision from enjoying it.

At the same time, the opposite stance—that it never pays for a company to invest in improving its environmental performance—is also incorrect. The reason is that, while externalities and public goods are examples of market failure (or, put another way, of departures from the assumptions of perfectly competitive markets), those market failures don't exist in isolation. The same markets that are imperfect because of externalities are also imperfect because economies of scale and scope limit the number of producers, because not all market players have access to the same information, and so on. (Market failures involving incomplete information are obviously of particular interest in the context of this chapter and others in this book.) These other market failures can intensify the tension between environmental and profit objectives, but they may also ease this tension, creating opportunities for firms to enhance shareholder value while helping to solve environmental problems.

How might this occur? There are five approaches that companies can take to reconcile shareholder value and environmental quality. Some companies may be able to improve their environmental performance and reduce their own private costs simultaneously. Others can distance themselves from competitors by differentiating their products and commanding higher prices for them. Still others may be able to "manage" their competitors by imposing a set of private regulations or by helping to shape the rules written by government officials. And some companies may even be able to make systemic changes that will redefine competition in their markets to their own benefit and that of the environment. Finally, almost all firms can learn to improve their management of risk and thus reduce the outlays associated with accidents, lawsuits, or boycotts. In the long run, these improvements in environmental risk management will prove a source of competitive advantage.

The degree to which any of these approaches is feasible depends on the structure of the industry in which the firm competes and on the firm's competitive position within that industry. Industries with differentiated products, price-insensitive customers, or small numbers of producers are more

likely to be favorable settings for reconciling environmental protection and shareholder value. Similarly, market leaders by definition have more power to alter the rules of competition in an industry than followers in the same industry. Regulatory structures obviously affect firms' latitude as well. Tighter regulation forces firms' performance to converge to the regulatory norm; this imposes costs on all of them but, at the same time, reduces the latitude for firms to differentiate themselves along environmental lines. Similarly, firms in industries in which regulatory expectations are rapidly changing may find more opportunities for environmental investments that deliver value to shareholders. This is true even for global environmental problems, such as climate change, for which regulatory frameworks have not yet been developed.

The appeal of any of the five approaches depends on the time horizon over which they are evaluated. "Proactive" environmental strategies often involve the purchase of real options: capping a contingent liability, for example, or staking out a position in a market niche that might grow rapidly in response to regulatory pressure or changing customer demands. Hence the environmental policies that maximize short-run cash flow may not be the ones that position the company optimally for the long run. This is true of business strategies in general, of course, but it especially applies to the environmental arena because benefits from environmental investments are often realized over long periods.

Put another way, environmental problems emerge because private costs diverge from social costs. Over the long term, on average, one would expect private and social costs to converge as societies create more efficient institutions for environmental management (the manner in which they can do so is the topic of much of the rest of this book). But in the short term, firms that try voluntarily to bring private costs in line with social costs (for example, by reducing pollution more than is required by law) may confront problems in the marketplace if their competitors don't follow their example. Successful environmental strategy involves the identification and exploitation of other (nonenvironmental) market imperfections, so that the company can weather any short-term dislocations in pursuit of long-term competitive and environmental benefits.

Thinking about the five approaches discussed in this paper should help managers apply mainstream business principles to environmental problems in their own industries systematically and realistically. These approaches can enable some companies to deliver increased value to shareholders while improving their environmental performance. But to evaluate the likelihood that they can profit from one or more of these approaches, managers need better information of two sorts: on the environmental costs and benefits of current and potential future practices, and on the relative private costs, industry economics, and competitive dynamics that are the bread

45

and butter of traditional strategists. In other words, since they are trying to bridge the gap between private and social costs, they need information on costs of both types.

Reducing Costs

One widely discussed approach to reconciling shareholder value with environmental management focuses on internal cost reductions. Some companies are able to cut costs and improve environmental performance simultaneously. The idea that such opportunities are widespread is often called "the Porter hypothesis."[4]

Consider Xerox's efforts. After nearly three decades of market dominance, the company found its traditional markets crowded in the late 1980s with well-funded new entrants. Xerox's market share declined, and its margins eroded precipitously. In 1990, the company's managers responded with a new environmental management initiative—the Environmental Leadership Program—that eventually included waste reduction efforts, product "takeback" schemes, and design-for-environment initiatives. By the mid-1990s, Xerox's large manufacturing complex in Webster, New York, was sending only 2 percent of its hazardous waste to landfills and saving substantial amounts in both waste management and raw materials purchasing. Even in the early 1990s, before the more ambitious components of their program had a chance to bear much fruit, Xerox's executives were already labeling the program an unqualified success.

Xerox's story illustrates a common pattern: dramatic cost savings are often found when a company is under tremendous pressure. As long as Xerox was the unchallenged market leader, it could afford to be easygoing about cost management—and it was. When the going got rough, it rose to the occasion with creative initiatives.

Observers of this pattern have wondered whether stringent environmental regulation could put the same kind of pressure on companies that market competition does. They argue that "free" opportunities to improve environmental performance—for which the benefits to the firm exceed the costs, even if the environmental benefits are ignored—are ubiquitous, and that stricter regulatory requirements or changes in the tax code could force companies to uncover them. Others disagree. They point out that managers are paid to minimize costs and wonder how adding new regulatory constraints could possibly reduce costs. Economists call this dispute the "free lunch" debate. The underlying issue is the appropriate level of government regulation. (See Porter and van der Linde 1995, Palmer et al. 1995.)

The "free lunch" advocates overstate their case. Even low-hanging fruit can only be gathered after an investment of management time, and that resource is hardly free. Investments in environmental improvement, like all other investments, are worthwhile only if they deliver value even after all the management costs have been included. Often, in fact, the supposed "free lunches" discovered by managers arise only because the government has altered, through regulation, the private cost of certain activities like waste disposal. Obviously, the government can increase the cost of any private activity it chooses to, but this doesn't mean that those cost increases are welfare-enhancing. They will enhance welfare only if the unregulated cost of the activity is lower than its social cost.

Fortunately, companies can remain agnostic on the question of whether free opportunities to improve environmental performance are widespread. From a business point of view, even if such opportunities are rare, managers should look for them as long as the search doesn't cost too much in time or other resources.

To evaluate the degree to which environmental cost-cutting strategies are likely to be successful economically, firms need to assess the private costs and benefits of their current activities and possible alternatives. In addition, note that many environmental cost-cutting initiatives turn out, on closer inspection, to be driven by regulatory compliance. This implies that firms also need to analyze the social costs and benefits of their activities, and the ways in which regulation might alter the magnitude and distribution of those benefits and costs, to determine where political incentives for further government intervention are greatest. This risk management problem is discussed in more detail below.

As environmental costs increase as a fraction of total costs, and as improvements in technology drive down the cost of collecting and analyzing information, it will pay to track environmental costs in a more disaggregated way. Instead of lumping them into an overhead category to be allocated on some arbitrary basis, firms can isolate the environmental cost impacts of particular processes and assign those costs to particular products. This information is obviously useful in identifying priorities for cost reduction. It is also likely to be useful in determining whether to introduce new products and discontinue existing ones.

Internal systems of tradable environmental permits are likely to be especially useful as instruments for generating information on private environmental costs. BP has established such a system for carbon dioxide, which is traded among BP business units like any input to production processes. The costs in different businesses of reducing carbon dioxide emissions determine the transfer price at which the businesses buy and sell permits. The system makes every business unit manager aware of the cost of reducing carbon dioxide within his or her unit, information of enormous value to BP in devising corporate environmental policy (Reinhardt, 2000b).

Differentiating Products

A second approach to reconciling shareholder value with environmental management focuses on environmental product differentiation. The underlying idea is straightforward: find customers willing to pay for products that provide greater environmental benefits or impose smaller environmental costs than those of competitors. Such efforts may raise the business's costs, but they may also enable it to command higher prices, capture additional market share, or both.

Consider an example from the textile industry. When textile manufacturers dye cotton or rayon fabric, they immerse the material in a bath containing dyes dissolved in water and then add salt to push the dyes out of the solution and into the cloth. Ciba Specialty Chemicals, a Swiss manufacturer of textile dyes, has introduced dyes that fix more readily to the fabric and, therefore, require less salt.

The new dyes help Ciba's customers in three ways. First, they lower the outlays for salt: textile companies using Ciba's new dyes can reduce their costs for salt by 0.5 percent to 2 percent of revenues—a significant drop in an industry with razor-thin profit margins. Second, they reduce manufacturers' costs for water treatment. Used bathwater—full of salt and unfixed dye—must be treated before it is released into rivers or streams, even in low-income countries where environmental standards may be relatively lax. Third, the new dyes' higher fixation rates make quality control easier, thus lowering the costs of rework.

Ciba's dyes are protected against imitation by patents and by the unpatentable but complicated chemistry that goes into making them. For those reasons, Ciba can charge more for its dyes and capture some of the value it is creating for customers.

Note that this resembles any other story about successful industrial marketing: the firm adds value to its customers' activities and then captures some of that value. Lowering a customer's environmental costs adds value to its operations just as surely as a new machine that enhances labor productivity.

Three conditions are required for success in environmental product differentiation, and Ciba's approach satisfies all three. First, the company has identified customers who are willing to pay more for an environmentally friendly product. Second, it has been able to communicate credibly its product's environmental benefits. And third, it has been able to protect itself from imitators for a long enough period to profit on its investment.

If any of those three conditions is not met, the environmental differentiation approach is unsuccessful. Consider, for example, the experience of H.J. Heinz's StarKist subsidiary when it decided to market dolphin-safe tuna.

Over the years, traditional techniques for catching tuna caused the death of millions of dolphins. The yellowfin tuna of the eastern tropical Pacific—the staple of tuna canners—often swim underneath schools of dolphins. A boat's crew would locate and chase a school of dolphins, drop a basket-like net under the school when the chase was over, and then haul in the tuna and the dolphins, often killing the dolphins in the process. Criticism of this practice, dating from the 1970s, intensified dramatically in 1989, when an environmental activist group released gruesome video footage of dolphins dying in the course of tuna-fishing operations. In April 1990, StarKist announced that it would only sell tuna from the western Pacific, where tuna do not swim beneath dolphins.

But the company ran into problems with all three conditions for success. First, contrary to the company's survey findings that people would pay significantly more for dolphin-safe tuna, consumers proved unwilling to pay a premium for a cheap, unglamorous source of protein. Second, although StarKist made known its efforts to protect dolphins, it turned out that the fishing techniques practiced in the western Pacific were no environmental bargain. For each dolphin saved in the eastern Pacific, thousands of immature tuna and dozens of sharks, turtles, and other marine animals died in the western part of the ocean. Finally, the company had no protection from imitators. The main competing brands, BumbleBee and Chicken of the Sea, matched StarKist's move almost at once.

It would be easy to take from this story a universally gloomy message about the prospects for environmental product differentiation in consumer markets. Environmental quality, after all, is ordinarily a public good: if it exists, everyone gets to enjoy it, regardless of who pays for it. From the standpoint of economic self-interest, one might wonder why any individual would be willing to pay for a public good.

But that view is too narrow. People willingly pay for public goods all the time: sometimes in cash, when they contribute to charities, and often in time, when they give blood, clean up litter from parks or highways, or rinse their soda bottles for recycling. Companies need to find the right public good or offer an imaginative bundle of public and private goods that will appeal to a targeted market.

For example, sellers of "designer beef"—meat from cattle that have not been exposed to herbicides or hormones—offer consumers potential health benefits (a private good) in addition to a more environmentally friendly product (a public good). Patagonia, a California maker of recreational clothing, has developed a loyal base of high-income customers partly because its brand identity includes a commitment to conservation. Patagonia and the beef marketers have not only cleared the willingness-to-pay hurdle, but they also have found ways to communicate credibly about the products and to protect themselves from imitators through branding.

Managers who are thinking about differentiating products along environmental lines need information of several kinds, each of which relates to one of the three necessary conditions discussed above. They need to understand their customers' willingness to pay for environmental characteristics in their particular markets. To reach this understanding, they need to analyze the real environmental costs and benefits of their own products compared to those of their competitors. But while necessary, this information is not sufficient. Especially in consumer goods markets, environmental product differentiation is often driven, at least in the short run, as much by perception as by scientific reality. So conventional market research, as well as engineering-based studies of environmental impacts, is required. Information on consumer attitudes is also necessary for the company to assess its prospects of credibly informing its customers about the environmental attributes of its goods. Finally, the firm needs to be able to anticipate the competitive response. This implies not only knowledge of its rivals' cost structures, but also of their assumptions and attitudes toward environmental markets. In summary, a firm contemplating environmental differentiation needs information of the sort that would be useful to governments (on actual environmental costs and benefits), as well as business-oriented information on the tastes and perceptions of its customers and rivals.

Managing Competitors

Not all companies will be able to increase their profits through environmental product differentiation. But some may be able to derive environmental and business benefits in a different way: by working to change the rules of the game so the playing field tilts in their favor. A company may need to incur higher costs to respond to environmental pressure, but it can still come out ahead if competitors must raise their costs even more. It may be able to accomplish this by uniting with similarly positioned companies within an industry to set private standards, or by convincing governments to create regulations that favor its own product.

Firms in the chemicals industry have applied this approach with some success. In 1984, after toxic gas escaped from a Union Carbide subsidiary's plant in Bhopal, India, and killed more than 2,000 people, the industry's image was severely tarnished and it faced the threat of punitive government regulation. In response, the leading companies in the Chemical Manufacturers Association (CMA) created an initiative called Responsible Care, and developed a set of private regulations adopted by the association's members in 1988. By writing their own rules, the companies hoped to forestall punitive regulation and to improve their safety records without incurring unreasonable costs.

The U.S. companies that make up the CMA must comply with six management codes that cover such areas as pollution prevention, process safety, and emergency response. If they cannot show good-faith efforts to comply, their membership will be terminated. The initiative has enhanced the association's environmental reputation by producing results. Between 1988 and 1994, for example, U.S. chemical firms reduced their environmental releases of toxic materials by almost 50 percent. Although other industries were effecting significant reductions during this period as well, the chemical industry's reductions were steeper than the national average.

Moreover, the big companies that organized Responsible Care have improved their competitive positions. They spend a lower percentage of their revenue to accomplish those results than do smaller competitors in the CMA; similarly, they spend a lower percentage of revenues on the monitoring, reporting, and administrative costs of the regulations. Further, because the association's big companies do a great deal of business abroad, they have worked with the CMA's foreign counterparts to persuade them to initiate their own Responsible Care programs—even in developing countries where one might expect little enthusiasm for tough environmental policies.

The prerequisites for the success of private regulatory programs like Responsible Care are the same as they would be for government regulatory programs. They must have measurable performance standards, information to verify compliance, and credible enforcement mechanisms. Private programs also need at least the tacit approval of government: if they are found incompatible with other rules, such as antitrust laws, the private regulations won't hold up. Private regulations must cover all relevant competitors: it is no use for one group of companies to tie the hands of another group if a third group can undercut them both.

The commodity chemicals business is better suited than most to private regulatory initiatives. Performance standards are comparatively easy to define because, for example, a perchloroethylene plant in Louisiana looks much like a perchloroethylene plant in New Jersey or Italy. Verifying compliance is not a problem either, because the companies constantly sell products to one another and thus can examine competitors' plants. Companies found in violation of the rules can be ousted from the association—even though it is illegal under antitrust law for the CMA to make compliance with Responsible Care a prerequisite for doing business with association members.

As an alternative to private regulation, companies that want to tie their competitors' hands can work with government regulators. DuPont's activities in the markets for chlorofluorocarbons (CFCs) and their substitutes provide an example. DuPont invented CFCs in the 1930s. Unknown in nature, the simple organic molecules became widely used in air conditioning and refrigeration equipment, in the manufacture of foam products, and as solvents and aerosol propellants. They were thought to be wonderful compounds for these purposes:

they were nontoxic, not flammable, and not explosive. The hitch, first postulated in 1974 and confirmed by the collaborative efforts of myriad atmospheric scientists over the ensuing decade and a half, was that CFCs migrated to the stratosphere, where they broke down and catalyzed the widespread depletion of the stratospheric ozone that shields the earth from ultraviolet radiation.

In response to the early evidence, American authorities in 1978 banned CFCs as aerosol propellants in the United States. European and other regulatory agencies failed to follow suit, but the American action left the industry with excess production capacity, leading to considerably lower profit margins in most segments of the business. DuPont resisted further regulation in the early 1980s on the grounds that the science was inadequate. But in 1988, after the publication of a governmental study fingering CFCs as the culprit in widespread ozone depletion, DuPont announced, unilaterally, that it was getting out of the CFC business.

DuPont's dramatic announcement—and the explicit accompanying statements that it was confident that its research efforts on substitutes would be sufficiently successful to ensure a seamless transition to the next generation of compounds—gave regulators the confidence to impose a worldwide ban on CFC production. DuPont benefited from an accelerated transition to the substitutes, which, unlike the CFCs themselves, were protected both by patents and by complicated manufacturing chemistry that kept small firms out of the business.

From DuPont's perspective, the story has not been an unequivocally happy one, for two reasons. The first is that the worldwide ban is phased across countries, with CFC production currently illegal in the rich world but still sanctioned in developing countries. CFC smuggling has become a big business, undercutting the markets for DuPont's CFC substitutes in industrial nations. Further, DuPont's former CFC customers have been widely successful in identifying completely different ways of serving the purposes of the original products, so that, for instance, instead of cleaning printed circuit boards with DuPont's expensive substitutes, they just use pressurized carbon dioxide. Both forms of substitution for the CFC substitutes have eroded profit margins in DuPont's business.

Note that the approach of forcing rivals to match one's own behavior is fundamentally different from that of environmental product differentiation, discussed immediately above. A manager thinking about the choice between the two approaches needs to ask, am I better off if my competitors match my investment, or if they don't? If a company's customers are willing to reward it for improved environmental performance, the company will want to forestall imitation by competitors. But if its customers cannot be induced to pay a premium for an environmentally preferable good, then it may want its competitors to have to match its behavior.

In designing approaches involving the management of competitors within an industry, a firm needs to understand the costs and benefits of the status quo, both for the environment and for the economy at large. It needs to understand, as well, the costs and benefits of its proposed solution. But equally important, it needs to understand how those costs and benefits are distributed across companies and consumers. Notice that in both the CMA and DuPont cases the strategy relied for success on particular actions of the government: regulatory forbearance in the CMA case, regulatory initiative in the case of DuPont. Government officials will always worry about the distribution of costs and benefits, so firms must analyze that distribution to gauge the political feasibility of their plans.

Redefining Markets

Some companies are trying to follow more than one of these approaches at once. That is, they are trying to reduce costs and to increase their customers' willingness to pay simultaneously. In the process, they are rewriting the competitive rules in their markets.

As we've seen, Xerox has been a leader in searching for cost reductions. More dramatically, it has also attempted to redefine its business model. Rather than simply selling office equipment, it retains responsibility for the equipment's disposal, and takes its products back from customers when they are superseded by new technology. The machines are then disassembled, remanufactured to incorporate new technology, and sold again for the same price as new machines. This practice enables Xerox to reduce its overall costs and to make life difficult for competitors that lack similar capabilities. Customers benefit, too, because they no longer have to worry about disposing of cumbersome machinery.

Rethinking traditional notions about property rights, as Xerox has done, is a useful way of determining corporate opportunities to redefine markets based on environmental challenges. Instead of transferring all rights and responsibilities of ownership to their customers, Xerox and other manufacturers are retaining some of the obligations in return for control of the product at the end of its useful life.

Xerox reportedly saved $50 million in the first year of its initiative to reclaim, reuse, and recycle its copiers and their components. A drop in raw material purchases was the most significant component of the cost savings; fewer natural resources were used to make new products. In 1995, Xerox estimated that it was saving several hundred million dollars annually by taking back used machines. Other manufacturers of electronic equipment, such as Kodak, IBM, Canon, and Hewlett-Packard, have undertaken similar initiatives.

The lesson: companies that combine innovations in property rights and technology may be able to create very strong competitive positions, using environmental concern to redefine their markets. To cite another example, Monsanto, DuPont, Syngenta, Aventis, and other companies are using conceptually similar approaches in an attempt to redefine the crop protection industry. Instead of making traditional insecticides for crop pests, the companies transfer genetic material from naturally occurring bacteria to seeds so that the plants themselves will be inedible to insects. These new seeds are highly profitable and avoid the financial and environmental costs of making, transporting, and applying insecticides.

Like Xerox, Monsanto also redefined the property rights that go with its product. To recover its investment in seed technology, Monsanto needs yearly repeat customers. But farmers commonly engage in a practice known as "brown bagging"— they save seeds left over from one year's crop to plant the following year. In return for the right to use the new type of seeds, Monsanto requires farmers to stop brown bagging and to submit to inspections to ensure compliance with the ban.

As is well known, agricultural biotechnology companies have not been unequivocally successful, especially in Europe. European environmental groups and consumers have protested the sale of genetically engineered products in their markets, abetted by those seeking trade protection for farmers. At the same time, however, the engineered seeds have increased market share in the United States, Canada, and Latin America because of the private economic benefits (reduced spending on other agricultural inputs, reduced erosion, and higher yields) that they confer on the farmer.

Obviously, ambitious strategies of the sort followed by Monsanto and Xerox entail significant market, regulatory, and scientific risks; they're not for every company—or even for every industry. The companies that appear to be succeeding are leaders in industries that face intensifying environmental pressure. Those companies have the research capabilities to develop new ways of delivering valuable services to their customers, the staying power to impose their vision of the future on their markets, and the resources to manage the inevitable risks. Moreover, by creating an appealing vision of a more profitable and environmentally responsible future, they may be better able to attract and retain the managers, scientists, and engineers who will enable them to build on their initial success.

Since these strategies combine product differentiation or strategic behavior with cost cutting, it follows that their successful implementation depends on the firm's ability to gather and analyze information of the sort discussed in each of this paper's earlier sections. To assess the chance that market redefinition will succeed, firms need to understand the magnitude and distribution of costs and benefits of the current market arrangements, and they need to understand how the redefinition that they hope to impose will

alter those costs and benefits. In addition, they need to understand their ability to communicate credibly the social and private benefits of their approaches to customers, regulators, and the public; clearly this is a type of information in which Monsanto and its competitors did not invest sufficiently. More research on the attitudes and opinions of European consumers and citizens, and not just on the costs and benefits to agricultural growers, might have prevented the setbacks that the agricultural biotechnologists have suffered in Europe.

Managing Environmental Risk More Effectively

For many business people, environmental management means risk management. Their primary environmental objective is to avoid the costs associated with an industrial accident, consumer boycott, or environmental lawsuit. Fortunately, effective management of the business risk stemming from environmental problems can itself be a source of competitive advantage.

Note that most companies don't try to manage environmental risk in the same ways they manage other business risks. In many companies, environmental risk is handled by the department that deals with environmental, health, and safety issues, while management of currency and other financial risk is centralized under the treasurer or financial officers. Those different parts of the organization usually have widely varying approaches to risk management and may even be ignorant of each other's activities.

Further, in managing environmental risk internally, most companies rely heavily on command-and-control mechanisms—in the form of procedural manuals and rules—to govern line managers' behavior. That approach impedes flexibility and fails to tap the expertise of individual line managers— the same problems that arise when government imposes command-and-control regulations.

There are legitimate reasons for managing environmental risk differently from other risks. Environmental risk is exceedingly difficult to assess quantitatively: no one can really know the probability of an accident occurring at a particular factory. By contrast, it's easier, say, to assess the probability that the dollar will move up or down against the yen, and market instruments exist that allow companies to hedge against such a risk. In the environmental arena, some use of command-and-control rules within the firm is probably necessary, particularly since government regulators rely on them so heavily.

As information about environmental risks and their effects on a company's financials improves, it will make sense to handle environmental risk more like other risks within the organization. For example, companies typically buy insurance against environmental liability at the corporate level but don't charge operating managers for their unit's portion of the premiums. As information on unit-specific risks improves, it will be easier to impose charges like these on the individual business units, aligning the managers' incentives more closely with those of the company. Similarly, in many companies, managers' environmental performance is already considered in performance reviews and in the promotions process, but in a manner that relies heavily on intuition and judgment rather than data. Again, as firms invest in information about environmental activities and costs, environmental concerns can be factored into performance evaluations and promotion decisions in a more systematic way.

The return on investments in environmental risk management will continue to be difficult to measure. One can never be sure, even long after the fact, that investments designed to prevent an accident or a lawsuit made sense. Hence even sensible investments in risk management are extremely vulnerable to cost-cutting pressure. At the same time, the inability to determine measurable results can also lead to overspending on risk reduction and empire building in the environmental office. An increasingly data-driven approach can alleviate such pressures, but should not be expected to eliminate them.

Senior managers need to ensure that those responsible for environmental risk are clear about the potential benefits of their investments. Environmental risk managers should be pushed to articulate why the level and type of investments they have chosen are appropriate. Furthermore, they need to communicate with those responsible for other sorts of business risk so that their approaches are consistent. Again, that doesn't mean the approaches should be identical. Until managers have the same information about environmental risk as they have about currency risk, it won't make sense to manage the two in the same way—and that day is a long way off. But better data can accelerate the convergence. In the interim, environmental risk management should not be shoved off to one side of the organization chart and managed as a special case. Integrating it into the company's overall risk management approaches will yield better decisions over the long term.

To frame environmental risk management in an analytically rigorous way, notice that a company can take any combination of four categories of steps to reduce financial risk to its shareholders. It can reduce the probability of a bad event (in the case of oil spills from ocean-going tankers, for instance, it can use double-hulled vessels). It can reduce the total social costs of the bad event should one occur (to continue with the oil spill example, an oil company consortium now maintains rapid-response cleanup vessels in ports along the most heavily traveled routes). It can transfer risk to other parties to reduce the share of total social costs for which the company will be responsible if the bad event occurs (for

example by purchasing insurance policies). Or it can acquire more information that allows managers to analyze the costs and benefits of different options in the first, second, and third categories.

As noted above, the cost of information about environmental costs and benefits is falling, and the benefits of maintaining and using that information are increasing; and so it is with information on environmental risk. As companies understand the costs and benefits of environmental risk reduction better, they will be able to reduce their reliance on internal command-and-control rules and increase the degree to which they use other, more delicate tools.

As a starting point, managers ought to evaluate their current environmental risk reduction efforts, which may be carried out in different parts of the company; and they ought to try to develop at least a crude understanding of the relative costs and benefits of various investments in risk reduction. For example, they should ask whether their external insurance policies (which are basically risk transfer mechanisms) are cost-effective compared to a combination of self-insurance and internal risk reduction. They should make sure that when they retain risk they are doing so by choice rather than by default, and that they are compensated for bearing this risk. This exercise itself will identify the most important gaps in the firm's own knowledge about its environmental risk management, and suggest where it should start obtaining more information.

Beyond All Or Nothing

All-or-nothing arguments have dominated thinking about business and the environment. But this unprofitable polarization need not continue. Consider how ideas about product quality have changed. At first, conventional wisdom held that improvements in quality had to be purchased at a cost of extra dollars and management attention. Then assertions were made that "quality is free:" new savings would always pay for investments in improved quality. Now companies have arrived at a more nuanced view. They recognize that improving quality can sometimes lead to cost reductions, but they acknowledge that the right strategy depends on the company and its customers' requirements. It is time for business thinking on the environment to reach a similar middle ground.

As we've seen, environmental problems can be analyzed usefully as business problems. This does not mean, of course, that questions of social responsibility can be ignored. It does mean, however, that they are only part of the equation. Investments in environmental improvement, whether motivated by notions of social responsibility, appeals to sustainability or the importance of the "triple bottom line," or some other consideration, must ordinarily be justified both to executives within the firm and to the firm's shareholders on the basis of private value. Not all companies can profit from concern about the environment: there is no universal free lunch, nor is there a "one size fits all" environmental policy that can be imposed without regard to the particulars of competition in a given industry. But some firms will be able to return extra value to their shareholders while contributing to the solution of environmental problems.

Improved information on environmental risks and costs is critical in helping firms find such policies. From the point of view of the managers of a firm, the central question in evaluating an environmental policy is whether the benefits to the firm outweigh the private costs to the firm of internalizing environmental costs. To answer this question, they need information about the environmental costs of their current activities, the private costs of reducing their adverse environmental impacts, and their prospects for recovering some of those private costs from customers or other economic factors.

How can governments help firms contribute to maintaining environmental quality? Not, in general, by trying to create explicit subsidies for technologies that are thought to be environmentally friendly, and not simply by exhorting companies to provide more environmental public goods. The first strategy is unstable, the second inefficacious. Instead, governments should work to create the expectation of a steady convergence of private and social costs. They should shift their focus away from command-and-control rules that lock in technology, and toward market-based instruments that internalize environmental costs, creating the incentive for the discovery and exploitation of better information about environmental issues—and the costs and benefits of addressing them.

Firms should make environmental investments for the same reasons they make other investments: because they expect the investments to deliver positive returns or to reduce risks. They need to go beyond the question, "Does it pay to be green?" and ask instead, "Under what circumstances do particular kinds of environmental investments deliver benefits to shareholders?" Any answer to the first question will be driven by ideology and inevitably will prove unenlightening. Answers to the second question will be far more useful, and improved data about the costs and benefits of environmental quality are essential to provide those answers.[5]

References

Esty, Daniel C. and Michael E. Porter, "Industrial Ecology and Competitiveness: Strategic Implications for the Firm," *Journal of Industrial Ecology* vol. 2, no. 1 (winter 1998), 35-43.

Gabel, H. Landis and Bernard Sinclair-Desgagné, "The Firm, Its Routines and the Environment," in Richard Starkey and Richard Welford, eds., *The Earthscan Reader in Business and Sustainable Development* (London: Earthscan, 2001).

Ghemawat, Pankaj, *Strategy and the Business Landscape* (Upper Saddle River, NJ: Prentice Hall, 2001).

Hartwick, John, "Intergenerational Equity and the Investing of Rents from Exhaustible Resources," *American Economic Review* vol. 67, no. 5 (1977), 972–974.

Palmer, Karen, Wallace E. Oates, and Paul R. Portney, "Tightening Environmental Standards: The Benefit-Cost or the No-Cost Paradigm?" *Journal of Economic Perspectives* 9, no. 4 (Fall 1995): 119–132.

Porter, Michael E., "America's Green Strategy," *Scientific American*, April 1991, 168.

Porter, Michael E. and Claas van der Linde, "Toward a New Conception of the Environment-Competitiveness Relationship," *Journal of Economic Perspectives* 9, no. 4 (fall 1995): 97–118.

Reinhardt, Forest, "Bringing the Environment Down to Earth," *Harvard Business Review* vol. 77, no. 4 (July-August 1999a), 149–157.

—, *Down to Earth: Applying Business Principles to Environmental Management* (Boston: Harvard Business School Press, 2000a).

—, "Global Climate Change and BP Amoco," Harvard Business School case number 700-106. (Boston: Harvard Business School, 2000b).

—, "Market Failure and the Environmental Policies of Firms: Economic Rationales for 'Beyond Compliance' Behavior," *Journal of Industrial Ecology* vol. 3, no. 1 (winter 1999b).

—, "Sustainability and the Firm," *Interfaces* vol. 30, no. 3 (May-June 2000c), 26–41.

—, "Tensions in the Environment," in "Mastering Risk" series, June 13, 2000d, reprinted in James Pickford, ed., *Financial Times Mastering Risk; Volume 1: Concepts* (Harlow, United Kingdom: Pearson Education Limited, 2001).

Solow, Robert. "Sustainability: an economist's perspective," J. Seward Johnson Lecture to the Marine Policy Center, Woods Hole Oceanographic Institution, 1991, reprinted in Robert N. Stavins, ed., *Economics of the Environment: Selected Readings* (fourth edition; New York: Norton, 2000).

World Commission on Environment and Development 1987, *Our Common Future*, Oxford University Press, Oxford, England.

Endnotes

1 For an early articulation of this important disparity, see Esty and Porter, 1998.

2 For a succinct treatment of the business strategist's perspective on sustainability, see Ghemawat, 2001. Despite their disparity, these two perspectives can be reconciled, at least conceptually. Both are related to the macroeconomic definition of sustainability developed by, among others, John Hartwick and Robert Solow (see Hartwick, 1977; Solow, 1991). Notice that the macroeconomic definition of Hartwick and Solow is focused more narrowly on environmental and natural resource considerations and, hence, is more amenable to testing and measurement than the Brundtland definition. In crude terms, the economic definition requires that per capita stock of total assets, natural and human-made, be kept constant. Notice, too, that this macroeconomic definition of sustainability has an obvious analogue at the firm level: a nondeclining total asset figure on the balance sheet, when assets are valued at social costs; see Reinhardt, 2000c.

3   The relationships between these perspectives on environmental risk can also be reconciled; see Reinhardt, 2000d.

4   For articulations of this view, see Porter, 1991, and Porter and van der Linde, 1995. Note that these papers do not include the term, "Porter hypothesis," but it is used, for example, in Gabel and Sinclair-Desgagné, 2001.

5   This paper draws on earlier research, including Reinhardt (1999a, 1999b, 2000c). I am grateful to Peter Cornelius and Dan Esty for their insightful comments and suggestions. Remaining errors are mine.

Chapter 5

Financial Markets and Corporate Environmental Results

Frank Dixon[1]

Environmental sustainability looms as one of the largest challenges ever faced by business. Financial impacts on companies are growing as customers, regulators, investors, and other stakeholders continue to press for improvements in corporate environmental performance and movement toward sustainability.[2] This drives increasing demand from the financial community for high quality analysis of corporate environmental performance. The sophistication of analysis has increased greatly over the past five years. However, limited data availability poses ongoing challenges for the capital markets, business, and other interested parties, such as non-governmental organizations (NGOs).

As environmental conditions continue to decline globally, and calls for action to address the situation increase, the need for performance data at the macro and micro levels grows. As noted elsewhere in this volume, the availability of transparent, high quality data spurs competition and performance improvement. At the macro level, data quality varies greatly by region. At the micro level, corporate environmental performance data are often unavailable, inconsistent, lagged, inaccurate, unverified, and biased. This lack of high quality data clouds relative corporate environmental performance,

reduces the efficiency of markets, and presents a significant barrier to meeting the sustainability challenge.

This chapter discusses the drivers of growing financial community interest in corporate environmental performance, methods of measuring such performance, and future measurement challenges. The EcoValue'21™ model, developed by Innovest Strategic Value Advisors, is presented to illustrate one method of addressing data challenges and to show how environmental performance ratings can be used increase investor returns.

Overall, the chapter argues that the financial relevance of corporate environmental performance is increasing and that this will drive expanded incorporation of environmental analysis into investment decisions. Ultimately, ongoing improvements in corporate environmental performance data will be needed to minimize investor risk and maximize financial returns.

Financial Community Interest in Corporate Environmental Performance

The corporate sector has a large impact on the environment through waste emissions to land, air, and water resulting from the production, use, and disposal of goods and services. This impact contributes to the ongoing decline in global environmental quality. Deteriorating environmental conditions translate into financial pressures on companies through increasing regulations; growing demand from consumers for environmentally responsible products, services and corporate policies; political pressures; rapidly expanding information transparency, largely through the internet; growing concerns among the general public about the state of the environment; increasing investor awareness of the financial benefits of improving corporate environmental performance; and growing competition among firms to improve their pollution control and natural resource management results. As environmental pressures continue to increase, companies that improve environmental performance relative to peers are likely to achieve superior financial returns and competitive positioning over the mid- to long-term. In addition, corporate environmental leaders frequently report achieving enhanced profitability in the short-term.

In spite of this growing financial relevance, mainstream investors have not traditionally incorporated environmental analysis into investment decisions. The financial community can thus be segregated into two groups: mainstream and socially responsible investors (those basing investments on environmental, social, and financial considerations). Traditionally, mainstream investors (representing about 85% of invested assets in the U.S.) considered environmental issues only to a limited degree. The environment was seen as a potential liability (e.g., Superfund sites), risk (e.g., catastrophic event), and/or cost item (e.g., regulatory compliance). It was not traditionally seen as a source of competitive advantage or investment out-performance. The concept of fiduciary responsibility has limited the use of environmental analysis in the investment process. This concept holds that the duty to maximize investor returns precludes consideration of issues that do not seem to be financially relevant to the company, such as the longer-term impacts of pollution on the environment and society.

More recently, mainstream investors, such as Dreyfus and Mellon Capital Management, have been using relative corporate environmental performance to differentiate companies and increase investment returns. Several factors drive growing financial community interest in corporate environmental performance. First, most studies of links between environmental and financial performance find positive correlations. Second, a reasonable theory exists for explaining correlations and postulating causation—environmental performance is a strong proxy for management quality. Third, socially responsible investment funds (which include environmentally screened funds) frequently outperform nonscreened funds. Fourth, growing pressure for corporations to assume fuller responsibility for their environmental and social impacts increases investor risk exposure. And, finally, recent regulatory requirements and evolving legal views encourage or mandate increased investor focus on corporate environmental performance.

Research Findings

Since 1990, most studies of links between corporate environmental and financial performance found positive correlations.[3] A U.S. Environmental Protection Agency study (2000) found, "A significant body of academic research relates measures of corporate environmental performance to measures of financial performance. The most striking aspect of this research is that most of it shows a moderate positive relationship between the two kinds of performance—regardless of the variables use to represent each kind of performance, the technique used to analyze the relationship, or the date of the study. In fact, the empirical evidence is of sufficient consistency and scale to embolden some to argue that a positive relationship between environmental and financial performance is without doubt."

A recent review (Cram and Koehler, 2001) of over 40 studies in this area found: "The 1990s has seen an expanding series of studies probing at the association between a firm's environmental and financial performance with the general conclusion that it is statistically significant." Another study by Bank Sarasin (Butz and Plattner, 1999) concluded that "there is definitely a statistically significant positive correlation between environmental performance and the financial return on equities in sectors where environmental performance is relevant in the public perception."

Positive correlations are also often found in the broader socially responsible investing category, which analyzes social (e.g., labor, supply chain, product safety, international, community development, etc.) as well as environmental metrics. For example, Pava and Krausz (1995) examined 21 empirical studies and found that twelve had positive correlations, eight had no statistically significant correlation, and only one had a negative correlation.

The EPA and Cram/Koehler studies point out various gaps and errors in some of the previous research which could produce misleading results. Potential flaws in some of the studies include poor data quality, small sample size, short time frame, inadequate control of non-environmental factors influencing financial performance, inappropriate cross-sectoral comparisons, emphasis on backward-looking data, failure to establish causation, vulnerability to greenwashing,[4] difficulty in assessing impacts on human health and the environment, and inconsistent definitions of environmental performance. To illustrate the complexity of definition problems, Cram/Koehler notes that performance based on emissions will improve if a company divests a polluting facility. But if the company continues to buy materials from the same facility, its overall environmental performance may not actually improve.

The Cram/Koehler paper concludes that study flaws "raise considerable doubt on the findings described in the literature thus far" and that this literature "offers no conclusive guidance on managerial and investor decision-making with respect to firm environmental performance." However, the implication that positive correlations found in the bulk of previous research may not be valid due to study flaws is probably too aggressive for several reasons. First, most of the alleged flaws have about equal probabilities of producing false negative and positive correlations. In other words, most of the potential flaws do not bias the studies toward finding positive correlations. The fact that positive correlations are nevertheless consistently found indicates the overall findings are probably valid.

Second, Cram and Koehler argue that as more sophisticated modeling techniques are employed to overcome earlier flaws, positive correlations are still found. Third, measuring highly complex environmental interactions with frequently poor data sometimes requires simplifying assumptions and use of suboptimal data sets. These approaches are open to challenge, but without them research in this area could hardly be done. Fourth, failure to show causation does not invalidate the findings since investors frequently make investment decisions based on factors, such as earnings, that are known to be correlated, but may not be causally linked, to stock price. (In the strictest sense, it is impossible to prove direct causation between stock price and anything since stock price is a collective market opinion.)

Fifth, the authors imply investors would use analysis of corporate environmental performance exclusively to choose stocks. Except for eco-enhanced index strategies, this rarely occurs. Environmental analysis is typically used in combination with traditional financial analysis to identify well-managed companies and stock market outperformance potential (discussed below under performance measurement). And finally, Cram/Koehler does not adequately address the fact that there is a logical and intuitively obvious explanation for the existence of positive correlations—the proxy value for management quality.

More research is needed in this area to test the existence of positive correlations between environmental and financial performance. Cram/Koehler's implication that the existing research is flawed, and investors cannot rely upon it, is probably overly critical. However, the authors add significant value by pointing out methodological weaknesses and suggesting "future directions for this body of research."

Management Quality and Other Explanatory Factors

Causation is notoriously difficult to prove since so many factors influence financial performance. As a result, studies often focus on explaining correlations rather than attempting to prove causation. Several of the studies using historical data to identify correlations between environmental and financial performance cite management quality as a primary explanatory factor. Future-oriented assessments of corporate environmental performance, such as those done by Innovest, find positive correlations more frequently and also attribute the findings largely to management quality. This factor probably has a greater influence on financial performance than any other. Management quality influences all aspects of business performance, largely determining success in key functional areas such as product development, marketing, and production. One could say that most factors enabling firms to outperform peers, such as lower costs or strong patent protection, originate from superior management.

Nearly every financial analyst would agree that management quality is one of the primary determinates of financial performance. However, very few have an objective way of measuring it. Assessments are usually subjective, based on opinion. Management quality is difficult to quantify since it involves assessing intangible factors such as the intelligence and business savvy of corporate leaders. Assessing management quality can be done by using proxies, for example by measuring some aspect of management performance, rather than trying to measure management quality directly. This, in effect, is done by analyzing earnings and other financial measures. However, there are so many other internal and external influences that the pure assessment of management quality is clouded. A clearer assessment can be had by narrowing the focus down to one issue. In other words, test how well management performs in one business area, preferably a highly complex area. Success here implies management will be effective at dealing with less complex business issues and will therefore produce superior returns.

It turns out that the environment is one of the most complex challenges facing management, especially in resource intensive sectors. There is a high degree of technical, regulatory, and market uncertainty. There are many internal and external stakeholders to deal with, requiring sophisticated communication skills. There are many complex issues to address such as global warming. And there are many nontraditional, often nonfinancial metrics to track. It is implied that companies dealing well with this high level of complexity have the sophistication to succeed in other parts of the business. The relationship between good environmental management and good general management may explain why companies recognized as environmental leaders frequently earn superior financial and stock market returns.

Another factor explaining correlations between environmental and financial performance is the direct impact improved environmental performance can have on profitability. Specifically, improving environmental performance can lower risk exposure and costs as well as increase revenues. Improving performance can lower risk exposure in areas including damage to corporate image and reputation, loss of market share, vulnerability to increasing regulations, product obsolescence, impairment of property values, and delay or cancellation of mergers and acquisitions. On the upside, reported financial benefits of improved environmental performance include reduced materials, energy, and waste disposal costs; enhanced product quality, market image, and market share; lower regulatory, insurance, and financing costs; enhanced innovation capacity; improved stakeholder relations; and enhanced employee morale and productivity. For example, 3M claims to have saved $810 million between 1975 and 1997 through pollution prevention initiatives (3M, 1998).

Companies pursue environmental performance improvements when the perceived benefits of doing so exceed the costs. However, benefits can be intangible, longer-term, and sometimes simply difficult to quantify. As a result, it may be hard to justify financially investments focused on improving environmental performance. Our research suggests that leading companies tend to have greater ability to deal with intangibles, such as future market perceptions or the indirect financial impacts of pollution. These firms will, at times, make marginal improvements in environmental performance that may not make sense on a strict internal rate of return basis. They perceive financial benefits that less sophisticated companies may not be able to see. The complexity of environmental decision-making further indicates why environmental performance can be used to differentiate management quality and stock market potential.

Some studies suggest positive correlations may also occur since more profitable companies may be able to spend more on the environment. Across the spectrum of companies, it is likely that all three explanations are valid—proxy value for management quality, direct impacts on profitability, and superior profits driving superior environmental performance. More research is needed to determine the relative importance of each factor in producing positive correlations between environmental and financial performance.

Fund Performance

Traditionally, most investment advisors believed that screening investments for environmental, social, or any other so-called nonfinancial criteria reduced financial returns, even though most studies have shown the opposite. Inappropriate fund comparisons have contributed to the idea that socially responsible investment funds underperform. For example, some studies have inappropriately grouped different types of funds (e.g., value, growth, etc.) together and compared them to other fund classes. However, socially responsible investing is a discipline that can be applied to all fund classes, and not a separate fund class.

Earlier approaches that exclusively used negative environmental and social screening did underperform the market at times.[5] Newer fund strategies using best-in-class approaches, that maintain diversity and shift investments toward presumably better managed companies, have frequently outperformed non-screened (mainstream) funds. More recently, both negatively screened and best-in-class (or positively screened) funds have often matched or exceeded the performance of mainstream funds. For example, 14 of the 16 environmentally and socially screened funds with over $100 million in assets received top scores from the rating services Morningstar and Lipper Analytical Services in 2000. Overall, 65% of the funds with over a three-year performance record received top scores (Social Investment Forum, 2001).

The availability of better information on environmentally and socially screened fund performance and findings from the many studies noted above are causing a growing belief in the financial community that environmental and social screening does not lower returns, and in many cases may enhance them. This partly explains the rapid growth of these funds in North America, Europe, and Japan over the past five years. At the end of 1999, environmentally and socially screened investments represented over $2 trillion in U.S. assets (Social Investment Forum, 1999).

The Evolving Role of Corporations in Society

As environmental conditions continue to decline, companies may be called upon to take fuller responsibility for their negative environmental impacts and to adopt a systems view of the environment. As firms are held responsible for a wider range of impacts, investor risk exposure will likely increase. Before the industrial revolution, society's impact on the environment was insignificant in relation to the ability of the environment to absorb the impact. Driven largely by the use of fossil and nuclear energy, human impacts are now signif-

icant and may be approaching the point of overwhelming the environment's ability to regenerate itself. All aspects of the Earth and its atmosphere are one interconnected system. No part operates in isolation. Traditionally, the total Earth system was too complex to study as a whole. Through reductionism, the parts were studied in depth. However, the overall system and relationships between its parts were not studied nearly as well. As a result, solutions to problems in one area became problems in another, as in fossil fuel combustion causing global warming.

Economic and commercial systems developed when human impacts on the environment were relatively small did not hold companies fully accountable for their negative environmental impacts. For example, these systems considered clean air and water to be free goods because they were so plentiful. As a result, companies were not charged the full cost of consuming these and other resources. Their focus was narrowly on the efficient production of goods and services.

Now, as the negative environmental impacts of companies become more obvious, these costs, which have been largely externalized onto society, are being internalized through increasing regulations, customer demands, taxes, and other mechanisms. To an increasing degree, companies are being called upon to expand their operating focus to include being a responsible corporate citizen and minimizing their negative environmental and social impacts. Companies failing to move in this direction will likely face growing financial risks and penalties both directly from government-imposed regulations, and indirectly from customers and capital markets.

Regulatory and Legal Issues

Growing financial market interest in corporate environmental performance is being driven by regulatory requirements and evolving legal views on the fiduciary responsibilities of fund advisors. For example, the U.K. government implemented legislation in July 2000 which requires pension funds to disclose their methods of screening investments for environmental, social, and ethical factors. As a result, 22 of the 25 largest pension funds in the U.K. have adopted such screening methods.

In the legal area, a significant barrier to the use of environmental and social screening has been the belief that directors of pension and other funds have a fiduciary responsibility to maximize returns which precludes them from considering so-called nonfinancial factors, such as environmental and social performance. This view is changing as studies by organizations such as the law firm Baker & McKenzie (Gibson, Levitt, and Cargo, 2000) find that fiduciaries may consider environmental and social issues when making investment decisions, provided reasonable due diligence is conducted.[6] Beyond allowing environmental and social screening, other studies find that, in some cases, the Duty to Monitor (Koppes, 1995) and the Duty of Obligation (Cogan, 2000; McKeown, 1997; Solomon and Coe, 1997) place fiduciaries under a legal requirement to consider environmental and social issues.

Measuring Corporate Environmental Performance

How do the capital markets identify superior environmental performance? Corporate environmental performance remains hard to measure. Yet, new quantitative analytic tools are being developed. For example, Innovest's EcoValue'21 model mitigates data problems and provides a relevant and objective measurement tool for the financial community. This section discusses the current state of corporate environmental performance data, measurement options and complexity, analytic methods, measurement results, and application by the financial community.

Data Quality

Standardized public financial reporting has been required in industrialized nations for many years. As a result, it is relatively easy for stakeholders to compare firms on many financial metrics. The requirement for standardized public reporting is not nearly as well developed in the environmental area. While U.S. global leadership has diminished markedly in recent years, the United States government remains the world leader in providing publicly available data that permits analysis of corporate environmental performance.

The U.S. Environmental Protection Agency (EPA) provides several publicly available databases.[7] These include the:

- Toxic Release Inventory (TRI, provides data on releases and transfers of over 600 toxic chemicals from manufacturing facilities)

- Emergency Response Notification System (ERNS, tracks spills and releases of toxic substances)

- Accidental Release Information Program (updates the ERNS database)

- Water Permit Compliance System (tracks permit violations and penalties assessed under the Clean Water Act)

- RCRA Information System (tracks handlers of hazardous waste under the Resource Conservation and Recovery Act)

- RCRA Biennial Reporting System (tracks the generation and shipment of hazardous waste)

- EPA Legal Action Data (tracks civil cases filed on behalf of the EPA)

- CERCLA Information System (tracks hazardous waste sites under the Superfund Program)

While this information facilitates corporate environmental performance measurement, there are many improvements that could be made. For example, the TRI tracks emissions from domestic manufacturing facilities. Since data on the environmental impacts of nonmanufacturing and international operations are usually not available, it makes comparing companies with significant activities in these areas difficult. Also, there is often a long lag time before data is made public. In addition, information is reported at the facility level and tagging methods often make it difficult to attribute the data correctly to the ultimate parent company. Finally, data is often uploaded by different states and regional EPA offices, making the quality uneven. Through various initiatives, the EPA is attempting to remedy some of these issues.

Some European governments, such as those in the U.K. and Sweden, require limited disclosure of corporate environmental performance data. However, varying disclosure requirements in Europe make it difficult to compare companies, especially those with operations in many countries. The European Union is working on standardizing disclosure requirements, so it is likely that higher quality government supplied data will be available in Europe over the next few years. Beyond government databases, various types of corporate environmental performance data are available from NGOs and watchdog groups.

While government supplied data is useful, especially in the U.S., the largest source of corporate environmental performance data is usually the companies themselves. Government-supplied data in Europe is sparse compared to the U.S. However, European companies generally provide higher quality data on corporate environmental performance (with many exceptions in North America and Japan). Companies in Europe tend to report more consistent quantified data on environmental impacts. This type of information is more useful to those assessing relative performance than the anecdotal information usually provided by less proactive firms. The quality and quantity of data provided by European firms often more than compensates for the lack of government-supplied data in Europe.

However, corporate self-initiative data not reported under mandatory disclosure schemes remain open to reliability questions. To reduce the perception that company-supplied data is biased, many European firms are seeking third-party certification of their environmental reports. While third-party verification does not eliminate bias potential, since companies still choose which data to report, it is a step in the direction of providing investors with more reliable data. To further improve data quality, several firms in Europe and North America are beginning to report under standardized environmental and social reporting schemes, such as the Global Reporting Initiative. Expanded use of programs such as these has significant potential to facilitate corporate environmental performance measurement. However, as long as the programs remain voluntary, their usefulness will be limited.

In the U.S., accounting rules for contingent liabilities can result in significant underreporting of environmental liabilities. For example, Financial Accounting Standards Board Statement No. 5 states that contingent liabilities, such as environmental remediation liabilities, shall only be accrued if they are probable and can be reasonably estimated. If the estimated cost is within a range, only the minimum cost shall be accrued. Disclosure is only required if a claim has been made, or it is probable one will be made. As a result, financial statements may significantly understate environmental liabilities. This could have a growing negative impact on investors as environmental pressures on companies increase.

A recent study of 13 pulp and paper companies (Repetto, 2000) found that each company could expect negative financial impacts from environmental issues of at least five percent of total shareholder value (over ten percent for some companies), and that these highly probable impacts were not disclosed in financial statements. Several parties, including the investment firm Calvert Group, have asked the Securities and Exchange Commission to increase investor protection by expanding current disclosure requirements. The SEC has explored the issue, but has not acted yet. Mandatory disclosure of corporate environmental performance data probably holds the greatest potential for facilitating measurement and prompting competition among firms to improve performance.

Measurement Issues and Complexity

Corporate environmental performance assessments can be done in many ways. Options include absolute versus best-in-class ratings, one versus many scores, and historical versus forward-looking analysis. Absolute ratings potentially describe a company's ultimate environmental impact more accurately as well as facilitate cross-sector comparisons. However, investors primarily use corporate environmental performance analysis to choose between companies within sectors, so a best-in-class approach is generally most effective. Condensing many metrics into one score provides less information than several separate scores. However, having a bottom-line score helps investors choose between companies, so a single-score approach remains effective. Historical analysis provides a more accurate assessment of a company's actual environmental impact. However, investors are interested in future performance, so a forward-looking analysis is often more appropriate. Though forecasting future results is much more difficult than reporting on past performance.

Beyond numerous assessment options, difficulty in quantifying corporate environmental impacts further complicates performance measurement. As companies are held responsible for a wider range of environmental impacts, minimizing investor risk will require ongoing improvements in corporate impact data. Yet quantifying corporate impacts is difficult since the Earth's ecosystem is multifaceted and highly complex. The sinks into which companies emit waste (e.g., atmosphere, water bodies, land masses) are so large that there are often long feedback loops before impacts can be identified. It is difficult to quantify a firm's impact on the environment because of these long feedback loops and because impacts can take many forms (e.g., human health, ecosystem damage, biodiversity loss, etc.). In addition, it is often difficult to disentangle a firm's contribution to a specific impact.

By using proxies and impact indicators, the process of assessing environmental impact can be simplified, assuming the data are made available. To illustrate, many companies, mostly in Europe, are reporting impacts under standard categories, such as global warming, ozone depletion, acidifying emissions, smog-forming emissions, eutrophication, toxic wastes, water use, energy use, and heavy metals. This facilitates cross-company comparisons and helps to minimize investor risk exposure.

Analyzing Corporate Environmental Performance

Throughout the industrialized world, a mix of nonprofit and for-profit organizations analyze corporate environmental performance and provide this information to investors and other stakeholders. For the most part, this analysis involves gathering data through questionnaires, then summarizing anecdotal and quantitative data (when available). Some firms, such as Kinder Lydenberg and Domini, assign ratings in various environmental and social categories, but most organizations are in the business of providing summarized data to investors, who then do the analysis on their own. Providing financially oriented analysis of corporate environmental performance data is done by fewer firms.

The EcoValue'21 model, developed by Innovest Strategic Value Advisors, uses a multifactored approach to overcome data problems and more accurately assess corporate environmental performance. Using one or only a few data points to assess performance makes the analysis vulnerable to "outliers" and other data weaknesses, potentially providing misleading results since so many factors influence overall environmental performance. A multifactored analysis minimizes the impact of faulty data since, even if a few data points are inaccurate, companies will usually be placed in nearly the same relative order based on overall averages.

To assess corporate environmental performance, one must first define it. With many interpretations available, there is no one "correct" definition. Definitions vary based on the goals of the user. An investor seeking to gauge the financial impact of corporate environmental performance would likely want to assess the financially relevant aspects of it. These include risk exposure, upside potential (engagement in environmentally related business opportunities), and overall management of upside and downside environmental issues. The EcoValue'21 model analyzes about 60 metrics grouped into these three categories. While the model was developed for investors focused on financial performance, its comprehensive nature allows it to be used by other stakeholders who are purely interested in environmental performance.

Metrics and weightings for the EcoValue'21 model were selected using criteria including the quality of available data (with preference given to quantitative, third-party verified data), business judgment about the relative importance of the metric to overall corporate environmental performance, and, most importantly, correlations to stock returns, since the model was being used to project this. The model was developed with strategic partners including PricewaterhouseCoopers and Morgan Stanley. To guide selection of metrics and determination of metric weights, regression analysis was done to analyze correlations between many environmental metrics and stock returns for 350 S&P 500 companies over a five-year period.

In the risk area, the model analyzes factors including site liabilities, hazardous waste generation, toxic emissions, compliance violations, spills and releases, and other indicators of environmental burden, such as those noted above, frequently published by European firms. In addition, energy and resource use efficiency is assessed along with market and regulatory risk exposure. Beyond these general categories, sector-specific metrics are analyzed based on the relevant risk factors for each sector. Examples include fuel mix in the electric utility sector, exposure to EPA's Cluster Rule in the pulp and paper sector, and fuel efficiency in the automobile sector. A key element of the analysis is adjusting for risk exposure between companies in the same sector due to factors including product mix and geographic location. In addition, the model analyzes performance over time as well as static risk indicators. Especially in the risk area, preference is given to quantitative data, and supplemented with qualitative assessments when this is not available.

In the opportunity area, the model analyzes capacity to develop environmentally favorable products and services, market positioning to sell them effectively, and actual involvement in marketing them. Capacity indicators include resources devoted to this area, research and development focus, and strategic planning procedures. Market positioning indicators include geographic regions served, demographics of customer base, and vulnerability to substitution. Involvement indicators include assessing the extent to which companies are developing and/or marketing environmentally favorable products and services.

The EcoValue'21 model is strongly future-oriented. As a result, factors such as management of environmental issues and strategic positioning get higher weightings. Beyond risk assessment, quantitative assessments of historic and current environmental performance are used to authenticate stated commitments to improving corporate environmental performance. Or, in other words, to verify that the company is "walking its talk" and not "greenwashing" its image.

In the management area, the model analyzes many metrics in the areas of planning, environmental management systems (EMS), and governance. Planning indicators include the extent to which environmental issues are incorporated into the overall business strategy. This includes an assessment of the degree to which the environment is being used to build competitive advantage, versus being used only for public relations purposes. EMS indicators include the quality of the environmental policy, use of life-cycle analysis to assess impacts, evaluation of eco-efficiency initiatives, EMS quality including the use of third-party certification schemes such as ISO 14001, performance monitoring and accounting systems, training procedures, supplier screening and engagement programs, quality of public reporting, auditing procedures, and participation in voluntary programs and product labeling schemes. Governance indicators include board involvement, management structure, compensation programs, and more subjective indicators such as company culture and senior management commitment.

Implementing the EcoValue'21 process involves first assessing the key upside and downside environmental issues in each sector as well as analyzing the strategies of sector leaders. This information is used to build a template against which the sector will be analyzed. Then information is gathered from sources including government databases, NGOs, periodical and web searches, financial community reports, and company information such as annual reports, 10Ks, 10Qs, environmental and sustainability reports, websites and any other available publications. Following this, company executives (often Vice Presidents of Environment, Health & Safety) are interviewed to complete the data gathering process.

Once data is gathered for each of the roughly 60 EcoValue'21 metrics, weightings are assigned as described above. These weighted scores are added to produce bottom-line numeric scores for each company in a given sector. Then best-in-class letter scores, ranging from AAA to CCC, are assigned which are intended to project relative stock market performance. The box on the next page provides the rationale for one company's rating, FPL Group.

Measurement Results

The EcoValue'21 model was designed to estimate stock market potential over the mid- to longer-term. The model is not intended to forecast short-term stock market potential. Nevertheless, it has been successful in doing so. In every high environmental impact sector and nearly every other sector, the average total stock market return of companies with above average EcoValue'21 ratings exceeded the average

FPL Group

FPL Group, the parent of Florida Power & Light, received a AAA EcoValue'21 rating in the U.S. electric utility sector. The rating reflects FPL's moderate risk level, excellent risk management, and sector-leading development of environmentally favorable businesses. With a diversified fuel mix, the company has lower risk exposure than those relying more heavily on coal or nuclear power. Proactive risk management includes extensive training and performance measurement, linking compensation to environmental performance, and implementing a leading environmental management system. Aggressive efforts to improve eco-efficiency include improving power plant efficiency, extensive recycling, and replacing toxic materials with lower impact substitutes. FPL is also one of the largest developers of wind power in the U.S.

Table 1

EcoValue'21 Performance

Sector	Time Period	Top Half – Bottom Half Differential (basis points)
Electric Utilities (U.S.)	8/98 – 7/01	4000
Automobiles (Global)	3/99 – 3/01	2300
Mining (Global)	2/98 – 2/01	2200
Food (Global)	9/97 – 9/00	3500
Steel (Global)	8/97 – 8/00	2700

return of bottom half companies by 300 to 3000 basis points per year. (Established in 1996, Innovest has rated over 1,500 mostly large cap companies in North America, Europe and Asia.) The following table shows top half – bottom half differentials in a few sectors.[8]

A study by QED International Associates (Blank and Carty, 2001) further illustrates how environmental ratings can be used to enhance investor returns. In the study, three different tests of EcoValue'21 were conducted. In the first test, the stock market returns of two equally-weighted portfolios were compared, top rated companies (AAA and AA—27% of the rated universe) versus the universe of companies rated by Innovest for the years 1997 to 2000. Over the entire period, the top-rated companies returned 12.4% annually versus 8.9% for the universe of rated stocks. The top rated portfolio also had lower volatility than the total universe. (More detailed results and a fuller explanation of the study are available at www.innovestgroup.com.)

Figure 1

The Growth of a $10,000 Investment of Top-Rated Stocks Versus the Total Universe

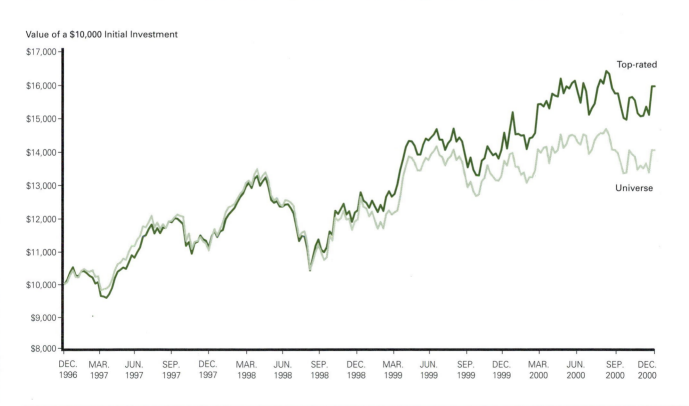

Value of a $10,000 Initial Investment

Source: Innovest and QED International Associates, Inc.

Figure 2

Growth of $10,000 in the Value of Portfolios Tilted to Top-Rated Stocks Versus the S&P 500

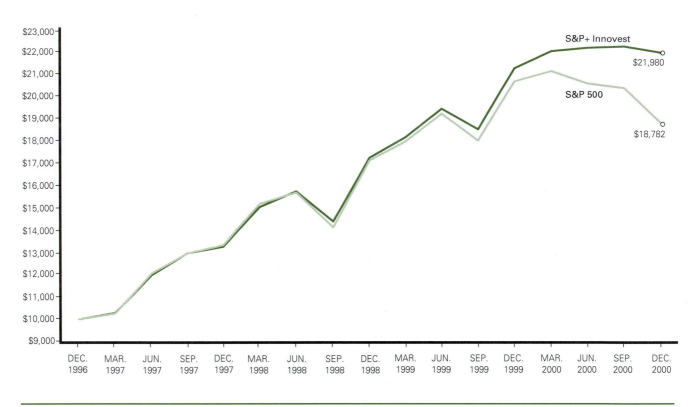

Source: Innovest and QED International Associates, Inc.

Figure 1 shows that $10,000 invested in the Innovest-ranked universe would have grown to $14,037 over four years as compared to $15,946 for the top-ranked companies.

To neutralize the effect of varying sector weights, which may have affected results in the first test, a second test was conducted. In this case, a portfolio was constructed that had the same risk profile as the S&P 500, but favored stocks that were highly rated by Innovest, subject to a 50 basis point tracking error. This portfolio was compared to the S&P 500. Figure 2 shows that the Innovest-enhanced portfolio grew from $10,000 to $21,980 as compared to $18,782 for the S&P 500 over the years 1997 to 2000. The enhanced portfolio also displayed lower volatility than the S&P 500 over the four-year time period. QED estimates roughly half of the widening performance differential in 2000 is due to the downturn in internet-related stocks.

The third test of the EcoValue'21 model compares top-rated companies (AAA and AA) to bottom-rated companies (B and CCC) in the most environmentally sensitive sectors (chemicals, electric utilities, forest products, mining, petroleum, and steel) over the four-year period 1997 to 2000. Figure 3 shows that the portfolio of top-ranked stocks grew to $17,844 in four years versus $14,043 for the bottom-ranked stocks. Over the entire period, the top-rated companies returned 18% annually versus 10.2% for the bottom-rated companies.

The primary reason for the short-term outperformance of environmental leaders found in Innovest's sector analyses and the QED study is most likely that the EcoValue'21 ratings are accurately gauging management quality. In effect, the ratings assess management's ability to deal with complexity, address forces acting upon the company, and modify strategies accordingly. As environmental pressures on firms continue to increase, it is likely that stock market performance differentials between environmental leaders and laggards will also grow. As a result, the importance of incorporating corporate environmental performance analysis into investment decisions will most likely increase over time.

Application by the Financial Community

Many pension funds, asset managers and other financial sector organizations use research on corporate environmental performance to help guide equity investments. The large majority of these investments are made through traditional negative screening approaches. However, a growing number of mainstream financial firms use relative environmental performance under a best-in-class, or positive screening, approach. For example, firms including ABN-AMRO, Bank Sarasin, Cambridge Associates, Neuberger Berman, Rockefeller & Co., Schroders, Société Générale, T. Rowe Price, and Zurich Scudder purchase environmental and social research and advisory services from Innovest. Overall,

Figure 3

A Comparison of the Growth $10,000 Investment in Top and Bottom Rated Stocks in Environmentally Intense Industry Groups

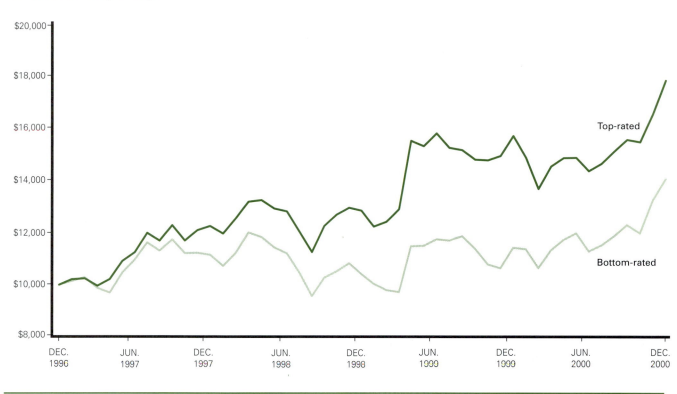

Source: Innovest and QED International Associates, Inc.

Innovest's research is used to positively screen and manage approximately $2.5 billion of invested assets.

Positive screening (e.g., investing in companies with superior environmental performance) can be used for stock picking and indexing investment approaches. To illustrate, the Dutch pension fund ABP (the world's largest pension fund with $175 billion in assets) is using positive screening to establish two $100 million portfolios, one focused on North American equities and the other on European equities. Innovest's environmental and social research will be combined with traditional financial analysis to guide investments toward environmental and social leaders. ABP's goals include reducing risk exposure, enhancing returns, and improving the environmental performance of its investments.

To illustrate indexing approaches, Dreyfus and Mellon Capital Management launched an eco-enhanced index fund strategy in early 2000. The fund maintains the same sector weightings as its benchmark, the S&P 500. But within sectors, Innovest research is used to overweight environmental leaders and underweight lower rated companies. By shifting investments toward presumably better managed companies, the fund is intended to outperform the S&P 500, which it has done. Figure 4 illustrates the structure of the Dreyfus-Mellon fund relative to its benchmark.

Figure 4

Dreyfus Eco-Enhanced Index

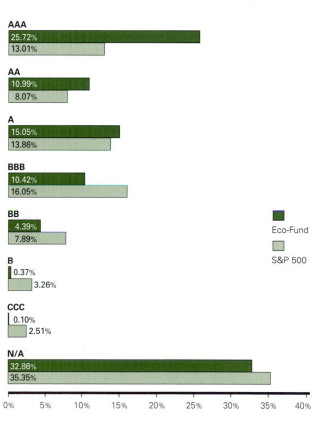

Future Directions

The main barrier to providing investors with corporate environmental performance information has been the lack of high quality data. As noted above, data quality is improving in industrialized nations. However, it remains poor in developing countries. The next barrier on the performance information front is likely to be finding the key drivers of environmental performance that differentiate companies. As more firms recognize the financial and strategic value of improving environmental performance, leading edge environmental management systems and strategies have been implemented more frequently. In some sectors, such as the European pharmaceutical sector, this has already occurred, raising standards broadly and making it difficult to differentiate firms on a corporate environmental performance basis. More refined data, increased disclosure of quantitative impacts, and new analytic tools may, however, continue to permit some differentiation among firms.

To date, corporate environmental performance analysis has usually been focused on assessing incremental performance improvements and systems to foster them. As companies adopt leading edge practices, differentiation on this basis will become more difficult. Given the growing importance of the broader concept of environmental sustainability, an alternative approach is to focus analysis on adoption of visionary strategies, pursuit of large-scale change and movement toward sustainability. To effectively analyze relative sustainability performance, researchers must understand both the concept of sustainability and the actions required to achieve it.

Sustainability is likely to be the greatest challenge ever faced by business because it implies management must address an extraordinarily complex set of interconnected issues and concepts. As environmental pressures and other forces compel companies to bear more complete responsibility for their negative environmental and social impacts, sustainability will likely become a central focus of management and a key driver of business success. This, in turn, will likely cause investment advisors to incorporate the issue into their investment decisions more fully. As evidence continues to mount that environmental performance is financially relevant, not screening for it could become a violation of fiduciary responsibility.

The role of corporations in addressing the challenge to become sustainable is likely to expand significantly for several reasons. Through globalization, privatization, and increasing influence over political processes, corporations are continuing to take on a larger role in society, which implies they will be called upon to play a more central role in sustainability. As pollution and other impacts increase and environmental systems approach limits, and as measurement technology continues to improve, negative corporate impacts will become more obvious, and pressures to mitigate those impacts will increase. And as the necessity to address the sustainability issue from a systems perspective becomes more apparent (meaning recognizing that no part of the Earth's system operates in isolation), calls for companies to more fully address their environmental and social impacts will increase.

Providing investors with high quality information on relative corporate sustainability performance is difficult in part because there is no agreed upon definition of environmental sustainability or the means to achieve it. One method of estimating performance is to develop a theoretical set of actions needed to achieve sustainability, then assess corporate engagement and effectiveness in taking these actions. This requires estimating the responsibility of the corporate sector in achieving sustainability. From a systems perspective, companies impact all aspects of society. As a result, corporations may increasingly be called upon to participate in societal arenas that have traditionally not been their focus.

It is unclear whether the bombing of the World Trade Center on September 11th will affect views on corporate responsibility or the role of corporations in achieving sustainability. However, concerns about security may prompt some companies to assess and mitigate their environmental and social impacts more fully. From an investor's perspective, growing pressures on companies from many different sources may cause sustainability analysis to become as important as more traditional areas of financial and strategic analysis.

Conclusion

The majority of existing research shows that environmental leaders tend to outperform in the stock market. This will likely be true to an even greater degree going forward as companies continue to face growing pressure to improve their environmental and social performance.

Businesses increasingly recognize the importance of environmental performance measurement. Companies across the world currently spend billions of dollars per year on reducing environmental impacts and moving toward sustainability—with growing investments in environmental metrics, accounting, and management systems. Attitudes toward environmental sustainability are diverse, ranging from reactive to proactive. For the most part, proactive companies are those that are best able to effectively address complex, uncertain, and intangible forces affecting their firms. They are usually also leaders in more traditional business areas. As a result, they tend to outperform their competitors in the stock market.

Analysis of corporate environmental performance is used widely in the socially responsible investing area, but not nearly as broadly by the mainstream investment community. This is changing as a growing number of mainstream firms, such as ABN-AMRO, Dreyfus, Mellon, and Schroders, use environmental performance as an additional factor in assessing stock market potential. As the financial impacts of environmental issues continues to increase, higher quality data on corporate environmental performance will be needed to minimize investor risk exposure and maximize returns.

References

3M, "3M Pollution Prevention Pays: Moving Toward Environmental Sustainability," St. Paul, Minnesota, 1998.

Blank, Herb D., and Michael C. Carty, "The Eco-Efficiency Anomaly," QED International Associates, Inc., April 2001.

Butz, Christoph, and Andreas Plattner, "Socially Responsible Investment: A Statistical Analysis of Returns," Bank Sarasin, 1999.

Clough, Richard, "Impact of an Environmental Screen on Portfolio Performance: A Comparative Analysis of S&P Stock Returns," Duke University, 1997.

Cogan, Douglas, "Tobacco Divestment and Fiduciary Responsibility, A Legal and Financial Analysis," Investor Responsibility Research Center, 2000.

Cohen M.A., S.A. Fenn, and J.S. Naimon, "Environmental and Financial Performance—Are they related?" Owen Graduate School of Management, Vanderbuilt University, 1995.

Cram, Don and Dinah Koehler, "The Financial Impact of Corporate Environmental Performance: Evidence of the Link between Environmental and Financial Performance," May 2001, Working Draft.

Dowell, G., S. Hart, and B. Yeung, "Do Corporate Global Environmental Standards Create or Destroy Market Value?," Management Science 46 no. 8, 2000.

European Federation of Financial Analysts, "Eco-Efficiency and Financial Analysis: The Financial Analysts View," Commission on Accounting, 1996.

Feldman, S.J., P.A. Soyka, and P. Ameer, "Does Improving a Firm's Environmental Management System and Environmental Performance Result in a Higher Stock Price?" Journal of Investing 6 no.4, 1997.

Gibson, Virginia L., Bonnie K. Levitt, and Karine H. Cargo, "Overview of Social Investments and Fiduciary Responsibility of County Employee Retirement System Board Members in California," Baker & McKenzie, 2000.

Hart, Stuart L. and Gautam Ahuja, "An Empirical Examination of the Relationship Between Pollution Prevention and Firm Performance," University of Michigan, School of Business Administration, 1994.

Johnson, M.F., M. Magnan, and C.H. Stinson, "Nonfinancial Measures of Environmental Performance as Proxies for Environmental Risks and Uncertainties," 1998, Working Paper.

Konar, S. and M.A. Cohen, "Does the Market Value Environmental Performance?" Review of Economics and Statistics, Volume 83, May 2001.

Koppes, Richard H. and Maureen L. Reilly, "An Ounce of Prevention: Meeting the Fiduciary Duty to Monitor an Index Fund through Relationship Investing," University of Iowa, The Journal of Corporation Law, 1995.

McKeown, William B., "On Being True to Your Mission: Social Investments for Foundations," Journal of Investing, Winter 1997.

Pava, Moses L., and Joshua Krausz, "Corporate Responsibility and Financial Performance: The Paradox of Social Cost," Quorum Books, 1995.

Repetto, Robert and Duncan Austin, "Coming Clean: Corporate Disclosure of Financially Significant Environmental Risks," World Resources Institute, 2000.

Russo, M.V. and P.A. Fouts, "A Resource-Based Perspective on Corporate Environmental Performance and Profitability," Academy of Management Journal 40 no.3, 1997.

Snyder, Jonathan, "The Performance Impact of an Environmental Screen," Winslow Management Company/Eaton Vance, 1993.

Social Investment Forum, "1999 Report on Socially Responsible Investing Trends in the United States," Social Investment Forum News, 1999.

Social Investment Forum, "7 Out of 8 Largest Social Funds Get Top Performance Marks for 2000," Social Investment Forum News, January 2001.

Solomon, Lewis D. and Karen Coe, "The Legal Aspects of Social Investing by Non-Profits," Journal of Investing, Winter 1997.

U.S. EPA, "Green Dividends? The Relationship Between Firms' Environmental Performance and Financial Performance," Office of Cooperative Environmental Management, 2000.

White, Mark A., "Corporate Environmental Performance and Shareholder Value," University of Virginia, McIntire School of Commerce, 1995.

Endnotes

1 The author wishes to thank Carla Tabossi for providing significant assistance in researching and writing this chapter. Sincere thanks are also given to Daniel Esty and Peter Cornelius for providing invaluable editorial feedback.

2 While there are many views on the meaning of environmental sustainability, the concept generally refers to not reducing the ability of the environment to meet the needs of future generations. The broader term, sustainability, addresses environmental and social issues, such as meeting basic human needs and respecting cultural diversity. The two terms are related, not only because they both refer to environmental issues, but also because environmental sustainability probably cannot be achieved without addressing social issues. For example, if basic human needs are not met, people might clear forests (needed for environmental sustainability) to survive in the short term.

3 Following are some of the studies in which positive correlations between environmental and financial performance were found: Konar and Cohen, 2001; Dowell, Hart, and Young, 2000; Johnson, Magnan, and Stinson, 1998; Russo and Fouts, 1997; Feldman, Soyka, and Ameer, 1997; Clough, 1997; European Federation of Financial Analysts, 1996; Cohen, Fenn, and Naimon, 1995; White, 1995; Hart and Ahuja, 1994; and Snyder, 1993.

4 Greenwashing generally refers to efforts by companies to portray themselves as being environmentally responsible, for the purpose of improving stakeholder relations, without significantly improving actual performance. An example of greenwashing might be when an electric utility focuses its corporate environmental report on building wildlife sanctuaries near its headquarters, but hardly addresses its power plants emissions. The sanctuaries are important, but probably not financially relevant to the firm, as emissions most likely would be.

5 Negative screening involves avoiding investments in certain industry sectors with high environmental risks, such as nuclear power, as well as avoiding investments in sectors such as alcohol and tobacco that investors oppose for ethical reasons. This approach can reduce diversity and increase portfolio risk.

6 For example, a letter from the U.S. Department of Labor to William M. Tartikoff, Senior Vice President and General Counsel of Calvert Group Ltd., dated May 28, 1998, stated that the Department of Labor "has expressed the view that the fiduciary standards of [ERISA] sections 403 and 404 do not preclude consideration of collateral benefits, such as those offered by a 'socially-responsible' fund, in a fiduciary's evaluation of a particular investment opportunity," provided that the investment is equal to or superior to alternative available investments on an economic basis.

7 Information from these databases can be downloaded from the nonprofit Right-To-Know network (www.RTK.net), which has been given authority by the U.S. government to disseminate these data to the general public.

8 More information about EcoValue'21 and Innovest in general is available at www.innovestgroup.com.

Chapter 6
Corporate Sustainability and Financial Indexes

Alois Flatz[1]

Introduction

Investors increasingly see forward-looking nonfinancial metrics as key to understanding a company's true value, while stakeholders demand ever greater transparency from multinationals on how they are managing their responsibilities to society at large and the environment. These trends are manifested in the increasing impact sustainability is having on mainstream investors, as illustrated by the following:

- Since its launch in 1999, the Dow Jones Sustainability World Index (DJSI World)[2] has benchmarked the performance of the world's leading sustainability companies, enabling investors to track and adapt their products along sustainable guidelines. As of August 2001, 33 licensees have created financial products based on the DJSI World, with total assets under management amounting to more than 2.2 billion Euro. These investors are offering sustainability-driven mutual funds, equity baskets, certificates, and segregated accounts.

- Other leading organizations on both sides of the Atlantic have also been involved in supplying the investment community with indexes around this theme. For example, since 1990, KLD (KinderLydenbergDomini) has managed the Domini 400 Social Index[3] with particular emphasis on the societal contribution of North American companies. Moreover, in July 2001, UK-based index provider FTSE (Financial Times Stock Exchange) launched the FTSE4Good[4] series of indexes to provide a benchmark to the SRI (Socially Responsible Investment) community.

- Recent legislative changes in Europe and Australia have also confirmed the trend toward incorporating sustainability criteria in investment decision-making. For example, both the UK[5] and Germany[6] have passed laws obliging pension funds to disclose their investment policy with regard to environmental and social criteria. In August 2001, Australia's Senate[7] passed an amendment to the Financial Services Reform Bill to require super funds and investment managers to disclose their policy on ethical investment.

- Significantly, the Swiss Federal Social Security Fund awarded State Street Global Advisors, the investment management arm of State Street Corporation, a 320 million Euro global equity index mandate based on the DJSI World in May 2001, the largest mandate of its kind.[8]

Previously, and despite high demand, the investment community was unable to rely on credible socioenvironmental data on which to base microeconomic decisions due to lack of availability of relevant data. Now, the ability to make macroeconomic decisions around socioenvironmental issues, which has been possible for many years, is complemented by a new ability to make associated microeconomic decisions due to recent developments in data provision around sustainability.

This paper outlines the challenges in assessing the sustainability performance of the world's largest companies and integrating the results into an index tracking the financial performance of leading companies. Following a definition of corporate sustainability, we focus on exploring the characteristics of traditional indexes and what implications these have on the challenge of indexing the performance of companies that embrace sustainability. Subsequently, we describe the process for developing the Dow Jones Sustainability World Index (DJSI World) as an introduction to the issue of how the specific challenges of assessing corporate sustainability have been addressed, with particular focus on the implications for indexes tracking the stock market performance of companies embracing sustainability. Finally, the risk and performance attributes of the DJSI World are explored to determine the value and validity of tracking the performance of corporate sustainability leaders in general.

Table 1

Sustainability Driving Forces

Ecological Forces	Sociocultural Forces	Economic Forces
Global climate changes and ecological instabilities	Global transparency in society through media and technological connectivity—corporate behavior is clear for all the world to see	Increasing speed-embracing innovation and product cycles, business relationships, and competition
Increasing ecological degradation with negative impact on human health and quality of life	Divergent demographic trends in developed and less-developed regions	Continuous scientific and technological progress
Loss of ecosystems and biodiversity	Wide social imbalances and inequalities in developed and less-developed regions (income, poverty, human rights, etc.)	Information is key factor
Lower capacity of natural sinks as carrying capacity is decreasing (soil, water, forests, etc.)	Urbanization and urban lifestyles	Technological connectivity and virtualization of (business) relationships
Scarcity of water in terms of both quality and volume	New lifestyles of self-organized groups with shared values	Globalization and liberalization of economic activities
	Consumer behavior changing due to awareness of inequalities, social imbalances, human rights, and unfulfilled development potential	Increasing power of multinational businesses compared to national states
	Consumer behavior changing due to awareness of ecological changes and social instabilities	Shift from supply-side to demand-side markets
	Healthy living as an important part of individual lifestyles	

Definition of Corporate Sustainability

One of the early attempts to define sustainability was presented by the Brundtland Commission[9] at the United Nations General Assembly in 1987: "Sustainable development is development that meets the needs of the present without compromising the ability of future generations to meet their own needs" (World Commission on Environment and Development, 1987).

If one considers businesses as living organisms, the need to behave in a way that avoids compromising societal and ecological systems is clear. Not compromising their ecological, societal, and economic environment in a wider sense is of vital interest to a company. As the environment in which economic activities are embedded becomes more unpredictable and fragile, higher-risk premiums are paid for increased volatility and uncertainty. Highly efficient economic systems (to ensure efficient resource allocation) require predictability and cannot afford to spend great resources on stabilizing the systems of the wider environment (for example, financing safety measures or remediation). Sustainability on a long-term basis requires businesses to satisfy the needs of their clients in a way that their products and the organization of their services (the value chain) follow the dynamic systems rules of ecological and sociocultural systems. Since societal, cultural, and ecological systems are not stable, their dynamic balances are in a state of permanent change.

Thus, businesses survive, not by maintaining a static business model, but by propagating their species, thereby indicating that sustainable business must be a dynamic state. Consistently reinventing a company in line with market, societal, and environmental realities and changes to ensure its own sustainability has proved the key to longevity and profitability of businesses (de Geus and Senge, 1997). The challenge of enabling sustainable economic growth, in the broad sense, offers new opportunities for companies to enhance their shareholder value by aligning with the emergent realities of the environment in which they operate. In line with this philosophy, *corporate sustainability* can be defined as:

a business approach to create long-term shareholder value by embracing opportunities and managing risks deriving from economic, ecological and social developments or changes.

These economic, ecological, and social developments or changes are trends that need to be considered and managed effectively, by maximizing the opportunities and minimizing the risks they present, if a company is to contribute to a sustainable future for itself and the parts of the systems in which the business is embedded. While a full discussion of these forces is beyond the scope of this paper, a summarized list of trends is illustrated in Table 1.

Increasing awareness of the trends highlighted above enhances the importance of a company's management of these developments. Companies that understand how these forces offer opportunities to enhance shareholder value, or pose risks to shareholder value, and adjust their corporate strategies and operational procedures accordingly, are on the road to achieving corporate sustainability. Probably one of the most important driving forces is the dramatic increase in transparency, caused by global media and connectivity. In turn, this increases the importance of brand integrity and the need to develop trust among all stakeholders.

Using Corporate Sustainability in the Investment Community

An index tracking the performance of companies addressing sustainability is valuable when it provides insight into the future financial prospects of a company or industry that conventional analysts are unlikely to incorporate, given a lack of focus on the potential for certain social, environmental, and economic issues in society to materialize and affect companies.

As traditional valuation metrics and historical corporate information increasingly concede importance to future-oriented, forward-looking indicators of the health of a company and its attractiveness to an investor, indexing the performance of companies addressing sustainability attempts to provide investors with the insights they are increasingly seeking (Funk, 2001). Thus, as an investment insight, equity research in relation to sustainability must be:

- forward-looking;

- based on industry-specific value drivers (as opposed to generic data);

- transparent and easily understood; and

- capable of adding value to existing valuation methods.

Assessing corporate sustainability aims to incorporate the characteristics mentioned above and offers insight across most equity asset classes and investment styles. The hypothesis for the business case for sustainability interpreted as a portfolio of stocks is that these stocks will be expected to outperform comparable portfolios, at least in the long run. The reasoning for this expectation is sound. Companies embracing global sustainability trends are likely to achieve a higher return on equity (ROE) and/or a lower required rate of return (RRR) than companies that ignore these trends. Higher ROE may result from a better understanding of investment opportunities or from lower non-operating cost, because of a better understanding of risks. Higher ROE may also result from social pressure groups channeling demand into sustainable products.

A lower RRR may result from a better understanding and management of risks. The RRR is a function of both operat-

ing and financial risks. Companies embracing sustainability trends may reduce their operating risks and, thereby, lower their equity costs. It presumably would also result in lower borrowing costs, leading to lower costs of capital, and, again, to higher ROE. Lower borrowing costs may also be the result of investors considering other parameters than just risk and return. High ROE and low RRR result in free cash flow that can be invested profitably when embracing sustainability trends. A portfolio, or an index composed of this type of company, thus will appreciate faster than a portfolio or an index of companies not embracing theoretically profitable investment opportunities. Investments in companies embracing sustainability thus promise higher returns and, due to lower business risk, better risk-return ratios. Based on this hypothesis, better performance can also be expected on a risk-adjusted basis.

Index Characteristics & Challenges for Sustainability

Characteristics of an Index

Security market indexes, or, more generally, security market indicator series, are designed to reflect valuation and changes in the valuation of a specific universe of securities (Reilly and Brown, 1997). They are intended to represent a universe or population of securities. Movements of the index's performance are supposed to allow inference to movements in the performance of the securities in the underlying universe. The universe targeted, together with the purpose of the indicator, determines which factors need to be considered in designing the index. Sample size, sampling method, and the weighting scheme are prominent examples of design elements.

The targeted universe can cover as much as the broad domestic economy or as little as some subset of a particular niche security market. The DJSI reflects changes in valuation of a universe of companies that are leaders in terms of corporate sustainability. The universe of companies embracing sustainability is broader than the DJSI, but the DJSI comprises the leaders in corporate sustainability, as judged by industry. Thus, the DJSI not only traces, but also implies, a universe of leading companies with regard to addressing sustainability.

Indexes are based on statistical information and calculation and, in general, have the following characteristics:[10]

- accurate and reliable data;

- clear, transparent, and replicable methodology;

- rules-based processes;

- objective and bias-free; and

- component data freely available.

An index tracking the performance of corporate sustainability leaders must have all the above-mentioned characteristics of a traditional index. In addition, it must be flexible enough to meet changing indexing trends and investor demands, such as the demand for broader benchmarks.

Challenges for an Index Tracking Corporate Sustainability

Over recent years, increasing numbers of investors have been aligning their investment strategies along sustainability criteria. This growing number of new financial products integrating sustainability in their core investment strategy provided the impetus for a neutral, rigorous, transparent, and easily replicable measurement of corporate sustainability. The challenge facing the indexing industry has been how to measure and quantify corporate sustainability, and how to integrate the results into an investable index that meets the needs of the investment industry.

By incorporating this type of equity research into an index, very specific new challenges arise—namely, and in no particular order:

Development of relevant assessment criteria (generic and industry-specific)

An index tracking the performance of corporate sustainability leaders first needs to define corporate sustainability and relevant assessment criteria. Criteria representing the challenges deriving from sustainability trends have to be developed and quantified in such a way that the best-positioned companies can be measured and identified.

Gathering of corporate information

An important challenge is how to gather the correct and relevant information to measure economic, environmental, and social performance dimensions. While some global companies publish corporate sustainability reports, the majority of companies are only beginning to understand and, hence, report on the concept of corporate sustainability. More important, not all data are consistent, relevant, or comparable.

Quantification of corporate sustainability

A key challenge in developing an index tracking corporate sustainability is how to quantify corporate sustainability. In most cases, sustainability developments are qualitative in nature, so they may lack easy quantification. While assessing companies' environmental performance and emission targets seems relatively straightforward, a consistent and equally quantifiable method is not readily available for many aspects of social and economic developments.

Identification of sustainability leaders in each industry group

Given that sustainability trends affect each industry differently, industry-specific challenges arise. As a result, industry

leaders need to be identified for each industry group, known as a "best-in-class" approach. Sustainability leaders within each industry group need to be ranked according to their corporate sustainability performance relative to one another. Individual industry groups should not be excluded based on the perceived sustainability of the particular sector.

Constructing an index with appropriate selection rules

A further challenge is how to construct the index. While most indexes represent a group of stocks with a specific goal, it is imperative to define clear selection rules reflecting the particular focus of the specific index. The number of companies considered corporate sustainability leaders and how their stock should be weighted is the critical consideration and challenge when selecting final components. When identifying leaders for each industry group, minimum sustainability standards need to be set to clarify at what threshold a company should no longer be considered a sustainability leader.

Fulfilling traditional index requirements

A corporate sustainability index needs to fulfill all the characteristics of a traditional index—it should be accurate, reliable, transparent, and consistent.

The Dow Jones Sustainability Indexes,[11] a collaboration of Dow Jones Indexes and Sustainable Asset Management (SAM), were developed as a response to these needs and challenges and, since its launch in 1999, have grown in number of licensees and assets under management. The next section of this paper explores the process used to develop the DJSI World as an example of how these challenges are being met.

Construction of the Dow Jones Sustainability Index

The index construction process of the Dow Jones Sustainability World Index (DJSI World) provides a good introduction to how index providers address the challenge to produce indexes tracking the financial performance of sustainability leaders. A description of the general challenges and how these were solved in the most important process steps of the DJSI World is described below (with the exception of the assessment step, which is described in detail in the next section).[12]

Investable Universe and Industry Allocations

In determining the initial investable universe from which the final components of the index are selected, key issues to consider are the core purpose of the index, what the index should represent, the acceptable level of liquidity of the stocks, the acceptable level of tradability, and the optimal level of convergence of the currencies represented. The Dow Jones World Index of the 2,500 largest companies in market capitalization, of the total universe of more than 5,000 companies, is used as

Figure 1

Overview of DJSI Construction Process

Dow Jones Sustainability Indexes

A cooperation of Dow Jones Indexes, STOXX Ltd. and SAM Group

Index Construction Process

- Investable Universe
- Allocation to Industry Groups
- Corporate Sustainability Assessment
- Ranking within Industry Groups
- Index Component Selection
- Index Calculation
- Ongoing Review

Annual Review

a basis for the DJSI World investable universe. The targeted universe can cover as much as the broad domestic economy or as little as some subset of some niche security market.

A further critical step is determining the allocation of industry groups for the investable universe. The homogeneity of the stocks allocated to each industry group must be fairly high because dependence on similar sustainability trends allows comparison among the relative performance of industry components, which are based on the 64 industry groups of the DJSI World.

Component Selection

Several challenges exist in the actual construction process. Regarding the ideal number of index components, the purpose of the index must be addressed. Decisions must be made about whether the index should provide an investment universe and benchmark for active asset managers or the direct basis for a financial product (such as an investment certificate). An active asset manager usually prefers to have a wide investment universe to have the possibility of choice in stock selections; however, a passive investor prefers a smaller universe to keep transaction costs low (for example, about 30 stocks for a certificate).

Moreover, the number of components has a major impact on the risk attribution of the index as a whole and depends on the overall sustainability score of companies that surpass a

minimum threshold of quality, the exact percentage of which is determined by specific assessor approaches. The DJSI decided to provide both a benchmark and an investment universe (DJSI in 2001–2002 contains approximately 300 components).

Setting the right threshold distinguishing the best-positioned companies from the others, and setting minimum standards to be applied should the overall level of quality within an industry be poor, is also a critical consideration regarding component selection. The threshold depends on the number of components needed per industry grouping because the "best-in-class" approach is not always applicable, given that there are industries in which few companies react to sustainability trends.

Component selection can be based on one of three possible approaches: market capitalization, numeric, or a mixture of both. Selection based on market capitalization has the major disadvantage of possibly being dominated by a single company with a very high market capitalization. Should the top-ranked company in terms of sustainability have a high market capitalization, no other companies embracing sustainability in that industry sector could qualify because the allocation of allowable market capitalization will already have been taken up by very highly capitalized stocks.

Selection based on the numeric approach would, for example, select the top 10 percent of companies based on the number of companies in the specific industry group. This method would select 250 companies from an industry group comprising 2,500 companies. This approach has the disadvantage of often not providing an ideal asset allocation per sector and, hence, a possibly limited risk spread.

Therefore, the DJSI World pursues a mixed approach incorporating both numeric and market capitalization-weighted elements,[13] which allows for a good representation of an industry's market capitalization while also assuring that the leading sustainability companies are included.

A further step in the selection of components is how the selected components should be weighted. With regard to traditional indexes, three methods are used to weight the components within indexes:

- market value-weighted (for example, NYSE Composite Index, Standard & Poor's 500 Index, Nasdaq Composite Index, Wilshire 5,000, London FTSE, MSCI Indexes);

- price-weighted (for example, Dow Jones Industrial Average, DAX 100); and

- equal-weighted.

Regional and sector allocation, currencies, and the method of stock weightings need to be similar. In the case of the DJSI World, free-float price-weighted market capitalization was selected to reflect the DJGI's move to a similar basis.

In the next section, we examine how the DJSI World specifically addresses the challenges to assessing corporate sustainability and to developing an index to track the performance of companies addressing this issue.

Challenges of Measuring (Assessing) Corporate Sustainability and Developing an Index

The assessment of corporate sustainability performance forms the basis for an index tracking the performance of companies embracing sustainability. In this section, four specific challenges of corporate sustainability assessment are addressed:

- defining corporate sustainability criteria;

- gathering corporate sustainability information;

- quantifying corporate sustainability assessment results; and

- meeting requirements of traditional indexes.

Defining Corporate Sustainability Criteria

Selecting relevant and quantifiable criteria to assess corporate sustainability is a major challenge because the quality of the index components depends heavily on this aspect of the assessment process. Assessment criteria should be easy to measure, understandable, clear, and precise. Corporate sustainability is widely based on qualitative criteria, so the most significant challenge is to develop quantitative proxies for qualitative data and integrate these into a system that meets the major requirements of indexing (for example, the need for replicability and objectivity).

However, even quantitative data can be difficult to access for many reasons. First, as there are no standards for sustainability accounting (of environmental and sociocultural issues) and no legal obligation for accounting along these issues, data are not readily available. Second, accessing and comparing companies based on environmental and social information is difficult because companies are active in very different business lines even if they are considered to be part of the same industry group (for example, IBM and Dell are not comparable). Third, system borders may be defined differently by different companies because most use production sites as system borders, although some may include transport and storage sites as well. The only viable approach to defining corporate borders, therefore, would be to consider the whole life cycle of a product. However, companies often do not have the data since they

do not control the whole value chain. Fourth, there are no specific clear indicators from the investment community, and only very divergent standards for normalization and categorization (such as aggregating emissions according to global warming potential). Finally, many companies simply do not have any data to report on many of the issues that corporate sustainability assessments must cover, and when they do, there is very little historical information.

In the case of qualitative criteria, integration of qualitative issues into criteria with precisely defined parameters (for example, closed-end or multiple-choice questions to which a score is attributed) is the approach used in most assessment methodologies. However, a major challenge is defining the criteria parameters. For example, how many environmental policies/charters are considered the ideal number to be signed—two, five, or ten—when the real issue is the quality of the implementation of these charters?

Moreover, maintaining relevance of the criteria is another important challenge. Trends and industry challenges are in constant flux, which means that criteria must be updated constantly. For the DJSI World, assessment criteria are updated annually to keep them relevant. Furthermore, criteria should represent not only challenges deriving from global trends, but also regional challenges where possible. SAM's approach is based on the hypothesis that large-cap companies face similar challenges based on global social, environmental, and economic trends. The same rigorous approach applied worldwide for the DJSI World allows for clear comparisons despite geographic spread, and worldwide relevance of criteria.

Given the wide range of sustainability trends and driving forces, criteria must be selected geared to distinguish between sustainability leaders and laggards. In addition, criteria need to be interdependent to represent the systemic nature of companies. Studies have shown that companies that are leaders in one criterion are usually leaders in others as well.[14] Therefore, it is important to select the right criteria, rather than a wide range of various unrelated criteria.

Table 2 provides an overview of the criteria used in the SAM Corporate Sustainability Assessment, which forms the basis for the DJSI World. This approach has been developed with the intention of addressing the challenges indicated above. In addition, criteria are derived from the sustainability trends highlighted in earlier sections and focus on factors such as industry value drivers and success factors in relation to managing the challenges of a changing environment. Each criterion listed is divided further into subcriteria that are not listed.

In essence, assessing a company by specific criteria incorporates evaluation of a number of issues, specifically:

- exposure of a company to a specific criterion (the challenge the company faces with regard to particular criterion);

- the company's policies and strategies to cover specific criteria;

- the management systems to implement policy/strategy;

- internal review processes to check progress in relation to specific criteria; and

- a company's track record—both quantitative (if released) and qualitative (documents, interviews, media, and stakeholders.

Criteria used in the DJSI World corporate sustainability assessment are reviewed annually, and external experts are asked to provide insight and recommendations to improve criteria and the entire assessment methodology. Moreover, efforts such as the Global Reporting Initiative are taken into account to ensure that the criteria used are fully aligned with efforts to standardize corporate sustainability reporting and assessment.

Gathering Corporate Sustainability Information

Standards for corporate sustainability assessment and reporting, unlike financial reporting and analysis, have not been defined or standardized so new methods to gather relevant and applicable information have to be created. Many sources of information exist, including company reports, questionnaires, industry studies, interviews with companies, third-party opinions (stakeholder inputs), and various media. Some sources are used to access basic information and insight; others are used to check the validity and truthfulness of company responses to questionnaires and interviews.

Because companies have differing interpretations of how to report sustainability performance, many environmental, social, and/or sustainability rating agencies have created questionnaires to fill this void. However, people may read and understand questions in an assessment questionnaire differently because of language barriers, lack of guidelines, or

Table 2

SAM Corporate Sustainability Assessment Criteria

	Opportunities	Risks
Economic	• Strategic and financial planning • Organizational development • Intellectual capital management • IT management and IT integration • Quality management • Customer relationship management • Branding and brand management • Reporting and accounting Industry specific (for example) • R&D programs	• Corporate governance • Risk and crisis management • Compliance systems • Corporate codes of conduct Industry specific (for example) • Specific risk management issues
Environmental	• Environmental policy and integration into overall strategy • Environmental charters • Responsible person for environmental issues • Environmental, health, and safety reporting Industry specific (for example) • Eco-efficient products and services (including strategy, R&D)	• Environmental management systems • Environmental input and output performance • Environmental profit and loss accounting Industry specific (for example) • Hazardous substances • Environmental liabilities
Social	• Social policy and integration into overall strategy • Responsible person for social issues • Stakeholder involvement • Social reporting • Employee benefits • Employee satisfaction • Remuneration systems Industry specific (for example) • Community programs	• Conflict resolution • Equal rights and nondiscrimination • Occupational health and safety standards • Freedom of employee organization • Standards for suppliers Industry specific (for example) • Personnel training in developing countries

cultural and regional differences. For example, there may be significant room for differing interpretations of questions by individuals within the company, and some companies may take an exaggerated rather than a more conservative approach to answering questions. Moreover, questionnaires are expensive for both parties, and feedback rates are usually low, leading to an inadequate assessment universe and information base.

Regarding documentation provided by the companies, language issues exist given the global coverage of the assessment. This issue is particularly acute with regard to Japanese companies, which often require the assessor to have all documentation translated from Japanese to English. Analysis of company documentation is also open to misinterpretation, and there is a high risk of error, exemplified by the fact that this part of the corporate sustainability assessment for the DJSI World has the highest potential for errors according to external reviews and auditor reports.[16] In addition, there are cultural differences in how much information actually is recorded in company documentation.

Furthermore, conducting company interviews is very costly, and their effectiveness depends heavily on the availability and seniority of the interviewee. Thus, replicating this approach across different companies is very challenging.

The wider media and the opinion of stakeholders are used primarily to verify the truthfulness of feedback provided by companies. This process is highly efficient, given the use of modern databases and the Internet. For example, for the DJSI World, the Dow Jones Interactive database, which covers more than 45,000 media sources worldwide, is tracked daily. An important consideration for the assessment is to avoid one-sided interpretations of data.

A further challenge in gathering company information for corporate sustainability assessments is the issue of fair disclosure to all shareholders. Companies are required by law to provide all shareholders with exactly the same information so as not to give one shareholder an advantage over another. Many corporate sustainability assessments ask companies for information that they are not willing to provide to the wider financial markets, and this creates gaps in much of the feedback provided to assessors and rating agencies.

For each company in the DJSI World assessment, the input sources of information for the corporate sustainability assessment consist of the responses to an online corporate sustainability assessment questionnaire, submitted documentation, publicly available information, analysts' direct contact with companies, and/or their main clients and the media. Questionnaires specific to each DJSI World industry group are distributed via www.sam-group.com to companies in the

DJSI World investable stocks universe. The questionnaire is designed to ensure objectivity by limiting qualitative answers through predefined multiple-choice questions. The completed company questionnaire, signed by two senior company representatives to ensure truthfulness, is the most important source of information for the assessment.

For the DJSI World, the questionnaire process has been streamlined to enable companies to answer the questions online. This interactive tool enables a company to update and change its most recent updates in the existing questionnaire and, in turn, facilitates efficient and accurate assessment by the analyst. The feedback rate of the corporate sustainability assessment questionnaire has improved from 15 percent in 1999 to 25 percent in 2001. Furthermore, the process helps to reduce the error margin of data input and interpretation by analysts.

Analyzed documents include sustainability reports, environmental, health and safety, social, and annual financial reports and all other sources of company information—for example, internal documentation, brochures, and websites. Analysts also review media, press releases, articles, and stakeholder commentary written about a company over the past year. Finally, each analyst personally contacts individual companies to clarify questions arising from analysis of the questionnaire and company documents.

Although corporate sustainability assessment information gathering depends heavily on a company's willingness to participate in the process (which is often determined by the incentive of being part of the index), the trend toward greater transparency in corporate reporting eventually will allow for more streamlined access to information and, hence, greater efficiency for both assessors and assessees.

Quantifying Corporate Sustainability Assessment Results

Quantification is clearly a major challenge in corporate sustainability assessment, the overall objective of which is to aggregate the performance of a company in terms of specific criteria into an overall sustainability performance score. Challenges exist regarding quantification of qualitative data, management of large volumes of company data, and consistent objective application of the assessment methodology

Qualitative criteria are measured via an ordinal metrics scheme, which allows the differences between companies' performances to be expressed in such a way that one company can be identified as better or worse than the next, rather than in absolute score terms. Subsequently, aggregation of the performance scores is done via the weighting of the criteria answers, the challenge being to bring the criteria into a meaningful relationship that represents the correct impor-

Table 3

Individual Criteria Weightings

Answer	Weight of Answer	Weight of Question (Question 45)	Weight of Theme (Environmental Charters)	Weight of Class Opportunities) (Strategic Sustainability Opportunities)
Member of more than 3 accepted charters	100	.40	.25	.15
Member of 2–3 accepted charters	.66			
Member of 1 accepted charter	.33			
No charters signed	0			
No answer	0			

tance of a criterion relative to the overall system. Regarding the potential for subjectivity in this quantification, this is addressed by adopting the Delphi approach of accessing expert input.

Moreover, the increased use of information technologies may increase the quality of data management, capacity, and efficiency. For the DJSI World, approximately 1,000 companies are assessed yearly and monitored constantly, which requires massive data management skill to ensure effective data capture and analysis to ensure, in turn, effective quantification of corporate sustainability performance scores. Use of an extensive database and introduction of an online questionnaire enhances data security and helps to increase accuracy and data replicability and comparability.

Corporate sustainability assessment data are open to wide misinterpretations from both the assessed company and the assessor. It is vital to ensure that subjectivity is minimized while developing a company's quantified score. Using the same analyst for an entire industry group facilitates a coherent application of views. Constant internal assessor training, clear working procedures, and cross-checking by different analysts foster objectivity and accuracy of results. External audit reviews also may be used to ensure objectivity and replicability, both of which are critical factors for an index.

The corporate sustainability assessment enables calculation of a sustainability performance score for each company. Reviewing, assessing, and scoring all available information in line with corporate sustainability criteria determines the overall sustainability score for each company in the DJSI World investable universe and, subsequently, allows comparison of performance and identification of leading companies to be included in the index. In the DJSI World, corporate sustainability assessment and subsequent quantification and scoring is conducted in three stages: questionnaire assessment, quality and public availability of

information, and verification of information. This process is described briefly below:

Stage 1: Questionnaire Assessment

All answers provided in the questionnaire receive a score. Each question has a predetermined weight for the answer, the question, and the theme and class within the question.[17] The total score for the question is the combination of these weights. For example (see also Table 3):

Question 45: Has your company signed environmental charters, or is it committed to the principles of sustainability councils/coalitions?

☐ Yes, please specify_____
☐ No charters signed

The weighting of each answer is automatically calculated by the SAM Information Management System and totaled to give a score for the questionnaire.

The weight of the questionnaire as an information source depends on whether the questionnaire and answers provided have been approved—in other words, signed by at least one senior management member of the company. If the questionnaire is not signed, less weight is given to the questionnaire and, accordingly, more weight is assigned to Stage 2.

Stage 2: Quality and Public Availability of Information

Company documents and publicly available information about a company are scored according to their scope, coverage, and ease of access.

The *scope and coverage* of a company's documentation is evaluated for each dimension: economic, environmental, and social. In this stage, the analyst assesses how well implementation of policies and management systems is documented across the entire company. The evaluation is scored based on the scale portrayed in Table 4.

Table 4

Level of Scope and Coverage

Level of Scope and Coverage	Score	Criteria
Low	0	No documentation *or* only few case studies without context or lack of worldwide coverage
Medium	1	Acceptable quality of documentation. Description of policy/management system/activity, and information about coverage in a systematic way
High	2	Good quality of documentation. Strategy, implementation of management systems, performance against targets described in detail

The *ease of access* to a company's publicly available information and documentation is evaluated for economic, social, and environmental dimensions (see Table 5). This evaluation covers all publicly available documentation as well as information provided to SAM Research for the assessment.

Stage 3: Verification of Information

Information provided in the corporate sustainability assessment questionnaire is verified to ensure that quantification of a company's sustainability performance reflects reality. The verification process assesses whether a company implements and commits to its stated policies and management practices. The verification process begins with cross-checking a company's answered questionnaire with documents provided by the company and publicly available documents, which are considered in addition to the issue of public availability of information mentioned above. In addition, the company's record is verified by reviewing media and stakeholder reports. If necessary, direct interaction and clarification with the company are also undertaken by the analyst to verify selected parts of the assessment.

For each dimension, a sample of company answers is analyzed in depth to verify truthfulness. If the answers provided by the company cannot be validated or are contradicted by information found in company documents or other publicly available documents, this is reflected in a lower truthfulness factor, which results in a lower allocation of performance scores.

Within this context, the consistency of a company's behavior and management of crisis situations is reviewed in line with its stated principles and policies. Issues such as commercial practices (for example, tax fraud, money laundering, antitrust issues, balance sheet fraud, and corruption cases), human rights abuses (for example, cases involving discrimination, forced resettlements, child labor, and discrimination against indigenous people), workforce conflicts (for example, exten-

Table 5

Level of Public Availability

Level of Public Availability	Score	Criteria
Low	0	No public information
Medium	1	Most information is for internal use, some public information exists
High	2	Most information is publicly available

sive layoffs and strikes), and catastrophic events or accidents (for example, fatalities, workplace safety issues, technical failures, ecological disasters, and product recalls) are included in this part of the corporate sustainability assessment.

In the internal review, SAM research weighs the severity of the crisis in relation to the company's reputation and crisis management quality. If the company fails to meet its stated policies and management practices as found through the review, scores may be reduced or deleted entirely for whole criteria or specific individual questions. In extreme cases, the verification process may exclude companies from the eligibility list (decided by the DJSI Index Design Committee).

Based on the three stages outlined above, a company's total corporate sustainability score is calculated. According to predefined scoring and weighting structures, answers provided on the questionnaire by the company are weighted against scores for quality and public availability and for truthfulness of information. The resulting corporate sustainability score is verified through review of a company's involvement in critical issues. A company's total corporate sustainability score at the highest aggregated level is then calculated.

Corporate Sustainability Assessment Score Formula:

- $TS = \sum[CLW*CRW*QUW*\sum AS*(QAW+DAW*DAS)] -QVS$ for all questions

- TS = Total Score

- CLW = Class Weight

- CRW = Criteria Weight

- QUW = Question Weight

- QAW = Questionnaire Assessment Weight

- DAW = Weight of Quality/Public Availability, and Truthfulness of Information

- DAS = Score for Quality/Public Availability, and Truthfulness of Information

- AS = Answer Score

- QVS = Questionnaire Verification Score

The results of the calculation outlined above are based on ordinal metrics, which means that the scores can only be used for comparison and ranking purposes, not to determine a company's absolute performance score. It is not possible furthermore to use the scoring for weighting purposes.

The calculation of the score for corporate sustainability assessment is the basis for selection of components for the DJSI World. A key focus in the future will be on how to reconcile this score with shareholder value creation or destruction.

Meeting Requirements of Traditional Indexes

An index tracking the performance of companies embracing sustainability has to meet the requirements of traditional financial indexes as far as possible, especially with regard to accuracy and reliability of data; transparency and replicability of the methodology; and processing based on rules, objectivity, and independence. In the following section, we discuss these requirements with respect to the sustainability component and information included in the DJSI World. The traditional index requirements (such as data quality for the index calculation) are not discussed at this stage because we assume they are subject to the same rigorous scrutiny as all other indexes of the DJGI index family.

To ensure *accurate and reliable data*, the DJSI World is reviewed quarterly and annually to ensure that the index composition accurately represents the top 10 percent of leading sustainability companies in each of the DJSI World industry groups. Various Information Technology (IT) systems help to increase data quality, verification systems are used as described earlier, and quantified proxies of qualitative information are designed to enable data accuracy and reliability.

Moreover, to ensure quality and objectivity, an external audit and internal quality assurance procedures, such as cross-checking information sources, are used to monitor and maintain the accuracy of the input data, assessment procedures, and results. In addition to quarterly and annual reviews, the DJSI World is continually reviewed for changes to the index composition necessitated by extraordinary corporate actions—for example, mergers, takeovers, spin-offs, initial public offerings, delistings, and bankruptcy—affecting the component companies and their corporate sustainability performance. Finally, corporate sustainability monitoring is part of the ongoing review process. Once a company is selected as a member of the DJSI World, its corporate sustainability performance is monitored continuously.

With regard to *transparency and replicability of the methodology*, each of the Dow Jones Sustainability Indexes is accompanied by publication of a guidebook[18] outlining all of the decisions that have been made in development of the DJSI, especially in terms of meeting all of the challenges mentioned in this paper. It includes the corporate sustainability methodology, index features and data dissemination, periodic and ongoing review, the calculation model, and the management and responsibilities of all parties involved. In addition, the primary research of the DJSI World is based on a consistent rule-based methodology, the details of which are transparent via web publication.

To ensure *objectivity and freedom of bias toward companies and investors* for an independent and accurate corporate sustainability assessment and index construction, all processes are reviewed by an external, independent auditor for the DJSI World. An average error margin for the corporate sustainability performance assessment is determined by reviewing a representative random sample of 25 companies among the upper half of the companies that were not selected because of relatively low corporate sustainability performance scores and the lower half of companies selected for the DJSI World. The average error margin in 1999 was 0.55 percent, 0.72 percent in 2000, and in 0.74 percent in 2001.

Free availability of data is assured by publication of all details related to the corporate sustainability questionnaires, criteria groups, weightings used for scoring aggregation purposes, overall results, and index components.

Finally, to ensure that the index assessment is *independent and objective*, the DJSI World has established two important committees. First, the DJSI World Index Design Committee is solely responsible for all decisions on the composition and accuracy of the DJSI World. In particular, the DJSI World Index Design Committee is solely responsible for all changes to the index methodology, which is detailed in the current DJSI World Index Guide. Second, the DJSI World Advisory Committee is composed of independent, third-party professionals from the financial sector and corporate sustainability performance experts. The Advisory Committee advises the DJSI World Index Design Committee on index composition, accuracy, transparency, methodology, and the corporate sustainability performance assessment in line with the latest DJSI World Index Guide.

Thus, there are many challenges in assessing, measuring, and quantifying corporate sustainability, and the DJSI World approach has been devised expressly to address these challenges.

In the next section, we examine the risk and performance attributes of the DJSI World to determine the value and viability of developing an index tracking the financial performance of sustainability leaders as a whole.

Figure 2

Performance of the DJSI World and the DJ Global

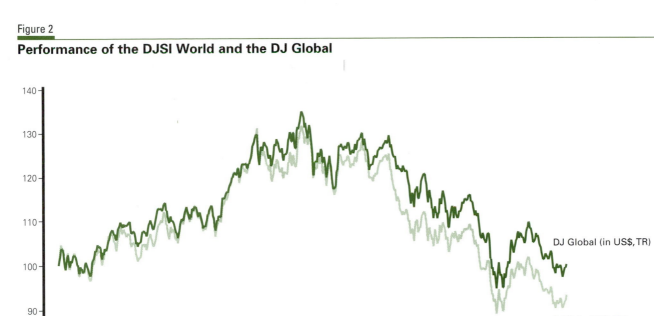

Risk and Performance Attribution of the Dow Jones Sustainability World Index

The Dow Jones Sustainability World Index was launched in 1999 to track the performance of companies that lead in corporate sustainability. According to the DJSI World, an investment in the companies leading in corporate sustainability would have yielded a cumulative negative return of –6.9 percent, or –2.7 percent per year in U.S. dollars. Volatility hovered at around 18 percent for the first 14 months, before dropping to some 16 percent for the last 10 months. As could have been expected, the sustainability leaders thus were not immune to the general downturn in equity prices the world has experienced since mid-2000.[19]

Aggregate performance is generally of less interest than a detailed analysis of its causes. After all, the negative performance may be due to factors unrelated to sustainability. Performance attribution, as it is called, is usually most revealing when compared to some other universe's indicator. It is tempting to compare the DJSI World to some general global indicator, such as the Dow Jones Global Index (DJ Global), a broad index comprising more than 5,000 companies around the world. We will use it for comparison here because the broader an index is, the less overlap there is. This may help in distinguishing sustainability factors from other factors. In addition, the DJ Global is calculated using free-

float adjustments that are comparable to the methods used for the DJSI, making it a more suitable partner than the MSCI World Indexes.

Comparing Performance

Figure 2 indicates how the two indexes have fared in comparison. From inception of the DJSI World until the end of July 2001, the DJ Global has fared better than the DJSI, losing all, but not more than, its interim gains.

Relative performance needs to be compared on an all-things-being-equal basis, if any specific claim is to be judged. Portfolio theory answers the question of what makes such comparisons by adjusting historic returns for the risks incurred. Since 1999, the DJSI World has displayed an annual volatility of 17.12 percent, while the DJ Global has had an annual volatility of 15.27 percent (see Table 6). The DJSI World has thus been more volatile. If this volatility is an unbiased reflection of the risks incurred, then the difference in volatility is big enough to warrant risk-adjusted return measurement. One risk-adjusted measure of performance is the Sharpe ratio, which relates the return a portfolio earned, above what an appropriate risk-free alternative could have earned, to the volatility experienced. Since 1999, the annual returns of the DJSI World and the DJ Global have been –2.7 percent and 0.0 percent, respectively. We assume the average

Table 6

Return and Risk Comparison

since 1999, total returns in US$

	DJSI	DJ Global	Difference
Volatility	17.1%	15.3%	1.8%
Return	-2.7%	0.0%	-2.7%
Risk-free rate*	4.0%	4.0%	—
Excess return	-6.7%	-4.0%	-2.7%
Sharpe ratio	-0.39	-0.26	-0.13

*Average of global eurocurrency rates.

Table 7

Country Allocation of Dow Jones Sustainability Index

Country/Currrency Allocation (without Emerging Markets)
as of July 30, 2001

	DJSI	DJ Global	Difference
Australia/New Zealand	1.03%	1.40%	-0.37%
Canada	2.24%	2.19%	0.05%
Denmark	0.38%	0.27%	0.11%
Euroland	22.78%	12.83%	9.95%
HK/Singapore	0.51%	1.14%	-0.63%
Japan	6.42%	10.02%	-3.60%
Norway	0.23%	0.13%	0.10%
Sweden	0.98%	0.83%	0.15%
Switzerland	8.13%	2.53%	5.60%
U.K.	18.91%	9.81%	9.10%
U.S.	38.38%	58.85%	-20.47%

risk-free rate to have been 4 percent, which yields a Sharpe ratio of –0.39 for the DJSI World and of –0.26 for the DJ Global.

Yet, calculating Sharpe ratios alone does not provide for a true all-things-being-equal basis. Nor is a comparison of Sharpe ratios sufficient to draw conclusions about the sustainability case. Three aspects need to be considered when dealing with relative performance. They are differences in exposure to common factors, differences in exposure to specific factors, and pure chance.

Exposure to Common Factors

Over any period, two indexes will be influenced to different degrees by common factors unless their composition is completely equal. Comparison of the composition of DJSI World with that of the DJ Global as of July 31, 2001, reveals that they were far from equal (see Table 7). First, companies in the Eurozone make up 23 percent of the DJSI World, compared with some 13 percent in the DJ Global. U.K. companies account for 19 percent of the DJSI World, but less than 10 percent of the DJ Global. In turn, Japanese companies account for only 6 percent in the DJSI World, while they account for 10 percent in the DJ Global. U.S. companies account for less than 40 percent in the DJSI World, but 60 percent in the DJ Global. There is thus a strong representation of European companies in the DJSI World, at the expense of Japanese and US companies.[20]

The excess weight of some 10 percent in Eurozone companies implies that the Euro has greater influence on DJSI World's return that on the DJ Global's return. Indeed, some 2 percent of the DJSI World's cumulative underperformance of 6.9 percent can be attributed to the Euro's 25 percent decline during that period.

Currency impacts can be controlled for, in part, by calculating indexes on a hedged basis. Other unequal impacts of common factors are much more difficult to control for. One example is the overweight in large-cap companies. The DJSI World has a much higher percentage of companies with large market capitalization than does the DJ Global. This difference in composition can hardly be controlled for. The significant overweight of large-cap stocks is caused by the assessment process, which starts from a universe of the 2,500 largest companies worldwide. Some 2,500 small caps that are members of the DJ Global are not assessed. Over periods during which large-cap companies do relatively better than small caps, the DJSI World does relatively better than the DJ Global. The opposite is true as well: there are periods where small caps do relatively better than large caps, as has been the case since January 1999, so the DJSI World will do worse than the DJ Global. At least 1.8 percent of the cumulative underperformance of the DJSI World since its inception can be attributed to this difference in composition.

The DJSI World also has an above average number of "growth" companies—that is, companies that tend to reimburse investors in capital gains rather than in dividends (see Table 8). Whether "growth" is a true characteristic of the sustainability universe is hard to assess. It is tempting to attribute the overweight in "growth" companies to the notion that sustainability companies are embracing global change and seeking opportunities to profit from it, so they find investment opportunities where companies not integrating sustainability considerations do not. This would justify an above average number of "growth" companies. Yet, this explanation is not necessarily the only one possible. In any case, in times when "value" companies outperform "growth" companies, the DJSI will lag the DJ Global. The small overweight in growth stocks has caused another 20 basis points of cumulative underperformance since inception.

Table 8

Risk Factors in the Index

Allocation According to Aegis Global Equity Risk Model™ Factors (without Emerging Markets) percent of standard deviations of developed markets universe, as of July 30, 2001

	DJSI	DJ Global	Difference
Size	19.80%	-39.80%	59.60%
Success	0.00%	5.20%	-5.20%
Value	2.60%	4.10%	-1.50%
Variability in Markets	9.80%	14.60%	-4.80%

Aegis Global Equity Risk Model™, Barra Inc.

Exposure to Specific Factors

The "best-in-class" approach applied to the sustainability universe leads to a small number of companies comprising the index. On July 31, 2001, the DJSI World comprised 225 companies, while the DJ Global comprised 5,029. Thus, each company included in the DJSI World carries considerably more weight than in the DJ Global. This make the DJSI World a less diversified universe and, thus, more vulnerable to specific factors.

Mispricing of companies and corrections of their mispricing is one such factor. The DJSI World is composed as a price index, as equity indexes usually are. Value-based indexes are rare because of the difficulty of assessing intrinsic value. Examples of value-based indexes are trade-weighted currency indexes based on purchasing power parities. Price indexes by definition cannot consider valuation. In addition, the DJSI World selects the leading companies with regard to their market capitalization. If the market capitalization is affected by severe mispricings, and there is no reason why leading companies embracing sustainability should be shielded from mispricings, the DJSI will tend to include the more severely mispriced companies within an industry, as long as the sustainability rating is comparable. The impact of a correction of the mispricing of these companies on the DJSI, therefore, is larger than the impact of the industry-wide correction on the broader index.

Mispricings and their subsequent corrections on the stock level have indeed been affecting the DJSI World. Lucent Technologies and EMC Corp. are two examples. In the 22 months of its membership in the DJSI World, Lucent alone accounted for a cumulative underperformance of 1.9 percent. EMC Corp. accounted for 2.9 percent of the relative underperformance during its nine months of membership alone. The performance of Lucent Technologies and EMC Corp. was not related to their sustainability rating, but to the recent IT bubble, so the DJSI World had no means of excluding them, regardless of their mispricing.[21]

Chance

As an unbiased measure of valuation and performance of the sustainability sector, the DJSI World is inevitably prone to be taken as "evidence" for or against the hypothesis of a sustainability business case. Interpreted as a portfolio, it will be expected to outperform comparable portfolios, at least in the long run. The reasoning behind this expectation is sound, as mentioned earlier. A portfolio or index of leading companies that embrace sustainability can be expected to appreciate faster than a portfolio or an index of companies that do not address profitable investment opportunities. Investments in such companies promise higher returns and, due to lower business risk, better risk-return ratios. Better performance can thus be expected on a risk-adjusted basis .

Unfortunately, the DJSI World will never be able to prove this reasonable expectation in relation to a comparable index. First, the issue of comparability will remain unsolvable. Different exposures to common factors can never be fully controlled for, and the influences exerted can take many years to even out. Second, specific factors come into play when index membership is restricted to the leaders within a universe. Third, the impact of pure chance, inherent in any investment, can override a sound investment case. A soon as chance has to be taken into account, statistical methods have to be used to decide for or against a case. The usual expectation is that the underlying value-added will persist sooner or later and will override all adverse chance influences, and perhaps even different exposures to common factors. Little consideration is given to the amount of time and the number of observations needed to make statistical methods applicable and enable them to detect significant differences in performance.

In fact, the number of observations needed for supporting any claim of "value-added" is uncomfortably large. For the moment at least, the index cannot help to determine whether investing in companies embracing sustainability is worthwhile. It may never do so. Nevertheless, the overall rationale behind this investment thesis is sound. Sustainability may be claimed to produce higher-than-average returns on a risk-adjusted basis. On a before-the-fact basis, portfolios comprising sustainable companies thus may be labeled "better" investments, with "better" defined in risk-return space, even when the index or the portfolio fails to deliver on the promise after a relatively short period of time.

Conclusion

To assess the importance and success or failure of sustainability indexing and investing, it is crucial to keep the primary goal of sustainability investing in mind—the identification of companies that are best positioned to profit from trends that are redefining the basis for business success. Clearly, the competitive rules affecting industries and businesses are shifting to the company's management of risk and opportunity derived from economic, social, and ecological trends. Leading companies addressing sustainability therefore derive competitive advantage while also reinforcing the likelihood of these trends becoming reality. Sustainability investing and indexing's "best-in-class" approach aim to identify the leaders. Promoting the fact that competition, in the Schumpeterian sense, is healthy and a driver for continuous innovation shows that this approach addresses one of the basic tenets of our market economy. Therefore, sustainability investing and indexes tracking the performance of leading companies in terms of sustainability are particularly appropriate as an investment hypothesis, compared with negative screening approaches designed to allow investors a direct expression of their personal ethical values.

Concerning the methodology for assessing corporate sustainability, there are a number of key challenges. Assessment of corporate sustainability is a very new concept, a lengthy history on which to base judgments is lacking. The assessment methodology described in this paper is a work in progress, and improvements must be made to align the content toward criteria that better reflect companies' performance and risk attributions. Furthermore, regional particularities will be given increasing prominence as specialized regional and emerging market assessment approaches are developed.

Moreover, there is a distinct dearth of scientific evidence to support the tenets of the sustainability investing hypothesis and approach. Increased collaboration among academia, science, and business should be promoted to close this gap. This cooperation between the sciences and the private sector will also provide the background for the much-needed standardization of corporate sustainability reporting.

Finally, it is premature to draw definitive conclusions regarding the business case for sustainability. As discussed, a much longer time frame is needed to attribute index or fund performance to particular sustainability criteria or strategies. Evidence regarding a positive impact of sustainability on outperformance of an index is not conclusive. However, available empirical evidence supports the view that sustainability investing has not led to a long-run-adjusted underperformance versus a conventional approach. Measured over shorter periods, risk-adjusted sustainability performance deviates from a conventional approach. Currently, investment biases (such as a regional bias or a slight growth tilt) are potentially the dominating driver of risk-performance differences. Therefore, it is possible to attest that companies that pursue sustainability do not underperform the wider market in the longer run, although they very clearly deliver major benefits to society and the environment as well as contributing positively to the vibrancy of the overall economic system. The fact that an index incorporating companies that lead in their approach to sustainability shows no negative effects on risk performance and may in fact, highlight hidden social, ecological, and economic value, explains the dramatic increase in interest in sustainability investing in recent years.

References

de Geus, Arie and Peter Senge, *The Living Company*, Boston: Harvard Business School Press, 1997.

Funk Karina, *Sustainability and Performance: Uncovering Opportunities for Value Creation* London, Cap Gemini Ernst & Young Center for Business Innovation, 2001.

Reilly F.K., and K.C. Brown, *Investment Analysis and Portfolio Management*, Fort Worth: Dryden Press, 1997.

World Commission on Environment and Development, *Our Common Future* (Brundtland report), Oxford: Oxford University Press, 1987.

Endnotes

1   With assistance from Erica Tucker-Bassin and Colin le Duc.

2   www.sustainability-index.com

3   www.kld.com/benchmarks/dsi.html

4   www.ftse4good.com

5   Under the terms of Pensions Act amendments that took effect on July 3, 2000, Britain's pension funds will be required to disclose whether they take full account of the environmental, social, and ethical impact of their investments.

6   As promulgated in the Federal Law Gazette, *Gesetz zur Reform der gesetzlichen Altersversicherung und zur Förderung eines kapitalgedeckten Altersvorsorgevermögens (Altersvermögensgesetz AvmG)*, Bundesgesetzblatt Teil I, 29.06.2001, 1310.

7   Australian Senate Table Office, *Senate Bills List 2001*, p. 40.

8   SSGA, *State Street Global Advisors announces CHF 500 million SRI mandate win from Swiss Federal Social Security Fund*, press release, May 21, 2001.

9   The United Nations General Assembly established the World Commission on Environment and Development in 1984, also known as the Brundtland Commission (after chair Gro Harlem Brundtland).

10 www.stoxx.com

11 The DJSI World is produced by SAM Indexes GmbH, a collaboration between Sustainable Asset Management and Dow Jones Indexes. The main function of SAM is to provide the research, while Dow Jones Indexes conducts the calculation and distribution functions for the DJSI.

12 www.sustainability-indexes.com/djsi_world/guidebook.html

13 www.sustainability-indexes.com

14 SAM internal studies to assess the efficiency of the corporate sustainability assessment methodology.

15 www.sustainability-indexes.com

16 PricewaterhouseCooper's audit letter of August 31, 2001, at www.sustainability-index.com

17 www.sustainability-indexes.com for full disclosure of questions and weightings in the assessment process.

18 www.sustainability-indexes.com/djsi_world/guidebook.html

19 This section has been contributed by Mr. Thilo Goodall.

20 It is unlikely, however, that U.S. companies lag others in sustainability to the degree indicated by these figures. Indeed, the construction process may account for part of the U.S. underweight.

The "best-in-class" approach, with "class" defined as an industry or industry group, is one of the characteristics distinguishing sustainability from other approaches. The approach clearly stresses industry specifics, allowing for a fair comparison within industries, while acknowledging the difficulties of comparing companies across industries. Focusing on industry strata can have an impact on country strata, if sustainability is distributed unevenly across industries within a country. Therefore, no direct inference should be made to the regional distribution of leading companies embracing sustainability.

21 Relative price movement across strata can thus shift the composition of the index if the relative movements are more pronounced in the index than in the entire universe. This has been the case and experience to date with the DJSI World. Large caps have corrected more than small caps since August 2000, and information technology and telecommunication companies have seen their prices slashed much more than those of companies from other sectors. U.S. companies provide large percentages of both large caps and of technology companies, so that about 6 percent of the current underweight in U.S. companies can be attributed to relative price movements since inception of the DJSI.

Chapter 7
Creating Green Markets: What Do the Trade Data Tell Us?

Peter K. Cornelius,
Friedrich von Kirchbach,
and Mondher Mimouni

Whether stringent environmental regulations undermine or promote competitiveness has remained the subject of controversial debate. While some (e.g., Siebert, 1977) have raised the concern that such regulations increase production costs and induce pollution-intensive industries to migrate to countries with less stringent frameworks ("pollution havens"), others, especially Porter (1991), Porter and van der Linde (1995), and Esty and Porter (1998), have argued that good environmental governance may actually foster higher productivity and economic growth. According to the latter view, environmental regulations provide important incentives to use input factors more efficiently, enhance total quality management, and, over the longer term, encourage firms to develop new and less-polluting products, capture new market niches, and thereby gain competitive advantage as environmental standards are tightened at home and abroad. Consistent with this argument, environmental regulations are believed to help create new markets for green products, environmental services, and pollution control technologies—a view that has become known as the "green-markets" hypothesis.

With public debate about the environment having intensified appreciably in recent years, an increasing number of studies have empirically tested competing hypotheses about the relationship between environmental governance and economic performance. The different approaches have been reviewed, for example, by Esty (1997), Panayotou and Vincent (1997), and, with more specific regard to the U.S. manufacturing sector, Jaffe et al. (1995). Focusing on both the micro- and macroeconomic evidence, these studies have generally rejected the hypothesis that environmental performance necessarily comes at the expense of economic competitiveness or standard of living. At the same time, however, empirical support for the much stronger claim that stringent environmental regulations may actually cause innovation and foster economic development has remained relatively rare.

Empirical studies about the nexus between environmental regulation and innovation typically have focused on the domestic evidence on R&D and productivity growth. In this paper, we take a different route. If environmental regulation indeed has a positive effect on the performance of domestic firms relative to their foreign competitors by stimulating domestic innovation, then one should be able to observe a competitive advantage first and foremost in green products themselves. Such a competitive advantage should be detectable in the foreign trade statistics on environment-friendly products, which already account for almost 5 percent of global trade. This paper, therefore, examines highly disaggregated trade data and investigates whether a relationship exists between trade performance and the stringency of a country's environmental regulations.

The rest of this chapter proceeds as follows. First, we describe the data set we are using and define what we mean by environment-friendly products. Moreover, we briefly describe how we measure environmental governance and trade performance. Second, we present and interpret the empirical evidence. Third, we focus on environmental regulations for air and water quality. The final part summarizes and draws some conclusions.

Innovation, Environment-Friendly Products, and Foreign Trade

There are a number of important reasons why poor regulatory regimes can be expected to prevent development and commercial use of environmental technologies. As Preston (1997, p. 140) argues: (i) government subsidies create incentives for overuse of certain resources; (ii) polluters are undercharged, leading to fewer incentives for innovators to develop or adopt cleaner technologies; (iii) innovation is not funded consistently through the different stages of technology development; and (iv) capital markets shy away from long-term risks because of regulatory uncertainty.

While, in theory, good environmental governance should foster innovation and productivity growth, the empirical evidence has remained inconclusive. A study by Lanjouw and Mody (1993), for example, finds that increases in environmental compliance costs lead to increases in the patenting of new environmental technologies with a one- to two-year lag. However, this study is inconclusive regarding whether compliance with more stringent environmental regulations actually leads to increases in inventive output.

In a similar study, Jaffe and Palmer (1997) examine the stylized facts regarding environmental expenditures and innovation in a panel of manufacturing industries. The authors find that lagged environmental compliance expenditures have a significant positive effect on R&D expenditures, when they control for unobserved industry-specific effects. They find little evidence, however, that an industry's inventive output—as measured by successful patent applications—is related to compliance costs.

Finally, in a recent study, Berman and Bui (2001) examine the effect of air quality regulation on productivity in some of the most heavily regulated manufacturing plants in the United States: the oil refineries of the Los Angeles Area (South Coast) Basin. Using refineries not subject to these regulations as a comparison group, this study finds that despite the high costs associated with intensification of local regulations, productivity in the Los Angeles Area Basin refineries rose sharply between 1987 and 1992. This leads the authors to conclude (p. 498) "that abatement cost measures may grossly overstate the economic costs of environmental regulation as abatement can increase productivity." However, whether such productivity gains actually overcompensated for compliance costs has remained unclear.

Similarly, it is still uncertain whether the market actually values good environmental performance. Konar and Cohen (2001) find that bad environmental performance is negatively correlated with the intangible asset value of firms, encompassing patents, trademarks, proprietary raw material sources, and brand names, as well as possible liabilities that detract from the earning power of the physical assets of a firm, such as consumer mistrust. The results suggest that firms have an incentive to overcomply with environmental regulations to portray an environmentally concerned image. While firms appear to be rewarded in the marketplace for their good environmental behavior, there is still uncertainty about whether this relationship is truly causal. Are highly reputable and profitable firms environmentally sound because they can afford to be, or does their environmental concern enhance their reputation and, thus, their market value?[1]

A similar degree of uncertainty persists at the macroeconomic level. Esty and Porter (2000; see also an update of this study in the present volume), for example, find a relatively close correlation between environmental performance and national competitiveness, suggesting that stringent reg-

ulation need not come at the expense of economic growth or standard of living. Specifically, the authors show that better antipollution performance and energy efficiency are strongly and positively associated with per capita income, especially for medium- and high-income groups, and that those nations with strong environmental regulatory regimes generally exhibit stronger competitiveness and GDP performance than their peer countries. While their findings are consistent with a variety of other analyses, the authors emphasize that their results do not establish causality.

As far as environmental regulation and foreign trade are concerned, an important part of the empirical literature has focused on the pollution-haven hypothesis and ecological dumping, arguably the single most important issue throughout the North American Free Trade Agreement (NAFTA) approval process (Esty, 1994). Grossman and Krueger (1993) and Tobey (1990), for instance, have analyzed whether net exports in industries subject to particularly stringent regulations are systematically lower. They found little empirical support for this hypothesis. Nor does there seem to be any evidence that an increasing share of trade in pollution-intensive products comes from developing countries, which generally have less stringent regulations (or are perceived to be more relaxed about their enforcement). Where such patterns can be observed, there is reason to assume that this is a reflection of increased demand for pollution-intensive products in these countries rather than a shift of pollution-intensive production to them.

While generally consistent with the weak hypothesis that there does not need to be a tradeoff between environmental and economic performance, these studies contain little information about the degree to which stringent regulatory regimes have actually helped specific industries to gain a competitive advantage. One industry of particular interest in this regard is, of course, environment-friendly products themselves, such as pollution-control technologies. However, notwithstanding the rapid growth of green markets in recent years, this industry has attracted surprisingly little attention from trade researchers.

An important reason for this disinterest may lie in the numerous empirical challenges one faces in investigating the green-markets hypothesis based on trade data. To begin with, environmental friendliness needs to be defined. Second, one has to be able to measure the stringency of environmental regulations across different areas and in particular sectors. And, finally, a country's sectoral trade performance needs to be quantified comprehensively. We address these challenges below.

First, as far as the definition of environment-friendly products is concerned, this paper follows an approach proposed recently by the Joint OECD-EUROSTAT Working Party on

Trade and Environment (OECD 2000). According to this approach, environment-related activities are first mapped into three groups: (1) pollution management group, (2) cleaner technologies and products group, and (3) resource management group. Then these activities are matched with the product codes for the corresponding outputs.

list issued by OECD and EUROSTAT has remained the only internationally recognized one. Box 1 provides examples for each category of green products; in Table A1 (see page 92) we present disaggregated export data of environment-friendly products for 1998–1999, the two most recent years for which trade data are available.[3]

Second, in assessing and ranking regulatory regimes, we adopt Esty and Porter's (2000) approach, which is based to a considerable extent on survey data from the World Economic Forum's *Global Competitiveness Report*. Their regulatory index measures the array of national regulatory standards, structures, information, and institutions, using common factor analysis. Specifically, this index reflects the stringency of a country's pollution standards; structure of regulatory standards and processes; strictness of regulatory enforcement; degree of subsidization of natural resources; extent of information available for policymaking and enforcement; and capacity of a country's environmental institutions.

Esty and Porter's rankings of regulatory regimes confirm a close correlation between the state of the environmental regime and the level of development as indicated by per capita income (on a purchasing power parity basis). Contrary to earlier concerns, governments in more advanced countries appear to have resisted pressures to reduce their environmental regulatory standards in a race to the bottom (Cooper, 1994; Esty, 1996). An important reason why developing countries generally suffer from inferior regulatory regimes often lies in their limited institutional capacity. However, doubts have been raised about whether the harmonization of standards at a higher level is desirable. As a matter of fact, it has been argued that poorer countries should resist adopting the same standards that prevail in the developed countries, allowing each country "to tailor its regulations to local economic circumstances and tastes. As countries' per capita incomes increase, so too will their demand for environmental quality" (Portney, 2000, p. 202). In light of this, we distinguish three groups of countries according to their level of economic prosperity to review their performance against their peers.

Third, our empirical assessment of national trade performance regarding environment-friendly products is based on United Nations COMTRADE data, the world's largest database of trade statistics. The data are derived from some 100 reporting countries and cover more than 90 percent of world trade. Using this database, we calculate a trade performance index (TPI) for countries' current trade competitiveness and for changes in sectoral trade competitiveness over time. These calculations are made for 184 countries, with exports and imports of nonreporting countries being estimated on the basis of partner-country data.

Of course, the OECD-EUROSTAT approach is far from perfect. Even at the six-digit level of the Harmonized System, products remain quite heterogeneous, with their classification based on a customs perspective according to their physical characteristics rather than their final use. Product categories do not necessarily distinguish between environmental and other purposes, and a given product generally has multiple uses. Conversely, the list does not encompass product groups, which do not differentiate environment-friendly products from other products. For instance, recycled paper, hydroelectric plants, wind turbines, sustainable forestry, fishery, or agriculture are not taken into account. Notwithstanding these limitations, the

Based on 1999 data, the *current TPI* reflects five complementary sector-specific trade indicators: (1) the share of a country's exports in world trade; (2) sectoral trade balance; (3) per-capita exports to control for the size of the economy; (4) degree of product diversification within a given category; and (5) degree of diversification across different export markets. From a dynamic perspective, the *change TPI* reflects (1) the change in the country's sector-specific share in world exports; (2) ability of exporters to increase their sectoral trade surplus or reduce their deficit; (3) degree to which exporters are specialized in particularly dynamic products; (4) changes in product diversification; and (5) changes in market diversification.[4] The *change TPI* captures the period from 1995 to 1999.

Environmental Regulation and Trade Performance

In 1999, world exports of environment-friendly products totaled almost US$200 billion, or around 5 percent of total exports. Wastewater management technologies account for the largest share of green products, representing more than 40 percent of world exports. The next two most important categories are environmental monitoring systems and solid waste management technologies, with shares of around 13 percent and 17 percent, respectively (Table A1). By comparison, exports of heat- and energy-saving technologies and air pollution control products have remained considerably lower, totaling less than US$10 billion each (less than 5 percent in terms of total exports of environment-friendly products).

Developed countries enjoy not only more advanced regulatory regimes, as indicated in Table 1; they also account for the bulk of exports of environment-friendly goods. As a matter of fact, in 1999, more than 90 percent of such exports originated from countries with per capita incomes of US$18,000 or more. The United States, Germany, and Japan alone accounted for around 45 percent. Compared to their peers, however, their regulatory regimes are perceived to be average. More advanced regimes were found, for example, in the four Scandinavian countries (Denmark, Finland, Norway, Sweden) and Switzerland. While exports from these top performers were substantially smaller in dollar terms, totaling less than 7 percent of world exports, environment-friendly goods represented a sizeable share of their total exports. This especially applies to Switzerland, where environment-friendly goods represent the highest percentage in terms of GNP (based on purchasing power parity (PPP)).

Middle-income and, especially, low-income countries account for a significantly smaller percentage of global markets for environment-friendly products. On average, countries with per capita incomes between US$5,000 and US$18,000 represent a share of less than half a percent, compared with an average of almost 5 percent in the case of high-income countries. At the same time, green goods represent a comparatively smaller share of middle- and low-income countries' total exports. Because these countries usually have less-developed regulatory regimes, it is not surprising that we find a relatively close correlation between the countries' exports of environment-friendly goods and the ranking of their regulatory regimes, as indicated by a correlation coefficient of −0.608.

There are important variations within the respective groups, however. Colombia and Venezuela, for example, are perceived to have substandard regulatory regimes, given their level of development; they perform poorly with regard to exports of environment-friendly goods. In turn, a number of low-income countries, notably Hungary and Mexico, appear to have developed comparatively sophisticated regulatory standards. Enjoying considerable foreign direct investment inflows, these countries show great success in penetrating foreign markets for environment-friendly goods. These observations are consistent with the green-markets hypothesis, although uncertainty remains about the extent to which good regulatory regimes actually *cause* such markets to develop. In this regard, China represents an important outlier, whose exports of environment-friendly goods in 1999 amounted to almost US$4.3 billion, or more than 2 percent of world exports, despite a weak regulatory regime (in line with that country's relatively low level of development). Again, foreign direct investment is likely to have played a considerable role for China's export success in these markets.

Taking into account additional trade indicators, such as sectoral trade balance and diversification of products and export markets, Germany, France, and the United Kingdom were the three top performers in 1999. By comparison, U.S. companies appeared less competitive, notwithstanding their leading position in terms of total exports of environment-friendly products. Overall, however, the trade performance index confirms that high-income countries enjoy an important competitive advantage in these markets. While it is difficult to quantify the extent to which good environmental governance actually has played a role in this regard, it is important to note that the correlation between the perceived stringency of environmental regulations and trade performance as measured by the TPI is even closer than in the case of the much narrower concept of export values or total world market shares (Figure 1; the correlation coefficient between the current TPI and environmental regulations totals 0.783).[5]

As far as changes in trade performance over the 1995–1999 period are concerned, seven of the top ten performers are low-income countries. Obviously, most of these countries

Table 1

Exports of Environment-Friendly Goods and Trade Performance, 1995–1999

	Country	Regulatory Index Overall ranking	Exports of Environment-Friendly Products				Trade Performance Index	
			Millions USD	% of world market share	% of total exports	% of GNP (PPP basis)	Current	Change
High-income Countries with per Capita Incomes of > US$ 18,000	**Finland**	**1**	**1279**	**0.64**	**3.02**	**1.22**	**10**	**18**
	Norway	2	802	0.40	2.02	0.75	13	39
	Switzerland	3	4824	2.42	6.13	2.55	7	41
	Netherlands	4	4695	2.36	2.37	1.38	9	27
	Denmark	5	3075	1.55	6.54	2.43	6	22
	Austria	6	3052	1.53	4.95	1.66	14	26
	Sweden	7	3447	1.73	4.08	1.99	5	19
	Germany	8	31487	15.83	5.83	1.84	1	40
	United States	9	35091	17.64	5.14	0.44	8	20
	United Kingdom	10	10414	5.23	3.82	0.85	3	32
	Canada	11	5830	2.93	2.72	0.79	23	27
	Japan	12	23890	12.01	6.16	0.82	11	46
	Belgium	13	5026	2.53	2.93	2.10	12	23
	Australia	14	622	0.31	1.11	0.16	17	50
	France	15	11583	5.82	3.77	0.88	2	34
	Ireland	17	725	0.36	1.15	1.07	16	8
	Italy	23	14756	7.42	6.13	1.27	4	45
	Hong Kong	26	175	0.09	…	…	29	51
	Average (unweighted)		8932	4.49	3.99	1.31		
Middle-income Countries with per Capita Incomes > US$ 5,000 < US$ 18,000	New Zealand	16	289	0.15	2.39	0.48	46	14
	Iceland	18	4	0.00	…	…	15	33
	Spain	19	3145	1.58	2.88	0.50	18	38
	Korea	20	2622	1.32	1.97	0.46	19	31
	Portugal	21	704	0.35	3.00	0.49	20	12
	Israel	22	894	0.45	3.84	0.86	26	13
	Malaysia	24	1311	0.66	1.79	0.85	37	24
	Brazil	25	1678	0.84	3.29	0.16	45	48
	South Africa	27	962	0.48	3.65	0.33	22	37
	Chile	31	204	0.10	1.37	0.11	42	35
	Argentina	32	338	0.17	1.34	0.09	36	29
	Czech Republic	33	1001	0.50	3.80	…	17	36
	Greece	35	187	0.09	1.93	0.14	25	10
	Poland	36	889	0.45	3.38	0.34	24	44
	Costa Rica	37	56	0.03	1.38	0.24	43	4
	Thailand	39	842	0.42	1.58	0.24	28	15
	Colombia	44	66	0.03	0.61	0.02	41	42
	Venezuela	51	93	0.05	0.54	0.05	50	52
	Average (unweighted)		849	0.43	2.28	0.34		
Low-income Countries with per Capita Incomes < US$ 5,000	Jordan	28	9	…	0.51	0.06	47	1
	Hungary	29	775	0.39	3.38	…	31	43
	Mexico	30	5433	2.73	4.62	0.69	34	2
	Slovak Republic	34	268	0.13	2.51	…	27	21
	Russia	38	824	0.41	1.12	0.14	21	11
	China	40	4292	2.16	2.33	0.03	35	3
	Peru	41	24	0.01	2.35	0.02	48	17
	Egypt	42	19	0.01	0.49	0.01	51	16
	India	43	294	0.15	0.89	0.02	38	30
	Indonesia	45	416	0.21	0.85	0.07	32	6
	Mauritius	46	9	0.00	…	…	30	49
	Philippines	47	494	0.25	1.68	0.19	40	5
	Ukraine	48	378	0.19	2.98	…	33	54
	Vietnam	49	35	0.02	0.39	0.03	49	53
	Zimbabwe	50	5	0.00	0.20	0.02	52	9
	Bulgaria	52	78	0.04	1.82	…	39	25
	El Salvador	53	29	0.01	2.30	0.17	44	7
	Average (unweighted)		787	0.42	1.78	0.12		

Figure 1

Environmental Regulations and Trade Performance

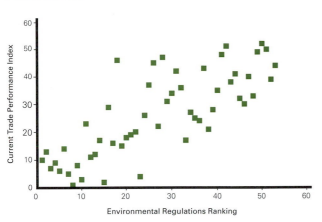

have benefited from a strong base effect. One important exception is Mexico, whose exports of environment-friendly products in 1999 already amounted to almost US$5.5 billion. That this export success is closely related to free trade within the space of NAFTA is clearly indicated by Mexico's low current TPI ranking, which suggests relatively poor market diversification.

By comparison, performance improvements in high-income countries are relatively smaller, as indicated by their TPI rankings. However, this finding is largely explained by a base effect, rather than poor environmental governance. Germany, for instance, the top performer in terms of the current TPI and the world's second-largest exporter of environment-friendly products, belongs in the bottom quartile in terms of improvements in trade performance.

Trade Performance in Specific Subsectors

The hypothesis that good environmental governance plays a role in gaining a competitive advantage in green markets may be examined further by looking at specific subsectors. Table 2 depicts exports of air pollution control products and an air pollution regulation index that is based on the *2000 Global Competitiveness Report* Survey. Again, industrialized countries generally have more stringent regulatory regimes than lower-income countries, with a relatively close correlation between regime stringency and export performance. With air pollution control products requiring a particularly high level of technological sophistication, it is not surprising that we find a very high degree of concentration of exporting by developed countries. In 1999, Germany was particularly successful in penetrating foreign markets, as evidenced by a world market share of more than 20 percent. Germany is perceived to have one of the most stringent regimes in this area.

However, that factors other than environmental governance have also played a role becomes evident when examining the export performance of other European countries, such as the United Kingdom, France, and Italy. Although their air pollution regulations appear to be comparatively less stringent, they have also been very successful in developing a competitive advantage in air pollution control products. This observation also applies to Japan, the third most important exporter of such goods.

Moreover, Table 2 shows a water pollution regulatory index and exports of potable water treatment technologies and wastewater management technologies, with the latter accounting for around 40 percent of all environment-friendly goods based on the OECD-EUROSTAT definition. Again, we observe a familiar pattern: the perceived stringency of regulations is closely correlated with the stage of development and export performance. Two countries represent important outliers, Mexico and China. Their combined share in world exports of wastewater management technologies amounts to almost 50 percent of the total export share of all middle- and low-income countries considered here. Individually, the two countries belong to the ten largest exporters in the world in this category, despite their overall stage of development and their comparatively low regulatory standards.

While these outliers do not change the overall picture materially, it is important to reemphasize that a country's export performance of environment-friendly goods appears to be driven by a multidimensional set of factors. For example, within the group of high-income countries, it is difficult to discriminate according to the perceived stringency of regulations. While those countries with a regulatory score of 6 or above (on a 1–7 scale) enjoyed, on average, a global export share of 4.4 percent, the average share of countries in this category with a score ranging from 5 to 5.9 amounted to only slightly less, that is, 4.2 percent.

Finally, we have calculated current and change TPIs for two (OECD-EUROSTAT) subcategories of environment-friendly products, namely: *pollution management and cleaner technologies* (Table 3). According to the rankings, Germany is the most competitive exporter in these product groups, followed by a number of other European economies. While a closer examination of the rankings within the different income groups suggests that good environmental governance may explain a country's trade performance only partly, it does appear that, in the absence of adequate environmental standards, firms have little incentive to develop new green products. Hong Kong represents an important example. At around US$22,500 per capita income in 1999 (PPP) Hong Kong is one of the richest economies in the world; however, its environmental regulations are perceived to be substandard. At the same time, Hong Kong underperforms considerably in terms of exporting pollution control products and cleaner technologies.

Table 2

Environmental Regulations and Exports of Specific Green Goods, 1999

	Country	Air Pollution Regulatory Index[1]	Exports of Air Pollution Control Goods		Country	Water Pollution Regulatory Index[1]	Exports of Potable Water Treatment Technologies		Exports of Waste Water Management Technologies	
			mn USD	% of world market share			mn USD	% of world market share	mn USD	% of world market share
High-income Countries with per Capita Incomes of > US$ 18,000	Finland	6.7	41.3	0.46	Finland	6.8	1.8	0.2	456	0.6
	Austria	6.6	116.9	1.29	Denmark	6.6	1	0.1	1802	2.2
	Germany	6.5	1917.2	21.16	Austria	6.6	0.8	0.1	8	0
	Denmark	6.4	85.4	0.94	Switzerland	6.5	8.3	0.7	1880	2.3
	Switzerland	6.3	367.5	4.06	Germany	6.5	659.7	57.6	11873	14.7
	Netherlands	6.3	154.5	1.70	Sweden	6.4	33.5	2.9	1234	1.
	Sweden	6.3	123.7	1.36	Netherlands	6.3	12.4	1.1	1884	2.3
	United States	6.2	1296.8	14.31	United States	6.3	93.6	8.2	12969	16
	Norway	5.9	27.8	0.31	Canada	5.9	41.8	3.6	2364	2.9
	Japan	5.9	1047.5	11.56	Japan	5.9	56.5	4.9	6413	7.9
	Canada	5.6	183	2.02	Australia	5.8	1.9	16.6	1189	1.5
	Australia	5.6	11	0.12	Norway	5.7	0.3	0.1	356	0.4
	United Kingdom	5.4	637.2	7.03	United Kingdom	5.6	15.7	1.4	4171	5.2
	France	5.4	717.4	7.92	Belgium	5.6	22.2	1.9	2313	2.9
	Belgium	5.2	452	4.99	France	5.6	13.8	1.2	5710	7.1
	Italy	4.9	557	6.15	Italy	4.9	26.7	2.3	7907	9.8
	Ireland	4.8	10.2	0.11	Ireland	4.8	2.3	0.2	412	0.5
Middle-income Countries with per Capita Incomes >US$ 5,000 < US$ 18,000	Iceland	5.7	0.01	0.00	New Zealand	5.6	0.7	0.1	92	0.1
	New Zealand	5.3	6.9	0.08	Iceland	5.6	...	...	3.7	0.0
	Spain	4.6	53.1	0.59	Spain	4.8	3	0.3	1552	1.9
	Czech Republic	4.6	32.2	0.36	Czech Republic	4.5	1.6	0.1	584	0.7
	Portugal	4.3	3.2	0.04	Israel	4.4	70.6	6.2	487	0.6
	Israel	4.2	5.9	0.07	Korea	4.3	1.8	0.2	1293	1.6
	Korea	4.1	103.6	1.14	Portugal	4.2	1.2	0.1	314	0.4
	Brazil	3.9	24.3	0.27	Costa Rica	4.1	0.03	0.0	21.3	0.0
	Chile	3.9	2	0.02	Malaysia	4.0	0.9	0.1	419	0.5
	Malaysia	3.8	12.6	0.14	South Africa	4.0	1.9	0.2	257	0.3
	South Africa	3.7	427.6	4.72	Chile	3.9	0.4	0.0	26.6	0.0
	Hong Kong	3.0	1.7	0.02	Greece	3.9	0.09	0.0	80.6	0.1
	Argentina	3.0	47.9	0.53	Brazil	3.7	0.04	0.0	662	0.8
	Thailand	3.0	5.5	0.06	Poland	3.5	1.2	0.1	396	0.5
	Colombia	3.0	1.1	0.01	Hong Kong	3.2	1.8	0.2	109	0.1
	Venezuela	2.9	0.6	0.01	Colombia	3.2	0.7	0.1	30.5	0.0
					Argentina	3.1	0.2	0.0	252.7	0.
					Venezuela	3.1	0.03	0.0	27	0.0
					Thailand	2.9	0.4	0.0	464	0.6
Low-income Countries with per Capita Incomes <US$ 5,000	Hungary	4.2	24.4	0.27	Hungary	4.0	2.2	0.2	218	0.3
	Mexico	3.8	283.2	3.13	Slovak Republic	3.9	0.07	0.0	137	0.2
	Slovak Republic	3.8	1.9	0.02	Jordan	3.6	...	...	3.3	0.0
	Jordan	3.4	0.07	0.00	Mexico	3.5	6.5	0.6	2198	2.7
	Russia	3.4	34	0.38	Egypt	3.4	0.5	0.0	15	0.0
	Bulgaria	3.1	0.2	0.00	Russia	3.3	3.1	0.3	389	0.5
	Ukraine	3.0	3.9	0.04	Bulgaria	3.2	0.1	0.0	63.3	0.1
	Egypt	2.9	...	...	Ukraine	3.0	6.5	0.6	295	0.4
	India	2.8	6.4	0.07	Mauritius	2.9	...	...	0.9	0.0
	Mauritius	2.7	...	...	Peru	2.7	1.1	0.1	17.2	0.0
	China	2.6	47.6	0.53	Zimbabwe	2.7	...	...	1.5	0.0
	Indonesia	2.6	4.2	0.05	China	2.6	4.8	0.4	2700	3.3
	Peru	2.5	0.5	0.01	India	2.6	4.9	0.4	151	0.2
	Zimbabwe	2.5	0.01	0.00	Indonesia	2.5	0.8	0.1	192	0.2
	Vietnam	2.4	0.2	0.00	Philippines	2.5	0.1	0.0	105	0.1
	Philippines	2.3	4.5	0.05	Vietnam	2.3	...	...	25.4	0.0
	El Salvador	2.1	...	...	El Salvador	2.2	0.008	0.0	7.2	0.0

[1] On a scale from 1–7.

Table 3

Trade Performance in Pollution Management and Cleaner Technologies Products

	Current Trade Performance		Change in Trade Performance	
Rank	Pollution management	Cleaner technologies	Pollution management	Cleaner technologies
1	Germany	Germany	Costa Rica	Indonesia
2	Italy	France	El Salvador	Mexico
3	United Kingdom	Italy	Philippines	Spain
4	France	Sweden	China	Ireland
5	Sweden	United Kingdom	Mexico	Peru
6	Switzerland	Belgium	New Zealand	Korea
7	Denmark	Netherlands	Ireland	Finland
8	United States	Switzerland	Portugal	Austria
9	Finland	Denmark	Greece	Russia
10	Japan	Japan	Russia	Thailand
11	Netherlands	Finland	Israel	Norway
12	Belgium	Hungary	Peru	Colombia
13	Norway	Austria	Indonesia	Hungary
14	Austria	Mexico	Malaysia	Switzerland
15	Ireland	Israel	Slovak Republic	Sweden
16	Spain	United States	Argentina	China
17	Czech Republic	Thailand	Sweden	Belgium
18	Australia	Spain	Finland	South Africa
19	Portugal	Portugal	Bulgaria	Greece
20	Korea	New Zealand	United States	Israel
21	Russia	Canada	Thailand	Canada
22	South Africa	Chile	Jordan	Germany
23	Malaysia	Korea	Denmark	Egypt
24	Israel	Australia	Canada	Argentina
25	Canada	Malaysia	Zimbabwe	Brazil
26	Poland	South Africa	Venezuela	Poland
27	Greece	Norway	Vietnam	United Kingdom
28	Slovak Republic	Czech Republic	Belgium	Denmark
29	Mauritius	Ireland	Austria	Philippines
30	New Zealand	Russia	Netherlands	United States
31	Thailand	China	United Kingdom	Netherlands
32	Ukraine	Slovak Republic	Chile	Bulgaria
33	Mexico	Poland	South Africa	Portugal
34	Indonesia	Indonesia	India	Chile
35	Argentina	India	France	Italy
36	Hungary	Argentina	Korea	Malaysia
37	China	Brazil	Czech Republic	Slovak Republic
38	Brazil	Greece	Iceland	Japan
39	India	Philippines	Egypt	El Salvador
40	Bulgaria	Venezuela	Germany	France
41	Chile	Vietnam	Norway	Czech Republic
42	Philippines	Hong Kong	Brazil	Zimbabwe
43	Colombia	Colombia	Colombia	India
44	Costa Rica	Costa Rica	Spain	New Zealand
45	Hong Kong	Ukraine	Switzerland	Jordan
46	Venezuela	El Salvador	Italy	Hong Kong
47	Vietnam	Bulgaria	Poland	Costa Rica
48	Iceland	Peru	Japan	Australia
49	Peru	Jordan	Mauritius	Ukraine
50	El Salvador	Egypt	Australia	Venezuela
51	Jordan	Zimbabwe	Hungary	Vietnam
52	Egypt		Hong Kong	
53	Zimbabwe		Ukraine	

It is interesting to note that Hong Kong also scores poorly on the change TPI, further evidence that the lack of good environmental governance may represent an important impediment to developing new technology because firms face fewer incentives to innovate to meet higher standards. As a result, such firms will be less likely to compete successfully abroad. Conversely, countries that set high standards relative to their stage of development appear well positioned to improve their trade performance. Costa Rica is a particularly interesting example of this.

Economic integration may provide an important stimulus to upgrade regulatory regimes. As far as the European Union is concerned, we anticipate that standards in the southern European countries—Greece, Portugal, and Spain—will gradually approach the level of stringency that exists in the core countries. The expected convergence will likely provide further impetus in innovation and development of environment-friendly goods, a process that has already begun, as evidenced by the change TPI rankings. Similarly, Mexico is expected to continue to implement more rigorous environmental standards as its integration into NAFTA gains further momentum. As a result, Mexico's trade performance in green products, which has benefited greatly from foreign direct investment (FDI) inflows and free trade within the NAFTA area, is likely to improve further.

Conclusions

This paper looks at disaggregated trade data with regard to environment-friendly products to examine whether empirical support exists for the green-markets hypothesis. The data provide some support for the hypothesis that strict environmental regulation can translate into a trade advantage. First, there is a close correlation between environmental governance and trade performance, measured narrowly by a country's exports or, more broadly, by a trade performance index, which also takes into account market and product diversification. Second, controlling for the stage of development of individual countries, it appears that good environmental governance can help to drive innovation, product development, and improved trade performance. Conversely, countries that lack adequate environmental regulatory regimes are less likely to provide a framework conducive to innovation and developing a competitive advantage. Foreign direct investment can play an important role in this context. While there is little evidence that foreign investors are deterred by stringent environmental regulations, it seems that foreign direct investment has helped to develop new products and gain a greater share in markets for environment-friendly products. Third, competing successfully abroad depends on a complex set of factors, and further research is required to quantify the effect of stringent regulations on competitive advantage.

Combined with earlier research on trade flows and environmental regulation, our findings have important policy implications. It appears that stringent regulations do not undermine a country's trade performance; in fact, they help to promote development of new technologies, which translates into trade share gains in a world market that can be expected to remain highly dynamic in the future. Based on the evidence presented here, it is hard to estimate the extent to which competitiveness and economic growth depend on an adequate environmental framework. It appears, however, that firms in countries with substandard regulations face fewer incentives to innovate and develop competitive advantages. Rather than participating in a regulatory race to the bottom, countries—especially developing nations—are thus well advised to upgrade their environmental standards continuously.

References

Berman, Eli, and Linda T.M. Bui, "Environmental Regulation and Productivity: Evidence from Oil Refineries," *The Review of Economics and Statistics* 83 (August 2001), pp. 498–510.

Cooper, Richard N., *Environment and Resource Policies for the World Economy*, Washington: Brookings Institution, 1994.

Cornelius, Peter, Friedrich von Kirchbach, Mondher Mimouni, Jean-Michel Pasteels, and Shilpa Phadke, "Sectoral Trade Performance," in *The Global Competitiveness Report 2001–2002*, Oxford: Oxford University Press, forthcoming.

Esty, Daniel C., "Making Trade and Environmental Policies Work Together: Lessons from NAFTA," *Aussenwirtschaft* 49 (January 1994), pp.59–79.

Esty, Daniel C., "Revitalizing Environmental Federalism," *Michigan Law Review* 95 (1996).

Esty, Daniel C., *Greening the GATT. Trade, Environment, and the Future*, Washington: Institute for International Economics, 1997.

Esty, Daniel C., and Michael E. Porter, "Industrial Ecology and Competitiveness: Strategic Implications for the Firms," *Journal of Industrial Ecology* 2 (Winter 1998), pp. 35–43.

Esty, Daniel C., and Michael E. Porter, "Measuring National Environmental Performance and Its Determinants," *Global Competitiveness Report 2000*, Oxford: Oxford University Press, 2000, pp. 60–75.

Grossman, Gene M., and A.B. Krueger, "Environmental Impacts of a North American Free Trade Agreement," in P. Garber (ed.), *The US-Mexico Free Trade Agreement*, Cambridge, Mass.: The MIT Press, 1993, pp. 13–56.

Hettige, H., R.E. Lucas, and D. Wheeler, "The Toxic Intensity of Industrial Production: Global Patterns, Trends, and Trade Policy," *The American Economic Review, Papers and Proceedings* 82, 1992, No. 2.

Jaffe, Adam B., S.R. Peterson, P.R. Portney, and R. Stavins, "Environmental Regulation and the Competitiveness of U.S. Manufacturing," *Journal of Economic Literature* 33 (1995), pp. 132–163.

Jaffe, Adam B., and Karen Palmer, "Environmental Regulation and Innovation: A Panel Data Study," *The Review of Economics and Statistics* 79 (November 1997), pp. 610–619.

Konar, Shameek, and Mark A. Cohen, "Does the Market Value Environmental Performance?" *The Review of Economics and Statistics* 83 (May 2001), pp. 281–289.

Lanjouw, Jean, and Ashoka Mody, "Stimulating Innovation and the International Diffusion of Environmentally-Responsive Technology: The Role of Expenditures and Institutions," The World Bank (mimeo), 1993.

Low, Patrick, and Alexander Yeats, "Do Dirty Industries Migrate?" in P. Low, *International Trade and the Environment*. World Bank Discussion Paper 159, Washington: World Bank, 1992.

OECD, "Environmental Goods and Services : An Assessment of the Environmental, Economic and Development Benefits of Further Global Trade Liberalisation," Joint Working Party on Trade and Environment, COM/TD/ENV/(2000)86/Final, 2000.

Panayotou, Theodore, and Jeffrey R. Vincent, "Environmental Regulation and Competitiveness," *Global Competitiveness Report 1997*, Geneva: World Economic Forum, 1997, pp. 64–73.

Porter, Michael E., "America's Green Strategy," *Scientific American* (April 1991), p. 168.

Porter, Michael E., and Claas van der Linde, "Towards a New Conception of the Environment-Competitiveness Relationship," *Journal of Economic Perspectives* 9 (1995), pp. 97–118.

Portney, P. R., "Environmental Problems and Policy: 2000–2050," *Journal of Economic Perspectives* 14 (Winter 2000).

Preston, J.T., "Technology Innovation and Environmental Progress," in M. R. Chertow and D. C. Esty (eds.), *Thinking Ecologically. The Next Generation of Environmental Policy*, New Haven and London: Yale University Press, 1997.

Siebert, Horst, "Environmental Quality and the Gains from Trade," *Kyklos* 30 (1977), pp. 657–673.

Tobey, J.A., "The Effects of Domestic Environmental Policies on Patterns of World Trade: An Empirical Test," *Kyklos* 43 (1990), pp. 191–209.

World Economic Forum, *Global Competitiveness Report 2001–2002*, Oxford: Oxford University Press, 2001.

Endnotes

1 In the case of the latter, environmental reputation and performance may be regarded as a proxy for good management, an issue that is discussed in greater detail by Dixon elsewhere in this volume.

2 This is consistent with the finding that toxic intensity increased more rapidly in inward-looking developing countries, while outward-oriented developing countries had a slowly increasing or declining toxic intensity of manufacturing; see Hettige, Lucas, and Wheeler (1992

3 Note that the calculations of the trade performance index are based on data that are even more disaggregated than in those in appendix Table A1.

4 For a detailed description of the TPI, see Cornelius et al. (forthcoming). See also www.intracen.org

5 The transition economies in central and eastern Europe represent interesting outliers in the sense that their trade performance appears considerably better than their poor regulatory regimes would suggest. Apart from the legacy of foreign trade relations under communism, foreign direct investment likely has played an important role in some countries, notably the Czech Republic and Poland.

Exports of Selected Environment-Friendly Products, 1998–1999 (in thousands of US dollars)

Country	Heat/energy savings and management 1998	1999	Renewable energy plant 1998	1999	Potable water treatment and distribution 1998	1999	Cleaner technologies and products 1998	1999	Environmental monitoring, analysis and assessment 1998	1999
Argentina	24,346	23,255	28,911	21,192	39	159	5,636	3,851	14,803	11,404
Australia	23,336	35,474	24,577	20,534	2,105	1,923	12,414	12,819	122,547	117,960
Austria	131,419	130,860	37,988	52,256	795	771	46,236	49,457	296,643	282,454
Belgium-Lux	266,447	311,456	58,962	74,999	23,990	22,235	95,156	92,255	298,174	329,235
Brazil	24,844	23,596	43,216	69,884	262	44	9,851	9,949	78,745	104,712
Bulgaria	1,610	892	2,964	3,793	68	118	266	210	901	1,387
Canada	342,606	354,717	178,588	155,928	62,317	41,836	54,854	57,604	675,264	751,144
Chile	7,096	6,508	137,007	144,952	48	409	220	203	1,251	877
China	184,441	275,597	131,854	152,892	6,131	4,760	22,490	14,987	187,123	198,735
Colombia	7,101	4,869	83	58	364	667	1,151	446	584	337
Costa Rica	92	455	14,848	13,130	569	35	1,539	1,952	4,893	3,558
Czech Rep	63,558	59,825	9,554	7,842	1,875	1,626	8,954	7,623	45,499	46,437
Denmark	257,696	277,188	40,927	28,837	821	1,023	36,336	40,168	431,000	417,242
Egypt	234	285	5	58	306	484	50	49	31	16
El Salvador	56	62	4,876	7,263	20	8	2,838	2,574	47	60
Finland	98,226	90,326	5,194	5,842	1,363	1,826	19,415	19,179	233,930	217,955
France	879,461	735,776	306,063	274,327	13,719	13,840	113,904	104,797	1,289,742	1,249,635
Germany	1,353,271	1,245,191	551,966	527,661	673,001	659,726	441,813	429,628	5,282,949	5,557,066
Greece	1,851	2,297	10,249	9,266	60	93	5,694	3,362	4,567	6,950
Hong Kong	14,896	6,649	7,442	5,042	1,924	1,806	441	238	22,059	14,488
Hungary	190,252	189,156	11,873	12,083	2,304	2,223	1,615	1,436	43,315	53,853
Iceland		18					28	70	238	123
India	12,664	6,949	7,496	4,062	5,271	4,914	1,327	764	11,235	6,953
Indonesia	14,972	37,698	32,423	61,278	314	803	393	490	2,364	7,037
Ireland	30,146	20,134	5,314	4,474	1,507	2,298	2,531	1,035	158,275	178,199
Israel	23,499	29,114	11,167	8,098	41,098	70,568	1,088	613	136,826	109,914
Italy	1,002,734	875,437	110,225	94,606	29,889	26,705	111,724	108,746	824,546	807,515
Japan	583,473	531,192	1,236,321	1,602,416	46,334	56,527	37,027	38,462	2,662,962	3,080,936
Jordan	51						778	1,521	500	799
Korea Rep.	161,961	182,929	110,586	113,918	1,820	1,821	9,706	11,751	95,023	90,096
Malaysia	69,586	98,081	393,217	390,345	1,335	941	10,105	9,314	130,712	139,634
Mexico	467,196	540,585	128,216	170,987	9,024	6,475	12,601	9,153	602,807	682,546
Netherlands	465,024	386,929	233,572	208,056	15,634	12,424	75,444	78,418	825,603	834,729
New Zealand	6,563	4,222	148,806	99,648	403	732	1,067	1,342	44,821	42,964
Norway	32,882	29,110	1,185	1,618	229	275	12,912	13,222	133,286	121,045
Peru	278	553		80	1,670	1,137	6	4	465	644
Philippines	6,436	7,856	175,735	172,825	132	147	16	140	143,442	146,754
Poland	130,689	138,810	4,449	6,693	1,263	1,222	4,739	5,612	22,297	28,149
Portugal	20,660	26,134	10,438	90,296	535	1,170	5,460	7,379	93,735	52,363
Russian Fed	25,793	32,783	42,886	73,992	9,623	3,144	1,733	1,803	144,824	161,042
Singapore	89,571	95,079	143,340	173,866	2,101	3,877	48,327	35,682	378,896	430,185
Slovakia	16,901	12,966	1,424	1,601	17	71	1,638	1,168	7,576	6,397
Spain	106,606	94,546	74,126	95,670	2,699	2,963	45,424	48,528	318,082	331,719
Sweden	190,242	412,435	53,006	59,306	31,754	33,537	64,306	74,786	597,035	641,145
Switzerland	285,165	259,549	54,336	97,192	5,840	8,313	22,840	26,439	856,056	869,692
Thailand	61,857	66,022	70,534	83,194	299	351	2,735	2,627	30,012	28,987
Ukraine	7,294	5,533	837	1,111	3,183	6,469	156	103	15,309	29,036
Untd.Kingdom	600,446	527,930	247,505	290,477	15,992	15,675	121,169	111,420	2,124,751	2,028,862
United States	1,338,354	1,195,298	1,010,864	979,322	79,692	93,631	284,535	303,488	6,283,385	6,660,318
Venezuela	919	817	108,135	52,581	41	35	239	182	738	644
Viet Nam	437	705	339	727				15	347	591
Zimbabwe			774	3,002					256	121
Rest Of World	351,891	317,048	989,340	1,005,655	31,255	33,622	53,296	50,586	457,980	409,532
Total	9,977,129	9,710,896	7,013,743	7,534,935	1,131,035	1,145,459	1,814,223	1,797,650	26,138,451	27,294,177

Exports of Selected Environment-Friendly Products, 1998–1999 (in thousands of US dollars)

Noise and vibration abatement		Remediation and cleanup		Solid waste management		Waste water management		Air pollution control		Total	
1998	1999	1998	1999	1998	1999	1998	1999	1998	1999	1998	1999
88,982	82,667	104	26	56,056	45,600	101,970	101,401	31,963	47,946	352,810	337,501
62,587	71,854	3,952	5,263	252,001	91,595	217,465	252,744	38,216	11,762	759,200	621,928
642,857	638,983	19,773	20,289	505,792	571,237	1,276,520	1,189,007	125,787	116,987	3,083,810	3,052,301
525,850	594,884	7,998	9,527	777,526	826,577	2,244,493	2,313,458	405,358	452,041	4,703,954	5,026,667
770,180	713,486	67	28	69,963	69,884	705,219	661,971	21,963	24,359	1,724,310	1,677,913
3,444	1,064	98	351	17,194	6,187	57,681	63,291	1,878	237	86,104	77,530
897,473	1,153,526	33,750	44,874	672,124	723,248	2,174,313	2,363,607	207,619	183,051	5,298,908	5,829,535
2,755	3,801	31	40	15,599	18,414	29,056	26,586	2,825	1,974	195,888	203,764
121,531	194,420	75,101	104,601	487,157	597,562	2,687,229	2,699,952	35,597	47,630	3,938,654	4,291,136
3,843	2,867	28	117	29,016	24,658	38,632	30,523	1,883	1,152	82,685	65,694
1	51			7,754	14,611	21,747	21,340	210	825	51,653	55,957
133,790	155,829	5,650	3,330	99,402	102,739	595,759	583,987	44,948	32,205	1,008,989	1,001,443
180,467	176,451	7,494	6,051	203,636	240,676	1,815,368	1,802,156	77,084	85,386	3,050,829	3,075,178
10	8	53		2,204	3,090	17,481	14,968			20,374	18,958
2			19	6,994	11,902	8,487	7,196			23,320	29,084
151,959	135,279	18,539	16,428	265,722	295,401	503,187	455,570	51,064	41,282	1,348,599	1,279,088
1,718,777	1,677,902	69,206	75,929	1,068,684	1,022,614	5,874,577	5,710,355	704,670	717,407	12,038,803	11,582,582
4,498,199	4,581,244	310,590	325,997	4,556,729	4,369,690	12,562,587	11,873,141	1,860,658	1,917,170	32,091,763	31,486,514
69,342	52,854	1,902	917	26,743	29,951	95,044	80,654	1,032	1,049	216,484	187,393
17,078	13,051	118	176	30,513	23,334	138,664	108,822	3,285	1,675	236,420	175,281
183,194	202,046	8,363	9,035	57,383	61,041	200,438	218,471	21,106	24,424	719,843	773,768
107	204			177	199	3,508	3,661	32	11	4,090	4,286
71,450	78,740	357		49,220	34,292	182,638	150,968	6,715	6,416	348,373	294,058
25,635	33,589	161	1,019	45,588	77,868	166,283	192,220	2,628	4,257	290,761	416,259
14,772	18,315	36,081	32,241	31,218	45,606	377,520	412,146	9,219	10,262	666,583	724,710
14,819	19,454	3,069	1,538	125,395	161,999	462,508	487,035	6,580	5,909	826,049	894,242
1,183,395	1,137,820	155,656	150,247	3,168,402	3,091,470	8,049,199	7,906,614	563,168	556,990	15,198,938	14,756,150
3,635,604	4,168,239	19,369	35,129	5,226,167	6,916,422	5,914,160	6,413,010	1,032,310	1,047,591	20,393,727	23,889,924
411	204		18	2,070	2,628	2,270	3,267	199	66	6,279	8,503
129,403	136,289	3,897	3,532	669,113	686,247	1,140,705	1,292,058	117,307	103,588	2,439,521	2,622,229
10,028	17,679	674	1,437	147,202	221,489	385,550	419,040	16,592	12,600	1,165,001	1,310,560
890,551	1,133,366	479	1,310	337,057	416,889	1,810,214	2,198,097	252,298	283,204	4,510,443	5,442,612
489,797	411,128	32,434	28,233	553,392	697,447	2,140,518	1,883,635	165,863	154,448	4,997,281	4,695,447
7,668	7,716	1,553	905	24,924	32,635	79,955	92,372	7,279	6,864	323,039	289,400
91,271	93,570	31,943	31,839	155,527	127,626	397,768	355,595	30,547	27,767	887,550	801,667
1,338	351			5,280	3,079	16,150	17,228	40	453	25,227	23,529
7,971	7,435	522	552	49,180	49,163	88,985	105,036	4,585	4,516	477,003	494,424
139,786	151,590	3,532	2,232	137,356	152,342	426,086	395,559	15,203	6,737	885,400	888,946
124,630	138,419	24,078	19,142	39,312	51,481	225,873	314,561	3,729	3,201	548,450	704,146
89,549	51,940	3,418	14,091	30,357	61,323	447,079	389,455	24,093	33,977	819,355	823,550
196,191	189,617	9,359	5,977	356,860	495,927	895,287	964,036	42,221	38,471	2,162,153	2,432,717
9,963	14,258	1,121	1,654	63,437	90,520	104,466	137,367	1,814	1,935	208,357	267,937
520,528	534,548	23,059	23,745	365,460	409,059	1,491,722	1,551,594	70,978	53,106	3,018,684	3,145,478
406,347	348,961	33,638	30,602	477,858	488,851	1,209,014	1,234,010	116,311	123,724	3,179,511	3,447,357
104,166	118,963	8,451	8,387	1,042,854	1,188,767	1,880,749	1,879,513	367,905	367,533	4,628,362	4,824,348
57,216	77,303	296	505	114,114	114,359	361,730	463,532	7,018	5,565	705,811	842,445
22,522	16,344	239	200	23,232	20,517	312,787	295,147	7,485	3,852	393,044	378,312
1,451,946	1,237,246	165,920	109,169	1,258,085	1,285,964	4,526,797	4,170,502	598,635	637,240	11,111,246	10,414,485
4,257,661	4,469,703	374,354	412,101	5,665,221	6,711,982	12,176,011	12,968,689	1,305,143	1,296,816	32,775,220	35,091,348
5,561	4,242	112		10,788	7,505	46,854	26,967	770	562		
125	63	62	10	4,222	7,161	14,537	25,388	14	190	20,083	34,850
291	367			28	155	1,601	1,523	30	5	2,980	5,173
716,960	774,121	75,442	53,613	1,235,910	1,226,516	3,616,297	3,640,377	386,938	555,976	7,862,944	8,035,543
24,749,983	25,844,051	1,572,093	1,592,426	30,623,198	34,027,499	80,320,738	80,969,402	8,800,795	9,062,394	191,914,866	198,853,850

Annexes

Annex 1
Making Environmental Data Comparable: The ESI Methodology

Marc A. Levy, Alex de Sherbinin, Francesca Pozzi, and Antoinette Wannebo

Environmental Performance Measurement:
The Global Report 2001–2002

In designing the Environmental Sustainability Index (ESI), we chose a methodology that was straightforward and transparent to facilitate understanding and reproducibility. The crux of the ESI methodology is the conversion of selected variables to relative performance benchmarks, which are in turn averaged. In this respect, the ESI can be understood as something like a grade point average. This annex describes how these comparative benchmarks are derived.[1]

Choosing Countries and Estimating Missing Data

As described in Chapter 2, the ESI is built from 22 indicators that are calculated by a total of 67 measured variables. We selected countries for inclusion in the Index based on the extent of their data coverage for these indicators and variables. We eliminated all countries for which data were insufficient to generate measures for at least 19 of the 22 indicators, and we included all countries for which the data permitted measurements of at least 20 indicators (94 countries). For those countries where the data permitted measurements of no more than 19 indicators (54 countries), we applied an additional criterion: if their overall data coverage included at least as many variables as the lowest number for countries missing two indicators, we included them in the Index (28 countries met this criterion). We ended up with 122 countries in the Index, each of which had data for at least 62 percent of the variables in our analysis.

The median country in the Index is missing 17 of the 67 variables. A quarter of the countries are missing between 22 and 26 variables, and a quarter are missing between one and seven. Altogether, this means that 22 percent of the 8,174 data points in our database are missing. If we left all of the missing values as they were, we would be limited in our ability to draw comparisons across countries. We also risked rewarding countries that may be withholding poor results. We therefore adopted the following data estimation procedure.

Table 1

Variable-Specific Imputations

Missing Value	Predicted by
Percent of population with access to improved drinking water supply	Human Development Index
Daily per capita calorie supply as a percent of total requirements	Human Development Index
Ecological footprint "deficit"	Population density
Expenditure for R&D as a percent of GNP	R&D scientists and engineers per million population
Fertilizer consumption per hectare of arable land	Consumption pressure per capita
Industrial organic pollutants per available fresh water	Suspended solids
NO_x emissions per populated land area	Coal consumption per populated land area
Percent of breeding birds threatened	Percent of mammals threatened
Percent of mammals threatened	Percent of breeding birds threatened
Pesticide use per hectare of crop land	Fertilizer consumption per hectare of arable land
Phosphorus concentration	Industrial organic pollutants per available fresh water
R&D scientists and engineers per million population	Expenditure for R&D as a percent of GNP
Scientific and technical articles per million population	R&D scientists and engineers per million population
SO_2 emissions per populated land area	Coal consumption per populated land area
Under-5 mortality rate	Human Development Index
Urban SO_2 concentration	SO_2 exports
Urban TSP concentration	Suspended solids
Vehicles per populated land area	Population density
VOC emissions per populated land area	Coal consumption per populated land area

First, we estimated missing values for 19 of the variables using other variables that were chosen specifically because of their direct, substantive correlation with the missing variable. We identified these variable-pairs by analyzing the correlation coefficients of all the variable-pairs, including a small number of external variables thought to be potentially useful in predicting missing values. For variable pairs that had both a high correlation coefficient and a direct, causal connection between the two variables, we used linear regression to calculate an equation that permitted us to estimate these variables. Table 1 lists variables that were estimated in this way.

Although some variables appear in both columns of this table, when estimates were calculated, only observed values were used as predictors.

One set of variables we chose not to impute at all. These are listed in Table 2. They included variables related to physical characteristics of a territory, such as hydrology or coal deposits, survey results, or other features for which there

was no sound basis to arrive at an estimate. Altogether, 13 variables were not estimated.

For all of the remaining variables, we estimated missing values using a set of three simple predictors that had high correlations with a large number of variables in the ESI. These were Gross Domestic Product (GDP) per capita; Human Development Index (HDI); and the level of corruption as measured by the World Bank's Aggregated Governance Indicators project. If our goal had been strictly to estimate the missing values as best we could, we would have used a multiple-regression model incorporating all of these variables. However, we also wished to avoid rewarding countries that were underreporting, so we calculated three separate estimates instead, using these predictor variables one at a time, and then adopted the estimate that rewarded the country the least.

Altogether just over 60 percent of the missing variables were estimated using one of the above techniques. The estimation protocol permitted us to generate a full set of 22 indicators for each of the countries in the Index. In reporting country-

Table 2

Variables Not Estimated When Missing

Coal consumption per populated land area

Compliance with environmental agreements (WEF Survey)

Degree to which environmental regulations promote innovation (WEF Survey)

FSC-accredited forest area as a percent of total forest area

Internal renewable water per capita

Number of ISO14001 certified companies per million dollars GDP

Per capita water inflow from other countries

Percent of country's territory under severe water stress

Percent of country with acidification exceedence

SO_2 exports

Spent nuclear fuel arisings per capita

Stringency and consistency of environmental regulations (WEF Survey)

Subsidies for energy or materials use (WEF Survey)

Table 3

Variables Transformed to Logarithmic Scale

CFC consumption (total times per capita)

CO_2 emissions (total times per capita)

Fertilizer consumption per hectare of arable land

FSC-accredited forest area as a percent of total forest area

Global environmental facility participation

Historic cumulative CO_2 emissions

Internal renewable water per capita

NO_x emissions per populated land area

Per capita water inflow from other countries

Percent change in forest cover, 1990–1995

Percent of mammals threatened

Renewable energy production as a percent of total energy consumption

Vehicles per populated land area

VOC emissions per populated land area

level ESI data, we explicitly listed the number of missing variables and the number of those estimated.

Making the variables comparable

After producing a complete working data set, we carried out operations to make the data as comparable as possible across variables and countries. The following section explains what we did and why.

We first denominated selected variables to facilitate fair comparison across countries. Some variables needed no change in denominator because they were already collected in a way that permitted international comparison. Variables having to do with national governance systems, for example, were already comparable. Most of the environmental stress variables, however, were not comparable as they were obtained. They typically reported the quantity of a particular pollutant, but provided no information about the likely impact of that pollutant on humans or ecosystems. We experimented with different ways to make such stress variables comparable and ended up in many cases with a denominator we called "populated land area." Populated land area in this case refers to the size of that portion of a country's territory where population density exceeds five persons per square kilometer. This measure avoids the mistake of considering countries with large sparsely inhabited land areas to "offset"

their pollution in one area by distributing the pollution allocation across the whole land area. It assumes that pollution and other stresses are highly correlated with the location of people, and that other things being equal, a given amount of pollution in a small area was worse than the same amount in a large area.

Other denominators included GDP and total population. The selection of the denominator is made explicit in each of the variable tables contained in Annex 4; the populated land area values appear in Annex 2.

We then converted variables whose distribution was highly skewed to a logarithmic scale. In the absence of such a transformation, the variable would convey less information and be less comparable to other variables. The variables listed in Table 3 were transformed in this way.

We next trimmed the tails of the variable distributions to avoid having extreme values overly dominate the aggregation algorithm, and to partially correct for the possibility of data quality problems in such extreme cases.[2] For any observed value greater than the 97.5 percentile, we lowered the value to equal the 97.5 percentile. For any observed value lower than the 2.5 percentile, we raised it to equal the 2.5 percentile. We did this for each variable, but the total number of affected values was very small.

On two variables we imposed an additional truncation procedure because we thought that there were natural upper or lower thresholds at work. For the population growth rate variable, we set all observed values lower than zero equal to zero. This reflects a view that negative growth rates are not any more sustainable than zero growth rates. Similarly, we cut off the calories per capita variable at 120 percent of daily requirements; sustainability does not increase as one moves beyond this level (it may even decrease).

Finally, we converted all the variables to a unitless scale. We chose the z-score, which has desirable characteristics when it comes to aggregation. In particular, the fact that the z-score always has an average of zero means that it avoids introducing aggregation distortions stemming from differences in variable means. The formula for z-score is the value of the variable minus the mean of the variable, divided by the standard deviation. For variables in which high observed values correspond to low values of environmental sustainability, we reversed the terms in the numerator to preserve this ordinal relationship. In other words, for variables such as "percentage of land area under protected status," we used the conventional z-score, whereas for variables such as "percentage of mammals threatened," we used the alternative specification.

Aggregating the indicators

We explored a number of different approaches to aggregating the variable measures to produce the ESI and, in the end, chose the simplest and most transparent approach. We calculated the 22 indicators by averaging the variables (after they are made comparable through the steps outlined above), and we calculated the ESI by averaging the 22 indicators. For each indicator, this approach gives equal weight to each of the variables within that indicator. If the underlying scientific processes were understood fully, they almost surely would support an algorithm of unequal weighting, with differential weights derived from the different degrees of impact on overall environmental sustainability. However, in our judgment, there was no firm basis for drawing such scientific conclusions with current knowledge, nor is there likely to be such a basis any time soon.

We tested our judgment by conducting a survey of environmental experts and members of the business community to determine their views on the relative importance of the indicators used in the Index. The survey asked respondents to rank, on a scale of 1–5, the relative importance of the indicators that comprise the Index. A total of 254 surveys were received, representing 73 countries. One major set of respondents was identified at the October 2000 meeting of the World Conservation Congress in Amman, Jordan (n=158); the survey was circulated in person at that meeting. The other was the World Economic Forum Global Leaders for Tomorrow (GLT) membership (n=58); GLT members were

Table 4

Survey Result

	N	Mean Score 1=low, (5=high)	Standard Deviation
Urban air quality	254	3.8	1.2
Water quantity	254	3.6	1.3
Water quality	254	3.9	1.1
Biodiversity	253	3.9	1.1
Land	251	3.7	1.1
Air pollution	253	3.8	1.0
Water pollution and consumption	251	3.9	1.0
Ecosystem stress	252	4.0	1.0
Waste production and consumption pressure	250	3.6	1.1
Population	250	3.5	1.2
Basic sustenance	250	3.5	1.2
Public health	252	3.5	1.3
Disasters exposure	251	2.8	1.3
Science and technical capacity	252	3.0	1.1
Capacity for policy debate	252	3.3	1.1
Environmental regulation and management	253	3.5	1.1
Tracking environmental conditions	251	3.3	1.2
Eco-efficiency	249	3.3	1.2
Public choice failures	251	3.6	1.2
Contribution to international cooperation	251	3.0	1.1
Impact on global commons	252	3.5	1.2

sent the survey by the WEF. A smaller number of questionnaires (n=36) was circulated at other meetings of environmental experts during the fall of 2000; each of these meetings was attended by recognized experts from a range of countries. (Note that the Private Sector Responsiveness indicator was not yet formulated at the time of this survey, so it is not included here.) Table 4 summarizes the results.

We drew two conclusions from this survey. First, we noted significantly lower importance scores for an indicator we had been developing on "exposure to environmental disasters." The opinion that this indicator should be lower in relative importance was observed across regions, across sectors, and across income levels of the respondents' home countries. In the end, we dropped the environmental disasters indicator. In addition to its being judged to be of lower importance, it had weak variables available to measure it. Second, the other variables were close together—in virtually all cases occupying overlapping 95 percent confidence intervals. Although the Environmental Sustainability Index scores are different if we apply these weights, in the end we decided not to use them because we were not confident that they reflected a meaningful set of differences.

A sensitivity analysis suggests further that the weighting methodology developed would not have changed the rankings appreciably. In particular, we calculated an Index score using the survey-generated weights (and applying the average weight for the two indicators that were not part of the survey). The average shift in rank was only 1.7 places out of 122.

Under the circumstances, then, the choice of weights is more a reflection of values than of science, and in this regard we felt strongly that the 22 environmental sustainability indicators ought to receive equal weight, at least as a baseline metric, because they correspond broadly to the entire set of environmental sustainability issues identified by the global community as important. We realize that there are alternative approaches to the weighting question, as discussed in Chapter 2.

In addition to calculating the overall ESI by averaging the 22 indicators, we computed separate values for each of the five components: Environmental systems, Reducing environmental stresses, Reducing human vulnerability, Social and institutional capacity, and Global stewardship. These component values were calculated by averaging the indicators that make up each component (because the components vary in the number of constituent indicators, the ESI is not an average of the components but can only be calculated by averaging the 22 individual indicators).

To make the ESI and component scores more intuitively understandable, we converted the z-score average (a typical range would be from about –2.5 to +1.8) to standard normal percentile. The standard normal percentile has a theoretical minimum of zero and a theoretical maximum of 100, but is calculated in such a way that the maximum and minimum values are realized only at observed values between about 2.5 and 3 standard deviations away from the mean. Values within that range receive scores in between the minimum and maximum, regardless of where other countries' values lie in comparison. Likewise, values that fall outside that range do not receive significantly better or worse scores than values that lie between 2.5 and 3 standard deviations from the mean. Therefore, the standard normal percentile comes closest to preserving the information contained in the original z-scores, while portraying them in a manner more graspable to a broad audience. When reporting the individual indicator values, we opted to report the original z-scores; this preserves more information from the underlying variable averages, because

for a handful of indicators observed, minimum and maximum values fall beyond the range a standard normal percentile assumes.

Data Reduction

Whether to rely on the 22 indicators as the primary unit of aggregation, or to seek a more spare structure with fewer elements, was a central methodological choice. Our final choice was influenced by the following analysis.

To explore whether there might be a way to reduce the number of dimensions in the ESI, we carried out principal components analysis, a technique that identifies distinct dimensions of a data set using purely statistical techniques. Using commonly applied thresholds of significance, we derived 11 principal components. While fewer than 22, this does not produce a radically more simple structure. More important, the structure it generates is not in conformity with any scientific or value-based sense of how variables ought to be clustered. For example, the first principal component assigns to participation in the Global Environment Facility one of the highest positive weights and assigns to participation in the Montreal Protocol Multilateral Fund one of the lowest negative weights. Our theory-driven structure is superior to the principal components.

Another test of whether the 22 indicators comprise a meaningful set is to look at the correlations among the indicator pairs; other things being equal, the lower such correlations were, the more appropriate our structure. As a group, the 22 indicators had an average bivariate correlation among themselves of only 0.11. Only 36 of the 231 possible pairs of indicators had correlation coefficients greater than 0.5. The highest such pairs were Basic Human Sustenance/Environmental Health (0.85), Environmental Health/Reducing Population Stress (0.82), Basic Human Sustenance/Reducing Population Stress (0.72), and Environmental Health/Science and Technology (0.69). Of these pairs, only the first lends itself to a plausible argument for combination. However, as long as the total number of highly correlated indicator pairs is relatively low, as is the case in the ESI, we think it is preferable to keep the indicators separate to permit investigation into potentially useful causal connections among the pairs, and to permit reporting of metrics that are relevant for discrete policy communities. For example, the most highly correlated indicator pair contains one indicator that is primarily relevant to the food security community and another that is primarily relevant to the public health community. Keeping the indicators separate lets us be relevant to both communities while permitting useful interactions between them.

Coping with Data Challenges

Finally, an important aspect of the ESI methodology has to do with the manner in which we responded to challenges posed by the data, in particular challenges having to do with scale differences and with substantive gaps in data availability.

Scale differences are very important, because environmental sustainability is a phenomenon that rarely unfolds at the level of a nation-state as a whole. It is observed more typically at a smaller scale—a river basin, a forest, an urban center. Yet for the most part, environmental data are reported at the national level. If a country's freshwater withdrawals are about equal to its freshwater availability, for example, then using only national-level data will lead one to an optimistic assessment. But if withdrawals are highly concentrated in one area, and availability is concentrated in a different area, these national figures are very misleading. We sought wherever feasible to incorporate data that were collected or reported at a more fine-grained resolution, and then to aggregate them up to national levels in a way that took into account the sustainability dynamics at the smallest relevant scale. We did this for measures of acidification damage, water stress, water quality, air quality, land degradation, and private-sector responsiveness.

It is noteworthy that almost all of these examples of data that were aggregated up from smaller scales came from sources outside the standard canon of international organization data products. For the most part, the standard sources of comparable national environmental data do not lend themselves to such analysis. Of the examples mentioned above, only water quality and air quality came from UN sources; the others were from national labs, university departments, or commercial firms. And these two UN sources were less than user friendly. The air quality measure was provided for specific cities, and had to be combined with separate data on city population to make it comparable across countries; even then, the measures were so spotty than such comparisons were problematic. The water quality data were even more difficult to work with. Although they are collected under the auspices of a UN effort, the UN Global Environmental Monitoring System, the data are not released in a usable format except through special arrangement that requires significant compensation to cover processing costs.

Substantive gaps in data coverage were even more problematic. Many important variables had shockingly poor country coverage. It was extremely frustrating to experience the gulf between statements by global bodies about the high priority of water quality and air quality as critical environmental concerns, and the reality that there is very little systematic global monitoring of these factors. Some variables were measured so poorly that we could not use any metric at all in the ESI. This was true for resource subsidies, wetland loss, nuclear reactor safety, and lead poisoning, for example.

One strategy we used to help deal with data gaps was utilization of modeled data. Increasingly, global environmental phenomena are the focus of intensive modeling efforts that take the best available empirical observations as inputs and add tested methods for generating global estimates of either individual variables or the interaction among variables. Such model data are typically far more sensitive to scale and place than conventional sources. The input data are harmonized to make them systematically comparable by teams of substantive experts publishing results in a peer-review process. This data harmonization task is of crucial importance, because to construct a measure relevant to environmental sustainability one must frequently combine information from disparate sources. Without researcher expertise in the subject area, errors are possible (for example, our first effort to measure the percent of mammals threatened had a maximum value of 150 percent because our data for number of mammals present and number of mammals threatened came from different sources; they used incompatible taxonomies, which we realized only because the error in this case was so obvious).

We used model data for water quantity, acidification damage, air pollution emissions, industrial organic pollution emissions, and population stress. We were selective in choosing modeled data; all the models we drew from had been subject to scientific peer review and/or endorsed by international organizations.

In a few select cases, we constructed our own data sets. We did this for environmental health, land area impacted by human activities, and membership in international environmental organizations. We also arranged with a few data holders to have custom data sets constructed for us; this was the case with our use of the Innovest EcoValue '21™ and Dow Jones Sustainability Group Index variables.

In the following sections, we report in more detail how we dealt with three sets of indicators that are often considered to be of high global priority: freshwater resources (water availability and quality), terrestrial ecosystems (anthropogenic impacts on land and acidification exceedance), and environmental health.

Water Availability

One of the problems we encountered with existing data sets on internal renewable water resources by country is that they are compiled from many different sources, and they sometimes include and sometimes exclude water flowing from and to other countries. This quotation from the data appendix to the World Resource Institute's (WRI) *World Resources 2000–2001* report illustrates the dilemma: "When data for annual river flows *from* and *to* other countries are not shown, the internal renewable water resources figure *may* include these flows. When such data are shown, they are *not* included in a country's total internal renewable water resources." Although the WRI report is one of the best compilations of water availability figures, the ambiguity of the data definition, and the fact that the data come from 11 different sources, renders them less useful for globally comparative analyses.

To address this problem, we worked with hydrological modelers at the Center for Environmental Systems Research at the University of Kassel in Germany to perform some special runs of its WaterGAP 2.1b model (WaterGAP stands for Water Global Assessment and Prognosis; Alcamo et al., 2000). WaterGAP belongs to a class of environmental models, called "integrated" models, that were first developed during the 1980s to study large-scale environmental problems. The advantage of the WaterGAP model is that it is based on a consistent set of methodologies using actual hydrological data on precipitation, evaporation, and river flows from 1961 to 1990. These data were converted by the modelers to a 0.5° by 0.5° latitude-longitude grid (approximately 50 km by 50 km at the equator, and 50 km from north to south and 25 km east to west at a latitude of 60 degrees). The model estimates the impact of evaporation, which greatly affects water availability and consumption in upstream nations. The internal renewable water resources data represent 1961–1990 average annual flow of rivers and recharge of groundwater generated from endogenous precipitation, taking into account evaporation losses from lakes and wetlands. The inflow data represent 1961–1990 average annual inflow of rivers flowing from other countries, taking into account the loss due to consumptive water use in those countries.

The disadvantage of using WaterGAP is that, owing to the grid cell size (as described above), the model does not easily accommodate "microstates" such as small islands or city-states. Where possible, we used alternative data for these countries.

Water Quality

We obtained original water quality data sets from the UNEP-Global Environmental Monitoring System/Water group (GEMS/Water) and analyzed them extensively over one summer. The GEMS yields a consistent data set for 45 countries for a wide range of water quality indicators. We selected from the GEMS/Water data set a smaller subset of variables based on the extent of country coverage for each variable and the degree to which the variable is recognized as an important measure of water quality. We arrived at the following four indicators:

Dissolved oxygen: This is a "headline" indicator for water. It tracks eutrophication levels and is positively related with stream flow and inversely related to nitrogen and phosphorous levels. The U.S. National Research Council report (2000) listed dissolved oxygen as one of four indicators that provide crucial measures of ecosystem health.

Suspended solids: This is a measure of turbidity and would be associated most closely with people's visual assessment of what clean water looks like. In heavily agricultural areas with high erosion levels, suspended solids levels can be quite high (for example, the Ganges or the Yellow River). There is a fairly high natural component to suspended solids in rivers, and the concentration of suspended solids tends to increase in proportion to discharge levels. However, when aggregated across water bodies at the national level, this measure remains an important means of assessing water quality.

Phosphorus concentration: This is a measure of the level of eutrophication: phosphorous is a limiting nutrient for plant photosynthesis in freshwater environments. Little natural component is measured by this indicator, so we can be reasonably certain that we are measuring anthropogenic impacts on water bodies.

Electrical conductivity: This is a bulk measure of the concentration of metals or salt in the water. Electrical conductivity is one of the most rapid and inexpensive measurements to assess water quality, so its measurement is precise and widespread. Conductivity of water is affected by the presence of inorganic dissolved solids such as chloride, nitrate, sulfate, and phosphate anions, or sodium, magnesium, calcium, iron, and aluminum cations. It is important to note that geology can have a major impact on electrical conductivity. Streams that run through bedrock areas tend to have lower conductivity, whereas streams that run through soils tend to have higher conductivity.

One limitation of the GEMS water quality data is that the participating countries provide data from monitoring stations that vary in number and may be located in quite different locations with respect to water quality stressors (for example, industry and agriculture) within the same country and from country to country. This makes it challenging to aggregate station data within countries, and also makes it somewhat difficult to compare the resulting measures across countries. Because there is no alternative, however, the GEMS data are what we used. In the interest of developing a more comprehensive worldwide water database with a carefully constructed analytical protocol, support needs to be provided to the GEMS/Water program to expand its country coverage.

Anthropogenic Impacts on Land

We created a new variable from selected global data sets that seeks to provide a measure of land affected by human activity within a country. The variable bundles several environmental stresses, including ecological impact of natural vegetation clearing, the environmental impact of specific land use activities, and the efficiency of land resource use by a country. The variable is currently experimental and in a developmental stage.

Clearing of natural vegetation results in habitat fragmentation and degradation. If the land is converted to agriculture, there can also be economic and ecological costs from increased soil salinity. If the land is converted to urban area, the change is generally irreversible, resulting in increases in the extent of impermeable surfaces (pavement) and, potentially, pollution-generating activities. There is a direct relationship between cleared natural vegetation and biodiversity loss, including species extinction (Brooks et al., 1999). The anthropogenic impact variable represents an inventory-based approach proxy measure of biodiversity loss to complement the current species-based ESI biodiversity variable.

Two types of anthropogenic impacts were identified: the built environment and the agricultural (including pasture) land. Two satellite-derived global data sets were combined to estimate the area of land in each country affected by anthropogenic activities. Estimates of built environment were derived from the Nighttime Lights data set and estimates of agricultural land from the Global Land Cover Characteristics (GLCC) database. An estimate of the proportion of built environment and agricultural land provides a proxy for the amount of natural vegetation cleared. Two complications are immediately obvious: plantation-style forested areas are not included, and some pastureland is natural grassland. Nonetheless, a relative measure of land cleared of its natural vegetation cover is possible.

The methodology was as follows. Version 2.0 of the GLCC database was downloaded for each region (North America, South America, Eurasia, Asia-Pacific, and Africa) in the Lambert Azimuthal Equal Area Project from an ftp site (edcftp.cr.usgs.gov). The data set documentation reported that land cover classes over 10 percent of the earth's surface were revised for version 2.0 based on user feedback (Brown et al., 1999) and broad lessons learned from the validation exercise of the IGBP DISCover land cover data (Scepan, 1999; Muchoney et al., 1999). This version of the database is still based on the 1992–1993 AVHRR time series, so it represents the land cover patterns for that period. The USGS Land Use/Land Cover System Legend (Modified Level 2) was selected for this application.

Urban areas for the GLCC product were extracted from the Digital Chart of the World (Defense Mapping Agency, 1992). A visual inspection of the urban class indicated that not all built environment areas were represented, so the Lights at Night data set was used as an alternative. The Lights at Night data set captures a wider range of human activity, including residential, commercial, industrial, public facilities and roadways (Elvidge et al., 1999).

Elvidge et al. (1997a, 1997b) have applied the time series data from the Defense Meteorological Satellite Program

(DMSP) Operational Linescan System (OLS), as processed by the NOAA National Geophysical Data Center in Colorado, to inventory human settlements. The data product selected for the ESI application was the stable lights for city areas. This data product has been filtered for clouds, gas flares, fishing lights, and fires and thresholded based on how often a particular grid cell is lit. This global data layer is at a nominal resolution of 1km and represents data for October 1994 to March 1995.

The two data sets were combined in the following fashion: A global binary version of the stable lights for city areas was created, resulting in a grid of lit or not lit areas. The global data set was then cut and projected to match the GLCC database regions. The lit areas from the lights at night data set were then "added" to the GLCC, replacing the previous classification.

To estimate anthropogenically affected areas, the areas for relevant classes in the composite data set were tabulated by country. The USGS Land Use/Land Cover System Legend has five classes that include cropland and/or pasture: dryland cropland and pasture, irrigated cropland and pasture, mixed dryland/irrigated cropland and pasture, cropland/grassland mosaic, and cropland/woodland mosaic. Areas for these five classes and the lights at night-derived lit area class were combined, resulting in the square kilometers of anthropogenic impact. The land area affected by human activity calculated from the composite data set was divided by the land area of the country, as reported in the ESRI global country data set, to calculate the proportion of land area affected by human activity.

Acidification Exceedence

The objective of this variable was to assess the degree to which terrestrial ecosystems were affected by acidification due to sulfur deposition from industrial air pollution. We calculated the proportion of each country at risk of acidification, based on the "Exceedance of Critical Loads for Terrestrial Ecosystems" map obtained from the Stockholm Environment Institute (SEI) at York (United Kingdom). This map was produced as follows (Kuylenstierna et al., 2000):

1. Creation of sulfur deposition map. The Global Emission Inventory Activity (GEIA) sulfur emission inventory for 1990 was used and integrated into the MOGUNTIA model to calculate sulfur deposition. In addition, a model for global emission, transfer, and deposition of soil dust (*base cation deposition*) was used. The base cation deposition, and particularly the calcium content, is a measure of the ability of the ecosystem to neutralize the acidifying depositions. Two deposition ranges (10 percent and 100 percent of calcium content) were used in the model. The acidic deposition derived from sulfur emissions is calculated as sulfur deposition minus the base cation deposition rate.

2. Creation of sensitivity to acidic deposition map. A method that combines three classes of Cation Exchange Capacity with five classes of base saturation to define five classes of sensitivity was implemented and applied to the digital FAO Soil Map of the World.

3. Conversion of sensitivity map to critical loads map. The conversion was made based on the assumption that the critical load is equal to the buffering (weathering) rate of the soil.

4. Production of the exceedance of critical loads map. This was obtained by combining the acidic deposition (1) and critical loads (3) maps.

For 1990, the Stockholm Environment Institute at York produced two maps of exceedance of critical loads:

- High Risk, obtained by using low critical loads and 10 percent calcium content of modeled dust, and

- Low Risk, obtained by using high critical loads and 100 percent calcium content in dust.

Given the small variability in exceedance values from the Low-Risk map, we decided to use only the High-Risk map for inclusion in the ESI. The areas at risk have been summed within each country, and then the percentage of a country at risk of exceedance was calculated.

The acidification exceedance variable is a good example of how a modeling exercise can generate a measure that is far more relevant to environmental sustainability than standard data sources provide. In this case, the relevance comes from combining information about the deposition of pollutants (international sources commonly report emissions but not depositions, which are affected by weather conditions), with information about ecosystem sensitivity, in way that preserves the variation in scale that both input variables exhibit.

Environmental Health

The concept of environmentally related diseases has begun to gain currency, but no one to date has produced indicators of diseases that are attributable to environmental conditions. The Global Burden of Disease (GBD) study was a step in this direction; it produced some measures of countries' burden of disease for 1990, and among the diseases included were a number that could be traced directly to environmental factors (see http://www.hsph.harvard.edu/organizations/bdu). Smith et al. (1999) analyzed the Global Burden of Disease numbers and demonstrated that, of all the diseases included in the GBD study, acute respiratory infections (ARIs) and diarrheal diseases were most linked to environmental conditions. They conclude that "25–33 percent of the global burden of disease can be attributed to environmental risk factors. Children under 5 years of age seem to bear the largest environmental burden, and the portion of disease due to environmental risks seems to decrease with economic development"(573).

Using a large data set from the World Health Organization, we extracted age- and sex-specific deaths by country for the most recent years available (we used no data older than 1990) for the two classes of disease mentioned above: ARI and intestinal infectious diseases. In the first case, we produced an indicator, called "Child Death Rate from Respiratory Diseases," that measures deaths from respiratory diseases (WHO classes B31 & B320 & B321) per 100,000 population aged 0–14 (using UN population data broken down by age). The diseases in this category included acute tonsillitis, acute laryngitis and tracheitis, other acute upper respiratory infections, deflected nasal septum and nasal polyps, chronic pharyngitis, nasopharyngitis and sinusitis, chronic diseases of tonsils and adenoids, acute bronchitis and bronchiolitis, pneumonia, and other.

For the intestinal infectious diseases, we followed a similar procedure, except that we calculated standardized death rates for each country's entire population that were comparable across countries. We used standard demographic techniques to impose a standard population structure to allow comparisons of mortality rates among countries. We then totaled the deaths this yielded for all age groups and divided by overall population to yield a standardized death rate per 100,000 population. The diseases in this category included cholera, typhoid fever, shigellosis, food poisoning, amoebiasis, intestinal infections due to other specified organism, ill-defined intestinal infections, and other.

Our broadest conclusion about the state of global data relevant to environmental sustainability is that the gap between what policymakers, activists, and scientists say is most important and what is actually getting measured is enormous. To create the ESI, we invested considerable resources in obtaining data from sources outside the channels that normally populate global data tables, including modeling groups, private firms, and national environmental labs. We also devoted significant effort to constructing new measures by processing data sets that were relevant to environmental sustainability, but not in a form to permit combination and comparison. Our efforts clearly were very limited, and much remains to be done at an overall level. The biggest gaps concern water quality, air quality, soil degradation, wetland loss, resource subsidies, and adequacy of domestic environmental policies.

We find it instructive that of all the criticisms engendered by the debate over the ESI, the most vociferous are from countries that argue that the data used in the ESI are not the most accurate portrayal of conditions in a particular country. One official from a central European country has told us that our air pollution estimates (from the World Bank) are off by a factor of two for his country. Murray Feshbach (2001) has pointed out that Russia's relatively benign score of 56.2 masks an ongoing environmental catastrophe of great magnitude. The reason for the discrepancy in Russia's ESI score and what experts such as Feshbach know to be true is that the available global data do not capture the dynamics underway in Russia, in particular the intense toxic contamination there. Belgium's low score in the ESI (the lowest of all the industrial countries) came under intense scrutiny within Belgium: although on balance this scrutiny heightened attention on the overall poor environmental conditions and performance within Belgium, it also generated a systematic comparison of Belgian data sources and ESI sources, with a strong prima facie case that overall results would have been better for Belgium had higher-quality data on the country been available through the global sources used in the ESI (whether this is true has not been tested). Similarly, when the Mexican government launched an investigation into the reasons for Mexico's relative low score in the 2000 Pilot ESI, the primary response of the Mexican officials was not to challenge the ESI methodology or analytical structure, but to complain that the major international data providers were not providing a complete picture of environmental conditions in Mexico.

Of course, any claim that "if you had used our own data, we would have come out better" deserves to be treated skeptically, and it is sensible to assume that, for every government that reports the existence of data putting it in a better light, there is another government sitting on data that puts it in a worse light. However, we are convinced that there are domestic and regional data collection efforts for a significant number of countries that are superior in quality and relevance to comparative national environmental sustainability measurements. Finding a way to exploit these information resources effectively is an important task for the future.

Conclusion

Relative performance benchmarks, of the sort used to calculate the ESI, are only one means of exploring national-level environmental sustainability. Clearly, there are alternatives to the steps described above that generate the specific benchmarks used in the ESI. We have tried to be explicit about the precise steps taken to generate the ESI to permit greater understanding of what the numbers mean, critical reflection on the choices taken, and experimentation with alternative approaches.

References

Alcamo, J., T. Henrichs, and T. Rosch, *World Water in 2025: Global Modeling and Scenario Analysis for the World Commission on Water for the 21st Century,* Kassel World Water Series Report No. 2, Kassel, Germany: Center for Environmental Systems Research, University of Kassel, February 2000.

Brown, J.F., Loveland, T.R., Ohlen, D.O., Zhu, Z., 1999, The Global Land-Cover Characteristics Database: The Users' Perspective. Photogrammetric Engineering and Remote Sensing, v. 65, no. 9, p. 1069–1074.

Brooks, T. M., S.L. Pimm, V. Kapos, and C. Ravilious, "Threat from Deforestation to Montane and Lowland Birds and Mammals in Insular South-east Asia," *Journal of Animal Ecology* 68 (1999), pp. 1061–1078.

Defense Mapping Agency, 1992, Development of the Digital Chart of the World: Washington, D.C., U.S. Government Printing Office.

Elvidge, C.D., K.E. Baugh, E.A. Kihn, H.W. Kroehl, and E.R. David, "Mapping City Lights with Nighttime Data from the DMSP Operational Linescan System," *Photogrammetric Engineering & Remote Sensing* 63 (6) (1997a), pp. 727–734.

Elvidge, C.D., K.E. Baugh, V.R. Hobson, E.A. Kihn, H. W. Kroehl, E.R. Davis, and D. Cocero, "Satellite Inventory of Human Settlements Using Nocturnal Radiation Emissions: A Contribution for the Global Toolchest," *Global Change Biology* 3 (1997b), pp. 387–395.

Elvidge, C.D., K.E. Baugh, J.B. Dietz, T. Bland, P.C. Sutton, and H.W. Kroehl, "Radiance Calibration of DMSP-OLS Low-light Imaging Data of Human Settlements," *Remote Sensing of the Environment* 68 (1999), pp. 77–88.

Feshbach, Murray, personal communication, 2001.

Kuylenstierna, J., H. Rodhe, S. Cinderby, and K. Hicks, *Acidification in Developing Countries: Ecosystem Sensitivity and the Critical Load Approach at the Global Scale,* York: Stockholm Environment Institute at York, 2000.

Muchoney, D., Strahler, A., Hodges, J., and LoCastro, J., The IGBP DISCover Confidence Sites and the System for Terrestrial Ecosystem Parameterization: Tools for Validating Global Land-Cover Data: Photogrammetric Engineering and Remote Sensing, v. 65, no. 9, 1999, p. 1061–1067.

Myers, N., R.A. Mittermeier, C.G. Mittermeier, G.A. B. da Fonseca, and J. Kent, "Biodiversity Hotspots for Conservation Priorities," Nature 403 (2000), pp. 853–858.

National Research Council, Ecological Indicators for the Nation, Washington: National Academy Press, 2000.

Pimm, S.L., and P. Raven, "Extinction by Numbers," Nature 403 (2000), pp. 843–844.

Scepan, J., Thematic Validation of High-Resolution Global Land-Cover Data Sets: Photogrammetric Engineering and Remote Sensing, v. 65, no. 9, 1999, p. 1051–1060.

Smith, K.R., C.F. Corvalán, and T. Kjellstrom, "How Much Global Ill Health Is Attributable to Environmental Factors?" Epidemiology 10 (5) (September 1999), pp.573–84.

World Conservation Monitoring Center (WCMC), Global Biodiversity: Status of Earth's Living Resources, Cambridge, UK: Chapman & Hall, 1992.

Endnotes

1   Numerous individuals participated in the development of the ESI methodology. Within the core ESI project team, critical contributions were made by Dan Esty at Yale, and by Kobi Abayomi and Bob Chen at CIESIN. We are also grateful to Frank Dixon, Bedrich Moldan, Tom Parris, Theo Panayoutou, and other experts who participated in a series of review workshops for sharing their views on the ESI methodology at crucial junctures.

2   There is good reason to believe that values extremely far from the mean are more likely to reflect data quality problems than other values, other things being equal.

Annex 2
Component and Indicator Scores

Environmental Performance Measurement:

The Global Report 2001–2002

This section provides tables summarizing the country scores for each of the ESI components and indicators, sorted in order from highest to lowest scores. Note that the component scores are presented as standardized normal distributions ranging from a theoretical low of 0 to a high of 100. The indicator scores are presented as z-scores, with zero indicating the mean for the 122 countries, +1 and −1 respectively representing one standard deviation above and below the mean, +2 and −2 respectively representing two standard deviations above and below the mean, and so on. More information on how z-scores are calculated can be found in Annex 1.

A reference ranking is also provided for the variable used to calculate "inhabited land area."

2001
Environmental Sustainability Index

		RANK
80.5	Finland	1
78.2	Norway	2
78.1	Canada	3
77.1	Sweden	4
74.6	Switzerland	5
71.3	New Zealand	6
70.9	Australia	7
68.2	Austria	8
67.3	Iceland	9
67.0	Denmark	10
66.1	United States	11
66.0	Netherlands	12
65.8	France	13
64.6	Uruguay	14
64.2	Germany	15
64.1	United Kingdom	16
64.0	Ireland	17
63.2	Slovak Republic	18
62.9	Argentina	19
61.4	Portugal	20
61.0	Hungary	21
60.6	Japan	22
60.3	Lithuania	23
59.9	Slovenia	24
59.5	Spain	25
58.8	Costa Rica	26
58.2	Bolivia	27
57.7	Estonia	28
57.4	Brazil	29
57.2	Czech Republic	30
56.6	Chile	31
56.3	Latvia	32
56.2	Russian Federation	33
55.9	Panama	34
54.9	Cuba	35
54.8	Colombia	36
54.3	Italy	37
54.3	Peru	38
54.1	Croatia	39
53.5	Botswana	40
53.1	Greece	41

		RANK
51.9	Nicaragua	42
51.9	Zimbabwe	43
51.8	Ecuador	44
51.2	Mauritius	45
51.2	South Africa	46
50.8	Venezuela	47
50.7	Armenia	48
50.2	Gabon	49
50.1	Mongolia	50
49.8	Malaysia	51
49.8	Sri Lanka	52
49.6	Israel	53
48.8	Paraguay	54
48.1	Belarus	55
48.0	Fiji	56
47.7	Central African Republic	57
47.6	Poland	58
47.4	Bulgaria	59
47.4	Moldova	60
47.2	Guatemala	61
47.1	Papua New Guinea	62
46.9	Ghana	63
46.9	Honduras	64
46.9	Singapore	65
46.5	Azerbaijan	66
46.5	Nepal	67
46.4	Egypt	68
46.4	Trinidad and Tobago	69
46.3	Bhutan	70
46.3	Turkey	71
46.1	Mali	72
45.3	Dominican Republic	73
45.3	Mexico	74
45.2	Thailand	75
45.1	Albania	76
44.7	Cameroon	77
44.4	Belgium	78
44.1	Romania	79
43.8	Mozambique	80
43.8	Uganda	81
43.7	El Salvador	82

		RANK
43.7	Kenya	83
43.7	Tunisia	84
43.4	Pakistan	85
42.5	Indonesia	86
42.3	Jamaica	87
42.3	Senegal	88
41.8	Morocco	89
41.6	Uzbekistan	90
41.5	Kazakhstan	91
41.0	Malawi	92
40.7	India	93
40.6	Algeria	94
40.4	Bangladesh	95
40.3	South Korea	96
40.1	Jordan	97
40.1	Tanzania	98
39.6	Kyrgyz Republic	99
39.5	Zambia	100
39.2	Benin	101
39.2	Macedonia	102
38.9	Togo	103
38.4	Iran	104
38.3	Burkina Faso	105
37.9	Syria	106
37.6	Sudan	107
37.5	China	108
37.5	Lebanon	109
36.8	Ukraine	110
36.7	Niger	111
35.6	Philippines	112
35.1	Madagascar	113
34.2	Vietnam	114
33.2	Rwanda	115
31.9	Kuwait	116
31.5	Nigeria	117
31.3	Libya	118
31.0	Ethiopia	119
29.8	Burundi	120
29.8	Saudi Arabia	121
24.5	Haiti	122

Environmental Systems

This component includes the following variables:

Air Quality
Water Quantity
Water Quality
Biodiversity
Terrestrial Systems

High numbers represent higher sustainability; zero represents the mean.

Score	Country	RANK
91.2	Canada	1
87.4	Norway	2
85.3	Finland	3
79.3	Sweden	4
79.1	Iceland	5
78.0	Gabon	6
72.6	Venezuela	7
71.2	Argentina	8
70.5	Colombia	9
70.1	Bolivia	10
69.7	Ireland	11
69.7	Uruguay	12
67.7	Central African Republic	13
66.3	Botswana	14
66.2	Nicaragua	15
66.1	Peru	16
65.8	Austria	17
65.7	Australia	18
65.6	Paraguay	19
65.4	Russian Federation	20
64.6	Mali	21
64.4	Papua New Guinea	22
63.8	Slovenia	23
63.1	United States	24
62.6	Ecuador	25
61.3	Mongolia	26
60.9	Slovak Republic	27
60.3	Switzerland	28
59.1	Estonia	29
58.8	France	30
58.8	Portugal	31
58.3	Latvia	32
58.2	Ghana	33
58.1	Zimbabwe	34
58.1	United Kingdom	35
58.0	Brazil	36
58.0	Netherlands	37
57.9	Lithuania	38
57.6	New Zealand	39
57.0	Croatia	40
57.0	Denmark	41

Score	Country	RANK
56.6	Trinidad and Tobago	42
56.5	Cameroon	43
55.8	Bhutan	44
55.0	Benin	45
54.5	Honduras	46
53.7	Zambia	47
53.6	Belarus	48
53.3	Chile	49
53.3	Czech Republic	50
52.9	Malaysia	51
51.6	Germany	52
51.2	Costa Rica	53
51.0	El Salvador	54
50.8	Panama	55
50.7	Guatemala	56
50.6	Togo	57
50.4	Hungary	58
50.4	Mozambique	59
50.3	Japan	60
50.3	Armenia	61
50.2	Malawi	62
49.9	Kenya	63
49.4	Moldova	64
48.8	Kazakhstan	65
48.0	Sudan	66
47.7	Libya	67
47.1	Senegal	68
46.9	Uzbekistan	69
46.8	Spain	70
46.1	Israel	71
46.0	Nepal	72
45.8	Cuba	73
45.6	Egypt	74
45.0	Niger	75
44.6	Singapore	76
44.6	Albania	77
44.2	Greece	78
44.2	Tanzania	79
43.9	Syria	80
43.4	Pakistan	81
43.4	South Africa	82

Score	Country	RANK
42.8	Kyrgyz Republic	83
42.7	Uganda	84
41.6	Nigeria	85
40.7	Algeria	86
40.1	Bangladesh	87
40.1	Fiji	88
39.9	Tunisia	89
39.8	Kuwait	90
39.0	Saudi Arabia	91
38.8	Azerbaijan	92
38.8	Lebanon	93
38.7	Macedonia	94
38.3	Mauritius	95
38.1	Turkey	96
37.4	Burkina Faso	97
37.1	Jordan	98
36.8	Romania	99
36.8	Italy	100
36.3	Thailand	101
35.1	South Korea	102
34.9	Iran	103
34.8	Rwanda	104
34.2	Poland	105
33.8	Jamaica	106
33.5	Indonesia	107
33.2	Vietnam	108
32.8	Ukraine	109
32.2	Dominican Republic	110
31.5	Burundi	111
31.5	Ethiopia	112
29.5	Sri Lanka	113
29.5	Morocco	114
25.7	Bulgaria	115
25.5	Belgium	116
25.0	Mexico	117
24.0	India	118
23.4	Madagascar	119
22.0	Philippines	120
20.8	China	121
12.2	Haiti	122

Reducing Environmental Stresses

This component includes the following variables:

Reducing Air Pollution
Reducing Water Stress
Reducing Ecosystem Stress
Reducing Waste and Consumption Pressures
Reducing Population Pressure

High numbers represent higher sustainability; zero represents the mean.

Score	Country	RANK
76.8	Kazakhstan	1
74.2	Armenia	2
73.8	Mongolia	3
71.2	Mozambique	4
69.8	Russian Federation	5
68.9	Cuba	6
68.8	Zimbabwe	7
68.7	Moldova	8
67.8	Kyrgyz Republic	9
67.5	Argentina	10
66.5	Estonia	11
66.0	Belarus	12
65.6	Central African Republic	13
65.4	Albania	14
65.2	Azerbaijan	15
64.8	Uzbekistan	16
64.5	Peru	17
64.4	Lithuania	18
64.1	Hungary	19
64.0	Bolivia	20
62.9	Bhutan	21
62.6	Brazil	22
62.1	Romania	23
62.0	Uruguay	24
60.9	Kenya	25
60.4	Rwanda	26
60.4	Colombia	27
60.1	Panama	28
59.9	Morocco	29
59.2	Bulgaria	30
59.1	Botswana	31
59.1	Croatia	32
58.9	Venezuela	33
58.9	Cameroon	34
58.6	Chile	35
58.4	Madagascar	36
58.1	Turkey	37
58.0	Finland	38
57.8	Dominican Republic	39
57.8	Indonesia	40
57.7	South Africa	41

Score	Country	RANK
57.5	Niger	42
57.2	Mexico	43
57.0	India	44
57.0	Sri Lanka	45
56.4	Sudan	46
56.4	Iran	47
56.3	New Zealand	48
56.3	Bangladesh	49
55.5	Ethiopia	50
55.3	Greece	51
55.2	Latvia	52
55.2	Gabon	53
54.9	Malawi	54
54.2	Ecuador	55
54.1	Mali	56
54.1	Fiji	57
54.0	Nicaragua	58
53.9	Sweden	59
53.6	Algeria	60
53.5	Ghana	61
52.6	Burkina Faso	62
52.6	Spain	63
52.6	China	64
52.3	Norway	65
52.2	Portugal	66
52.2	Papua New Guinea	67
52.1	Tunisia	68
51.9	Tanzania	69
51.9	Togo	70
51.8	Uganda	71
51.5	Senegal	72
51.2	Canada	73
50.8	Thailand	74
50.4	Australia	75
50.3	Nepal	76
49.5	Slovak Republic	77
49.5	Honduras	78
49.3	Haiti	79
49.3	Nigeria	80
48.5	Zambia	81
48.3	Egypt	82

Score	Country	RANK
47.9	Pakistan	83
45.8	Vietnam	84
45.7	Ukraine	85
45.5	Poland	86
44.8	Trinidad and Tobago	87
44.8	Switzerland	88
44.5	Jamaica	89
44.3	Syria	90
44.3	Burundi	91
44.2	Ireland	92
43.4	Slovenia	93
43.3	El Salvador	94
42.8	Guatemala	95
42.4	Benin	96
41.6	Libya	97
41.3	Mauritius	98
40.9	France	99
40.7	Italy	100
40.0	Paraguay	101
37.8	Macedonia	102
37.1	Austria	103
37.0	United States	104
36.8	Philippines	105
35.2	Germany	106
35.0	Saudi Arabia	107
34.5	Costa Rica	108
31.9	Malaysia	109
31.8	Jordan	110
31.0	Czech Republic	111
30.6	Denmark	112
27.9	Iceland	113
25.4	Japan	114
23.7	Netherlands	115
23.7	United Kingdom	116
21.3	Lebanon	117
20.0	Kuwait	118
17.8	Israel	119
16.8	Singapore	120
14.2	South Korea	121
10.0	Belgium	122

Reducing Human Vulnerability

This component includes the following variables:

Basic Human Sustenance
Environmental Health

*Note: These values differ somewhat from the ESI released January 2001
because of an error in the Environmental Health indicator which is corrected here.*

High numbers represent higher sustainability; zero represents the mean.

Score	Country	RANK		Score	Country	RANK		Score	Country	RANK
83.1	Japan	1		72.3	Latvia	42		43.2	Dominican Republic	83
83.1	Belgium	2		71.9	Lebanon	43		42.6	Ecuador	84
83.0	Denmark	3		70.9	Malaysia	44		42.5	Honduras	85
83.0	Switzerland	4		70.3	Argentina	45		39.9	Botswana	86
83.0	Germany	5		70.2	Saudi Arabia	46		37.3	Nicaragua	87
82.9	France	6		69.1	Trinidad and Tobago	47		35.6	Vietnam	88
82.9	Austria	7		68.0	Kazakhstan	48		33.3	El Salvador	89
82.9	Iceland	8		67.9	Iran	49		32.5	Zimbabwe	90
82.8	Slovenia	9		67.9	Ukraine	50		31.6	Peru	91
82.8	Canada	10		65.9	Macedonia	51		31.2	India	92
82.7	Italy	11		65.6	Uruguay	52		25.2	Bhutan	93
82.6	Australia	12		65.3	Chile	53		24.4	Pakistan	94
82.5	Ireland	13		64.9	Algeria	54		22.6	Gabon	95
82.5	Norway	14		63.7	Armenia	55		22.2	Nepal	96
82.5	United Kingdom	15		63.1	Colombia	56		22.1	Bolivia	97
82.5	New Zealand	16		62.3	Mexico	57		20.9	Bangladesh	98
82.4	Spain	17		61.9	Turkey	58		16.8	Papua New Guinea	99
82.4	United States	18		61.5	Paraguay	59		15.6	Senegal	100
82.3	Singapore	19		61.4	Jordan	60		15.3	Ghana	101
81.9	Israel	20		60.6	Brazil	61		14.0	Mongolia	102
81.7	Hungary	21		60.6	Azerbaijan	62		12.5	Sudan	103
81.7	Greece	22		59.3	Tunisia	63		11.6	Cameroon	104
81.5	Slovak Republic	23		58.3	Albania	64		10.7	Benin	105
81.1	Portugal	24		56.5	Syria	65		9.5	Togo	106
80.4	Mauritius	25		56.1	Libya	66		7.4	Kenya	107
80.4	Czech Republic	26		55.0	Uzbekistan	67		6.8	Tanzania	108
80.0	Bulgaria	27		54.8	South Africa	68		6.6	Madagascar	109
79.6	Kuwait	28		53.8	Jamaica	69		6.1	Mali	110
79.5	Netherlands	29		52.0	Kyrgyz Republic	70		5.9	Nigeria	111
79.1	Poland	30		51.9	Indonesia	71		5.7	Uganda	112
78.7	Finland	31		50.4	Romania	72		4.9	Haiti	113
78.5	Croatia	32		50.1	Egypt	73		4.8	Zambia	114
78.4	South Korea	33		49.9	Panama	74		3.7	Burkina Faso	115
77.7	Sweden	34		49.3	Sri Lanka	75		3.6	Niger	116
77.4	Estonia	35		49.0	Philippines	76		3.5	Burundi	117
77.2	Costa Rica	36		48.5	China	77		3.4	Malawi	118
77.1	Lithuania	37		48.1	Thailand	78		3.4	Central African Republic	119
76.7	Belarus	38		48.0	Morocco	79		2.4	Mozambique	120
76.5	Cuba	39		45.7	Venezuela	80		2.0	Rwanda	121
75.8	Russian Federation	40		44.7	Fiji	81		1.4	Ethiopia	122
73.1	Moldova	41		44.3	Guatemala	82				

Social and Institutional Capacity

This component includes the following variables:

Science/Technology
Capacity for Debate
Regulation and Management
Private Sector Responsiveness
Environmental Information
Eco-Efficiency
Reducing Public Choice Distortions

High numbers represent higher sustainability; zero represents the mean.

Score	Country	RANK
92.3	Switzerland	1
91.2	Finland	2
87.4	Denmark	3
87.1	Netherlands	4
86.6	United Kingdom	5
86.3	Sweden	6
85.3	Norway	7
84.1	Iceland	8
83.4	United States	9
83.3	New Zealand	10
83.2	Austria	11
82.8	Australia	12
82.8	Japan	13
82.5	Germany	14
82.5	Canada	15
80.7	France	16
72.9	Israel	17
72.5	Ireland	18
68.8	Costa Rica	19
68.2	Belgium	20
66.9	Spain	21
66.7	Italy	22
66.5	Portugal	23
66.2	Slovenia	24
65.2	Singapore	25
60.6	Chile	26
60.2	South Korea	27
60.0	Slovak Republic	28
60.0	Czech Republic	29
59.9	Uruguay	30
57.1	Fiji	31
56.6	Hungary	32
56.2	Argentina	33
54.0	Estonia	34
53.9	Sri Lanka	35
53.7	Panama	36
53.1	Brazil	37
51.7	Bolivia	38
50.7	Latvia	39
49.7	Nepal	40
49.7	South Africa	41

Score	Country	RANK
49.5	Botswana	42
49.3	Croatia	43
49.1	Lithuania	44
48.0	Mauritius	45
47.6	Thailand	46
47.1	Malaysia	47
46.6	Greece	48
46.2	Uganda	49
46.2	Cuba	50
45.8	Ecuador	51
45.8	Poland	52
45.6	Dominican Republic	53
45.1	Guatemala	54
44.6	Mexico	55
44.5	El Salvador	56
43.8	Paraguay	57
43.7	India	58
43.7	Peru	59
43.1	Honduras	60
42.5	Russian Federation	61
42.4	Turkey	62
42.1	Pakistan	63
41.7	Egypt	64
41.1	Tanzania	65
41.0	Colombia	66
40.8	Zimbabwe	67
40.4	China	68
40.4	Nicaragua	69
40.4	Jordan	70
39.9	Malawi	71
39.8	Ghana	72
39.6	Albania	73
39.3	Armenia	74
38.7	Burkina Faso	75
38.4	Macedonia	76
38.4	Jamaica	77
38.4	Romania	78
38.3	Bhutan	79
37.9	Morocco	80
37.8	Kenya	81
37.8	Philippines	82

Score	Country	RANK
37.8	Zambia	83
37.6	Lebanon	84
37.5	Mali	85
36.2	Central African Republic	86
36.0	Moldova	87
35.4	Rwanda	88
34.4	Senegal	89
34.3	Indonesia	90
34.3	Mongolia	91
34.2	Madagascar	92
34.1	Gabon	93
34.0	Mozambique	94
34.0	Trinidad and Tobago	95
33.5	Bulgaria	96
33.2	Bangladesh	97
33.1	Papua New Guinea	98
32.8	Venezuela	99
32.7	Burundi	100
32.1	Togo	101
31.6	Tunisia	102
31.4	Cameroon	103
30.6	Benin	104
29.6	Ethiopia	105
29.4	Kuwait	106
28.6	Belarus	107
28.5	Haiti	108
28.2	Ukraine	109
28.2	Nigeria	110
27.8	Azerbaijan	111
27.2	Iran	112
26.8	Kyrgyz Republic	113
25.5	Algeria	114
25.4	Sudan	115
25.2	Niger	116
24.9	Syria	117
23.9	Vietnam	118
21.5	Kazakhstan	119
20.5	Uzbekistan	120
18.1	Saudi Arabia	121
18.1	Libya	122

Global Stewardship

This component includes the following variables:

International Commitment
Global Scale Funding/Participation
Protecting International Commons

High numbers represent higher sustainability; zero represents the mean.

		RANK			RANK			RANK
80.6	Czech Republic	1	56.1	Niger	42	42.4	Vietnam	83
80.5	Sweden	2	55.9	Spain	43	42.2	Sudan	84
80.0	Slovak Republic	3	55.9	Guatemala	44	41.6	Honduras	85
75.6	Netherlands	4	55.5	Tunisia	45	41.5	Togo	86
75.3	Switzerland	5	55.4	Mongolia	46	41.4	Zambia	87
74.9	New Zealand	6	55.4	Jamaica	47	41.3	Algeria	88
74.3	Bulgaria	7	55.3	Poland	48	41.0	Botswana	89
73.9	Norway	8	55.2	Brazil	49	40.9	Uzbekistan	90
73.8	Mauritius	9	54.9	Benin	50	39.2	Turkey	91
72.7	Costa Rica	10	54.8	Italy	51	39.0	Morocco	92
72.1	Canada	11	54.6	South Africa	52	39.0	Singapore	93
69.9	Finland	12	54.2	Burkina Faso	53	38.6	Paraguay	94
69.8	Uruguay	13	53.0	Kenya	54	38.1	Syria	95
69.8	Lithuania	14	52.9	Egypt	55	37.4	El Salvador	96
69.5	Australia	15	52.9	Portugal	56	36.7	Bhutan	97
68.4	Denmark	16	52.5	Pakistan	57	36.4	Ethiopia	98
67.6	Austria	17	52.2	Mexico	58	36.3	Belarus	99
67.4	Belgium	18	51.6	Nepal	59	35.9	Romania	100
67.3	Bolivia	19	51.3	Zimbabwe	60	34.4	Croatia	101
67.3	Hungary	20	50.4	Madagascar	61	34.1	Israel	102
66.3	Malaysia	21	50.1	Gabon	62	33.8	Estonia	103
66.1	Papua New Guinea	22	50.1	Argentina	63	33.8	Russian Federation	104
66.0	Panama	23	50.0	Cuba	64	33.1	Fiji	105
66.0	Germany	24	49.8	Ireland	65	31.0	China	106
64.7	Azerbaijan	25	49.5	Ecuador	66	30.7	South Korea	107
64.2	Uganda	26	48.3	Iceland	67	30.2	Ukraine	108
63.9	Sri Lanka	27	48.1	Dominican Republic	68	28.6	Armenia	109
63.7	France	28	47.4	Malawi	69	27.9	Burundi	110
61.8	United Kingdom	29	47.4	Bangladesh	70	27.5	Macedonia	111
61.5	Cameroon	30	47.3	Slovenia	71	25.8	Haiti	112
60.6	Nicaragua	31	46.6	Trinidad and Tobago	72	24.9	Iran	113
60.1	Mali	32	46.4	Indonesia	73	23.5	Rwanda	114
59.9	Senegal	33	45.6	Philippines	74	22.7	Nigeria	115
58.3	Ghana	34	45.2	Venezuela	75	20.1	Moldova	116
58.3	Japan	35	44.9	Jordan	76	19.3	Albania	117
57.9	Mozambique	36	44.3	India	77	18.4	Kuwait	118
57.6	Central African Republic	37	44.1	Colombia	78	15.8	Saudi Arabia	119
57.6	Greece	38	43.7	Tanzania	79	15.7	Kyrgyz Republic	120
56.5	Latvia	39	43.3	Thailand	80	13.7	Libya	121
56.4	United States	40	43.2	Chile	81	11.4	Kazakhstan	122
56.3	Peru	41	42.6	Lebanon	82			

Air Quality

This indicator includes the following variables:

Urban Sulfur Dioxide (SO2) Concentration
Urban Nitrogen Dioxide (NO2) Concentration
Urban Total Suspended Particulates (TSP) Concentration

High numbers represent higher sustainability; zero represents the mean.

Score	Country	RANK		Score	Country	RANK		Score	Country	RANK
1.62	New Zealand	1		.20	Ecuador	42		-.34	Kazakhstan	83
1.58	Cuba	2		.19	Venezuela	43		-.34	Algeria	84
1.45	Sweden	3		.18	Romania	44		-.34	Libya	85
1.45	Australia	4		.17	Mongolia	45		-.38	Paraguay	86
1.36	Malaysia	5		.17	Kenya	46		-.38	Uzbekistan	87
1.28	Finland	6		.16	Moldova	47		-.40	Azerbaijan	88
1.13	Iceland	7		.15	Vietnam	48		-.40	Indonesia	89
1.10	Lithuania	8		.14	Armenia	49		-.41	Gabon	90
1.03	Spain	9		.12	Uruguay	50		-.42	Papua New Guinea	91
1.02	Norway	10		.11	Nicaragua	51		-.44	Madagascar	92
1.00	Slovak Republic	11		.11	Mauritius	52		-.45	Togo	93
.99	Switzerland	12		.07	Trinidad and Tobago	53		-.47	Costa Rica	94
.98	Canada	13		.06	Albania	54		-.49	Rwanda	95
.95	Germany	14		.05	Panama	55		-.49	Greece	96
.94	Austria	15		.02	Morocco	56		-.50	Cameroon	97
.93	Belarus	16		.00	Bhutan	57		-.53	Sudan	98
.91	Singapore	17		.00	Jordan	58		-.53	Haiti	99
.88	Czech Republic	18		.00	Colombia	59		-.55	El Salvador	100
.78	Portugal	19		-.03	Nepal	60		-.56	Senegal	101
.76	Argentina	20		-.03	Russian Federation	61		-.58	Zambia	102
.70	France	21		-.03	Tunisia	62		-.59	Uganda	103
.69	Denmark	22		-.06	India	63		-.60	Nigeria	104
.63	Netherlands	23		-.06	Ghana	64		-.61	Benin	105
.61	United States	24		-.08	Fiji	65		-.63	Malawi	106
.60	Thailand	25		-.10	Peru	66		-.63	Tanzania	107
.58	Ireland	26		-.13	Jamaica	67		-.64	Central African Republic	108
.52	Belgium	27		-.14	Lebanon	68		-.67	Brazil	109
.52	Slovenia	28		-.14	Pakistan	69		-.69	Chile	110
.48	Israel	29		-.14	Mali	70		-.74	Mozambique	111
.41	Turkey	30		-.16	Botswana	71		-.74	Italy	112
.41	United Kingdom	31		-.19	South Korea	72		-.75	Burundi	113
.40	Bangladesh	32		-.20	Saudi Arabia	73		-.79	Egypt	114
.36	Kuwait	33		-.21	Bolivia	74		-.80	Ethiopia	115
.35	Latvia	34		-.27	Ukraine	75		-.81	Burkina Faso	116
.32	Hungary	35		-.28	Philippines	76		-.98	Niger	117
.32	Estonia	36		-.29	Zimbabwe	77		-1.08	Guatemala	118
.29	South Africa	37		-.29	Kyrgyz Republic	78		-1.65	Iran	119
.28	Croatia	38		-.29	Dominican Republic	79		-1.87	Bulgaria	120
.28	Japan	39		-.30	Syria	80		-2.24	China	121
.22	Macedonia	40		-.32	Poland	81		-2.58	Mexico	122
.21	Sri Lanka	41		-.33	Honduras	82				

Water Quantity

This indicator includes the following variables:

Internal Renewable Water Per Capita
Water Inflow from Other Countries Per Capita

High numbers represent higher sustainability; zero represents the mean.

		RANK
2.37	Gabon	1
1.75	Bolivia	2
1.74	Colombia	3
1.73	Papua New Guinea	4
1.70	Canada	5
1.62	Peru	6
1.50	Central African Republic	7
1.49	Venezuela	8
1.45	Brazil	9
1.38	Uruguay	10
1.33	Norway	11
.99	Paraguay	12
.94	Nicaragua	13
.86	Iceland	14
.81	Ecuador	15
.79	Costa Rica	16
.73	Bhutan	17
.72	Argentina	18
.72	Croatia	19
.69	Honduras	20
.69	Russian Federation	21
.68	Cameroon	22
.66	Finland	23
.62	Mongolia	24
.58	Zambia	25
.56	Chile	26
.52	Slovenia	27
.48	Mozambique	28
.46	Latvia	29
.46	Estonia	30
.45	New Zealand	31
.44	Guatemala	32
.42	Bulgaria	33
.41	Sweden	34
.41	Malaysia	35
.39	Ireland	36
.38	Austria	37
.33	Slovak Republic	38
.25	Botswana	39
.23	Thailand	40
.22	Vietnam	41

		RANK
.22	Lithuania	42
.21	Hungary	43
.21	Benin	44
.21	Kazakhstan	45
.17	Romania	46
.17	Zimbabwe	47
.16	United States	48
.15	Albania	49
.14	Bangladesh	50
.08	Nepal	51
.06	Portugal	52
.05	Moldova	53
.04	Netherlands	54
.04	Indonesia	55
.03	Mali	56
.00	Belarus	57
-.03	Niger	58
-.05	Tanzania	59
-.09	Greece	60
-.10	Sudan	61
-.12	Panama	62
-.13	Azerbaijan	63
-.13	El Salvador	64
-.15	Uzbekistan	65
-.15	Togo	66
-.16	France	67
-.17	Senegal	68
-.17	Australia	69
-.18	Mexico	70
-.20	Ghana	71
-.21	Germany	72
-.21	Syria	73
-.21	Fiji	74
-.22	Nigeria	75
-.24	Uganda	76
-.28	Kenya	77
-.29	Madagascar	78
-.29	Rwanda	79
-.31	Burundi	80
-.34	Egypt	81
-.35	Czech Republic	82

		RANK
-.36	Belgium	83
-.37	Ukraine	84
-.38	Armenia	85
-.41	Malawi	86
-.42	India	87
-.42	Pakistan	88
-.48	Iran	89
-.50	Turkey	90
-.53	Poland	91
-.62	Dominican Republic	92
-.63	Libya	93
-.64	China	94
-.64	Tunisia	95
-.69	Haiti	96
-.70	South Africa	97
-.71	Jordan	98
-.75	South Korea	99
-.76	Burkina Faso	100
-.77	Spain	101
-.79	Italy	102
-.80	United Kingdom	103
-.85	Ethiopia	104
-.86	Switzerland	105
-.87	Singapore	106
-.87	Kyrgyz Republic	107
-.97	Philippines	108
-1.00	Algeria	109
-1.00	Jamaica	110
-1.05	Japan	111
-1.05	Macedonia	112
-1.05	Denmark	113
-1.09	Cuba	114
-1.12	Sri Lanka	115
-1.12	Trinidad and Tobago	116
-1.19	Lebanon	117
-1.21	Mauritius	118
-1.21	Morocco	119
-1.22	Israel	120
-1.23	Saudi Arabia	121
-1.27	Kuwait	122

Water Quality

This indicator includes the following variables:

Dissolved Oxygen Concentration
Phosphorus Concentration
Suspended Solids
Electrical Conductivity

High numbers represent higher sustainability; zero represents the mean.

		RANK			RANK			RANK
1.85	Finland	1	.11	Estonia	42	-.33	Kazakhstan	83
1.54	Canada	2	.11	Panama	43	-.33	China	84
1.53	New Zealand	3	.10	Slovak Republic	44	-.33	Libya	85
1.42	United Kingdom	4	.10	Turkey	45	-.35	Papua New Guinea	86
1.32	Japan	5	.10	Trinidad and Tobago	46	-.35	Malaysia	87
1.31	Norway	6	.09	South Africa	47	-.35	Israel	88
1.30	Russian Federation	7	.09	Croatia	48	-.36	Honduras	89
1.27	South Korea	8	.08	El Salvador	49	-.37	Paraguay	90
1.19	Sweden	9	.06	Fiji	50	-.37	Uzbekistan	91
1.13	France	10	.04	Bulgaria	51	-.39	Azerbaijan	92
1.09	Portugal	11	.04	Botswana	52	-.40	Gabon	93
1.04	United States	12	-.01	Venezuela	53	-.42	Senegal	94
1.03	Argentina	13	-.02	Lithuania	54	-.47	Ukraine	95
.93	Hungary	14	-.04	Jamaica	55	-.49	Bhutan	96
.91	Philippines	15	-.06	Ecuador	56	-.49	Madagascar	97
.87	Switzerland	16	-.06	Germany	57	-.53	Togo	98
.86	Ireland	17	-.08	Zimbabwe	58	-.54	Tunisia	99
.85	Austria	18	-.08	Peru	59	-.59	Thailand	100
.74	Iceland	19	-.11	Lebanon	60	-.61	Haiti	101
.73	Australia	20	-.13	Romania	61	-.62	Nigeria	102
.70	Netherlands	21	-.14	Albania	62	-.64	Mozambique	103
.66	Mali	22	-.15	Egypt	63	-.64	Algeria	104
.64	Brazil	23	-.16	Sri Lanka	64	-.67	Zambia	105
.63	Slovenia	24	-.18	Saudi Arabia	65	-.69	Mexico	106
.62	Singapore	25	-.19	Armenia	66	-.70	Benin	107
.61	Greece	26	-.20	Bolivia	67	-.70	Uganda	108
.60	Cuba	27	-.20	Cameroon	68	-.74	Ethiopia	109
.58	Spain	28	-.22	Moldova	69	-.77	Indonesia	110
.55	Denmark	29	-.22	Tanzania	70	-.77	Malawi	111
.52	Iran	30	-.22	Belarus	71	-.77	Mauritius	112
.47	Italy	31	-.23	Macedonia	72	-.78	Rwanda	113
.39	Uruguay	32	-.23	Vietnam	73	-.81	Central African Republic	114
.39	Kuwait	33	-.24	Mongolia	74	-.95	Burundi	115
.37	Poland	34	-.26	Kenya	75	-1.00	Burkina Faso	116
.27	Colombia	35	-.28	Dominican Republic	76	-1.04	Niger	117
.27	Czech Republic	36	-.28	Kyrgyz Republic	77	-1.06	Sudan	118
.23	Ghana	37	-.28	Nepal	78	-1.26	Jordan	119
.23	Costa Rica	38	-.29	Syria	79	-1.31	India	120
.19	Chile	39	-.30	Pakistan	80	-1.36	Morocco	121
.18	Bangladesh	40	-.30	Guatemala	81	-2.25	Belgium	122
.15	Latvia	41	-.32	Nicaragua	82			

Biodiversity

This indicator includes the following variables:

Percentage of Mammals Threatened
Percentage of Breeding Birds Threatened

High numbers represent higher sustainability; zero represents the mean.

		RANK			RANK			RANK
1.65	El Salvador	1	.42	Switzerland	42	-.26	Pakistan	83
1.60	Nicaragua	2	.40	Syria	43	-.28	Turkey	84
1.53	Trinidad and Tobago	3	.40	Nigeria	44	-.29	Mauritius	85
1.25	Guatemala	4	.39	Mozambique	45	-.29	Mexico	86
1.16	Togo	5	.37	Mali	46	-.31	Iran	87
1.15	Botswana	6	.35	Uzbekistan	47	-.36	United States	88
1.15	Burkina Faso	7	.35	Kyrgyz Republic	48	-.38	Nepal	89
1.12	Canada	8	.34	Austria	49	-.38	Algeria	90
1.11	Zimbabwe	9	.32	Slovak Republic	50	-.39	Russian Federation	91
1.07	Honduras	10	.30	Bolivia	51	-.40	Bhutan	92
1.05	Malawi	11	.29	Belgium	52	-.40	Sri Lanka	93
1.03	Benin	12	.26	Netherlands	53	-.45	Portugal	94
1.01	Albania	13	.26	Uruguay	54	-.49	Bulgaria	95
.96	Central African Republic	14	.23	Czech Republic	55	-.53	Romania	96
.91	Burundi	15	.22	Germany	56	-.54	Saudi Arabia	97
.89	Moldova	16	.19	Lebanon	57	-.56	Spain	98
.86	Zambia	17	.17	Mongolia	58	-.57	Morocco	99
.83	Uganda	18	.17	Macedonia	59	-.62	Malaysia	100
.80	Armenia	19	.16	Jordan	60	-.64	Thailand	101
.76	Estonia	20	.15	Kazakhstan	61	-.68	Jamaica	102
.76	Gabon	21	.13	Slovenia	62	-.70	Chile	103
.76	Rwanda	22	.10	Croatia	63	-.75	Egypt	104
.74	Belarus	23	.07	Ecuador	64	-.78	Brazil	105
.71	Paraguay	24	.06	Italy	65	-.79	Papua New Guinea	106
.69	Latvia	25	.05	Poland	66	-.97	Kuwait	107
.67	Ghana	26	.04	Colombia	67	-.99	Australia	108
.66	Denmark	27	.04	Argentina	68	-1.03	China	109
.61	Finland	28	.02	Azerbaijan	69	-1.07	Dominican Republic	110
.61	Senegal	29	.02	Kenya	70	-1.08	Vietnam	111
.61	Ireland	30	.01	Tanzania	71	-1.14	Indonesia	112
.59	Norway	31	-.01	South Africa	72	-1.16	India	113
.58	United Kingdom	32	-.03	Peru	73	-1.21	Bangladesh	114
.58	Niger	33	-.03	Libya	74	-1.53	Cuba	115
.58	Iceland	34	-.06	France	75	-1.58	Japan	116
.53	Panama	35	-.10	Hungary	76	-1.91	South Korea	117
.53	Sudan	36	-.13	Ethiopia	77	-2.21	Madagascar	118
.53	Venezuela	37	-.15	Israel	78	-2.58	Philippines	119
.52	Costa Rica	38	-.18	Tunisia	79	-2.73	Fiji	120
.49	Lithuania	39	-.22	Ukraine	80	-3.07	Haiti	121
.45	Sweden	40	-.23	Greece	81	-3.37	New Zealand	122
.44	Cameroon	41	-.25	Singapore	82			

Terrestrial Systems

This indicator includes the following variables:

Severity of Human Induced Soil Degradation

Land Area Impacted by Human Activities as a Percentage of Toal Land Area

High numbers represent higher sustainability; zero represents the mean.

Score	Country	RANK
1.70	Fiji	1
1.67	Papua New Guinea	2
1.54	Gabon	3
1.48	Egypt	4
1.47	Norway	5
1.45	Canada	6
1.29	Central African Republic	7
1.19	Algeria	8
1.07	Japan	9
1.06	Chile	10
1.06	Paraguay	11
1.04	Libya	12
1.00	Australia	13
.98	Bolivia	14
.95	Mali	15
.90	Sudan	16
.88	Bhutan	17
.85	Niger	18
.84	Finland	19
.82	Botswana	20
.81	Venezuela	21
.78	Malawi	22
.77	Saudi Arabia	23
.76	Israel	24
.75	Iceland	25
.72	New Zealand	26
.71	Mongolia	27
.69	Benin	28
.67	Mauritius	29
.67	Peru	30
.65	Colombia	31
.59	Ecuador	32
.59	Sweden	33
.56	Mozambique	34
.43	Morocco	35
.42	Uruguay	36
.41	Russian Federation	37
.41	Cameroon	38
.40	Ghana	39
.39	Brazil	40
.36	Mexico	41

Score	Country	RANK
.33	Kenya	42
.29	Pakistan	43
.27	Zambia	44
.26	Trinidad and Tobago	45
.25	Argentina	46
.22	United States	47
.20	Kuwait	48
.19	Kyrgyz Republic	49
.18	Senegal	50
.18	China	51
.16	Tanzania	52
.16	Uzbekistan	53
.15	Jordan	54
.15	Kazakhstan	55
.15	Ireland	56
.14	Indonesia	57
.12	Tunisia	58
.11	Nepal	59
.11	Zimbabwe	60
.10	Ethiopia	61
.04	Togo	62
.04	Denmark	63
-.02	Iran	64
-.02	Nigeria	65
-.04	Slovenia	66
-.05	Dominican Republic	67
-.09	Cuba	68
-.12	Switzerland	69
-.18	Lebanon	70
-.20	Burkina Faso	71
-.20	Madagascar	72
-.22	Uganda	73
-.22	Guatemala	74
-.23	Jamaica	75
-.25	Nicaragua	76
-.30	Croatia	77
-.33	Armenia	78
-.33	South Korea	79
-.37	Syria	80
-.37	Slovak Republic	81
-.37	Portugal	82

Score	Country	RANK
-.44	Malaysia	83
-.47	Panama	84
-.48	Austria	85
-.50	Estonia	86
-.50	South Africa	87
-.50	Honduras	88
-.50	France	89
-.52	Greece	90
-.52	Azerbaijan	91
-.55	Macedonia	92
-.58	India	93
-.60	United Kingdom	94
-.60	Latvia	95
-.62	Czech Republic	96
-.62	Netherlands	97
-.67	Spain	98
-.68	Italy	99
-.70	Germany	100
-.77	Bangladesh	101
-.79	Lithuania	102
-.90	Ukraine	103
-.91	El Salvador	104
-.92	Costa Rica	105
-.93	Haiti	106
-.94	Philippines	107
-.96	Moldova	108
-.99	Belarus	109
-1.08	Singapore	110
-1.16	Rwanda	111
-1.23	Sri Lanka	112
-1.23	Vietnam	113
-1.25	Turkey	114
-1.31	Burundi	115
-1.32	Hungary	116
-1.36	Thailand	117
-1.37	Bulgaria	118
-1.39	Romania	119
-1.48	Belgium	120
-1.60	Poland	121
-1.75	Albania	122

Reducing Air Pollution

This indicator includes the following variables:

Nitrogen Oxide (NOx) Emissions Per Populated Land Area
Sulfur Dioxide (SO2) Emissions Per Populated Land Area
Volatile Organic Compound (VOCs) Emissions Per Populated Land Area
Coal Consumption Per Populated Land Area
Vehicles Per Populated Land Area

High numbers represent higher sustainability; zero represents the mean.

		RANK			RANK			RANK
1.36	Bhutan	1	.42	Dominican Republic	42	-.24	Singapore	83
1.29	Madagascar	2	.39	Malawi	43	-.26	Portugal	84
1.25	Papua New Guinea	3	.38	Benin	44	-.28	Ukraine	85
1.20	Mali	4	.36	Syria	45	-.29	Jamaica	86
1.11	Mozambique	5	.34	Guatemala	46	-.31	Armenia	87
1.11	Niger	6	.34	New Zealand	47	-.35	Egypt	88
1.08	Peru	7	.33	Haiti	48	-.36	Macedonia	89
1.06	Ethiopia	8	.32	Nigeria	49	-.41	Bangladesh	90
.99	Burkina Faso	9	.32	Saudi Arabia	50	-.44	Rwanda	91
.95	Sudan	10	.27	Ecuador	51	-.45	Mauritius	92
.92	Fiji	11	.24	Mexico	52	-.49	Libya	93
.86	Argentina	12	.23	Sri Lanka	53	-.51	Moldova	94
.86	Tanzania	13	.23	Indonesia	54	-.51	Canada	95
.82	Bolivia	14	.22	Togo	55	-.51	France	96
.82	Mongolia	15	.22	Russian Federation	56	-.59	Hungary	97
.80	Nicaragua	16	.20	Turkey	57	-.60	Iceland	98
.80	Gabon	17	.19	Belarus	58	-.61	Bulgaria	99
.79	Morocco	18	.19	Venezuela	59	-.61	Italy	100
.79	Iran	19	.18	Estonia	60	-.64	United States	101
.78	Kazakhstan	20	.18	Sweden	61	-.66	Greece	102
.77	Cameroon	21	.17	Nepal	62	-.66	Burundi	103
.75	Zimbabwe	22	.17	Uzbekistan	63	-.69	Switzerland	104
.64	Honduras	23	.15	Latvia	64	-.70	Austria	105
.63	Panama	24	.09	China	65	-.76	Croatia	106
.60	Central African Republic	25	.08	Jordan	66	-.76	Poland	107
.60	Uruguay	26	.08	Thailand	67	-.81	Trinidad and Tobago	108
.59	Senegal	27	.08	Norway	68	-.90	Lebanon	109
.57	Zambia	28	.07	India	69	-.99	Slovenia	110
.57	Cuba	29	.06	Finland	70	-1.02	Australia	111
.57	Pakistan	30	.04	El Salvador	71	-1.06	Kuwait	112
.55	Ghana	31	.03	Ireland	72	-1.19	Slovak Republic	113
.55	Kyrgyz Republic	32	.02	Malaysia	73	-1.44	Germany	114
.54	Chile	33	-.06	Albania	74	-1.45	United Kingdom	115
.52	Costa Rica	34	-.09	Azerbaijan	75	-1.49	Japan	116
.52	Uganda	35	-.09	Vietnam	76	-1.54	Denmark	117
.52	Kenya	36	-.09	Spain	77	-1.72	Israel	118
.49	Algeria	37	-.10	Philippines	78	-2.42	Czech Republic	119
.48	Tunisia	38	-.13	Lithuania	79	-2.48	South Korea	120
.48	Colombia	39	-.13	South Africa	80	-2.88	Belgium	121
.48	Brazil	40	-.17	Romania	81	-2.92	Netherlands	122
.46	Paraguay	41	-.20	Botswana	82			

Reducing Water Stresses

This indicator includes the following variables:

Fertilizer Consumption Per Hectare of Arable Land
Pesticide Use Per Hectar of Crop Land
Industrial Organic Pollutants Per Available Fresh Water
Percentage of Country's Territory Under Severe Water Stress

High numbers represent higher sustainability; zero represents the mean.

Score	Country	RANK	Score	Country	RANK	Score	Country	RANK
1.06	Central African Republic	1	.36	Finland	42	-.17	India	83
1.05	Rwanda	2	.35	Kenya	43	-.18	Azerbaijan	84
1.05	Uganda	3	.35	Malawi	44	-.19	Malaysia	85
.98	Gabon	4	.31	Cuba	45	-.19	Denmark	86
.96	Bhutan	5	.31	Romania	46	-.25	Portugal	87
.93	Mozambique	6	.30	Norway	47	-.26	Vietnam	88
.90	Cameroon	7	.26	Hungary	48	-.26	Libya	89
.79	Mongolia	8	.25	Kazakhstan	49	-.31	Kyrgyz Republic	90
.74	Senegal	9	.20	Croatia	50	-.32	Pakistan	91
.74	Madagascar	10	.20	El Salvador	51	-.32	France	92
.71	Russian Federation	11	.20	Bulgaria	52	-.34	Greece	93
.70	Botswana	12	.20	Austria	53	-.38	United Kingdom	94
.68	Burundi	13	.19	Moldova	54	-.38	Algeria	95
.67	Albania	14	.17	Czech Republic	55	-.40	Ireland	96
.66	Mali	15	.17	Guatemala	56	-.43	Iran	97
.66	Latvia	16	.13	Jamaica	57	-.51	Saudi Arabia	98
.66	Togo	17	.13	New Zealand	58	-.53	Spain	99
.60	Ethiopia	18	.13	Australia	59	-.54	Syria	100
.57	Burkina Faso	19	.12	Fiji	60	-.54	Sri Lanka	101
.57	Zambia	20	.12	Peru	61	-.56	Morocco	102
.55	Nigeria	21	.11	Thailand	62	-.57	Japan	103
.53	Lithuania	22	.10	Dominican Republic	63	-.57	Uzbekistan	104
.52	Haiti	23	.09	Philippines	64	-.57	Sudan	105
.52	Niger	24	.07	United States	65	-.58	China	106
.52	Canada	25	.07	Belarus	66	-.64	Egypt	107
.50	Ghana	26	.05	Bangladesh	67	-.82	Kuwait	108
.48	Nicaragua	27	.05	Papua New Guinea	68	-.84	Netherlands	109
.47	Tanzania	28	.05	Poland	69	-.87	Tunisia	110
.47	Bolivia	29	.03	Ukraine	70	-.93	Lebanon	111
.46	Estonia	30	.03	Slovak Republic	71	-.94	Macedonia	112
.44	Paraguay	31	-.04	Turkey	72	-1.00	Costa Rica	113
.44	Panama	32	-.04	Colombia	73	-1.25	Trinidad and Tobago	114
.42	Indonesia	33	-.05	Armenia	74	-1.27	Jordan	115
.41	Uruguay	34	-.08	Nepal	75	-1.28	Italy	116
.41	Argentina	35	-.10	Mexico	76	-1.30	Iceland	117
.41	Benin	36	-.10	Honduras	77	-1.39	South Korea	118
.41	Sweden	37	-.12	Germany	78	-1.39	Mauritius	119
.40	Zimbabwe	38	-.13	Chile	79	-1.98	Singapore	120
.37	Venezuela	39	-.13	Switzerland	80	-2.13	Israel	121
.37	Brazil	40	-.15	Slovenia	81	-2.20	Belgium	122
.37	Ecuador	41	-.17	South Africa	82			

Reducing Ecosystem Stresses

This indicator includes the following variables:

Percentage Change in Forest Cover 1990-1995
Percentage of Country's Territory with Acidification Exceedence

High numbers represent higher sustainability; zero represents the mean.

		RANK			RANK			RANK
1.33	Armenia	1	.33	Papua New Guinea	42	-.08	Romania	83
1.33	Uzbekistan	2	.33	Chile	43	-.12	Togo	84
1.16	Greece	3	.32	France	44	-.14	Indonesia	85
1.12	Kazakhstan	4	.30	Central African Republic	45	-.16	Trinidad and Tobago	86
.83	Hungary	5	.30	Fiji	46	-.17	Malawi	87
.68	Portugal	6	.30	Burundi	47	-.19	Dominican Republic	88
.67	Lithuania	7	.29	Ethiopia	48	-.19	Ireland	89
.67	New Zealand	8	.29	Brazil	49	-.20	Ecuador	90
.67	Belarus	9	.28	Colombia	50	-.25	Slovak Republic	91
.63	Estonia	10	.27	Botswana	51	-.26	Iran	92
.48	Australia	11	.27	Gabon	52	-.39	Guatemala	93
.47	India	12	.26	Tunisia	53	-.44	Panama	94
.46	Azerbaijan	13	.25	Zimbabwe	54	-.49	Syria	95
.46	Egypt	14	.21	Cameroon	55	-.50	Sweden	96
.46	Iceland	15	.21	Senegal	56	-.53	Honduras	97
.46	Israel	16	.20	Mozambique	57	-.57	Switzerland	98
.46	Kuwait	17	.18	United States	58	-.58	Malaysia	99
.46	Kyrgyz Republic	18	.18	Burkina Faso	59	-.61	Jordan	100
.46	Libya	19	.16	Saudi Arabia	60	-.62	United Kingdom	101
.46	Mauritius	20	.14	Sudan	61	-.64	Nicaragua	102
.46	Moldova	21	.14	Norway	62	-.66	Slovenia	103
.46	Mongolia	22	.14	Madagascar	63	-.69	Paraguay	104
.46	Niger	23	.14	Japan	64	-.70	Thailand	105
.46	Singapore	24	.13	Bangladesh	65	-.76	Netherlands	106
.46	Turkey	25	.13	Nigeria	66	-.83	Pakistan	107
.46	Russian Federation	26	.10	Mexico	67	-.92	Costa Rica	108
.45	Uruguay	27	.09	Uganda	68	-.96	Austria	109
.41	Latvia	28	.09	Mali	69	-.98	Poland	110
.40	Rwanda	29	.09	Tanzania	70	-.99	Germany	111
.40	Finland	30	.08	Bulgaria	71	-1.01	Vietnam	112
.40	South Africa	31	.03	Sri Lanka	72	-1.06	El Salvador	113
.39	Albania	32	.03	Nepal	73	-1.07	Denmark	114
.37	Ukraine	33	.03	Venezuela	74	-1.16	Haiti	115
.37	Kenya	34	.00	Bolivia	75	-1.19	Jamaica	116
.37	Argentina	35	.00	China	76	-1.19	Lebanon	117
.36	Spain	36	.00	Zambia	77	-1.19	Philippines	118
.35	Morocco	37	-.01	Italy	78	-1.25	South Korea	119
.34	Peru	38	-.02	Algeria	79	-1.36	Czech Republic	120
.34	Canada	39	-.03	Cuba	80	-1.52	Belgium	121
.34	Croatia	40	-.03	Benin	81	-1.63	Macedonia	122
.34	Bhutan	41	-.04	Ghana	82			

Reducing Waste and Consumption Pressures

This indicator includes the following variables:

Consumption Pressure Per Capita
Radioactive Waste

High numbers represent higher sustainability; zero represents the mean.

Score	Country	RANK		Score	Country	RANK		Score	Country	RANK
1.31	Azerbaijan	1		.57	Peru	42		-.13	Ghana	83
1.31	Moldova	2		.56	Macedonia	43		-.18	Philippines	84
1.31	Pakistan	3		.53	Nigeria	44		-.20	Australia	85
1.25	Armenia	4		.50	Jordan	45		-.20	Gabon	86
1.23	Kyrgyz Republic	5		.49	Egypt	46		-.21	Estonia	87
1.12	Bangladesh	6		.49	Honduras	47		-.22	Switzerland	88
1.08	Sudan	7		.49	Mexico	48		-.24	Costa Rica	89
1.08	Bolivia	8		.45	Romania	49		-.25	Lithuania	90
1.05	Mongolia	9		.43	Cameroon	50		-.35	Mauritius	91
1.02	Niger	10		.43	Panama	51		-.35	Italy	92
.98	Rwanda	11		.42	Bulgaria	52		-.39	Germany	93
.97	Ethiopia	12		.40	Argentina	53		-.44	Uruguay	94
.94	Mozambique	13		.40	Jamaica	54		-.50	Spain	95
.92	Madagascar	14		.37	Guatemala	55		-.50	Papua New Guinea	96
.92	Haiti	15		.37	Croatia	56		-.51	Paraguay	97
.91	Syria	16		.37	Tanzania	57		-.51	Denmark	98
.90	Burundi	17		.36	Bhutan	58		-.54	Greece	99
.85	Nicaragua	18		.34	Slovak Republic	59		-.54	Sweden	100
.82	Uganda	19		.34	Kenya	60		-.73	Finland	101
.81	Dominican Republic	20		.31	Benin	61		-.75	Fiji	102
.81	Central African Republic	21		.30	South Africa	62		-.79	Belgium	103
.81	Trinidad and Tobago	22		.29	Tunisia	63		-.86	Ireland	104
.80	Burkina Faso	23		.28	Indonesia	64		-.88	Portugal	105
.80	Zimbabwe	24		.27	Hungary	65		-.92	Canada	106
.79	Algeria	25		.23	Ecuador	66		-1.01	Norway	107
.79	Colombia	26		.18	Turkey	67		-1.03	New Zealand	108
.78	Nepal	27		.15	Senegal	68		-1.14	Malaysia	109
.71	Malawi	28		.14	Botswana	69		-1.15	South Korea	110
.71	Uzbekistan	29		.13	Russian Federation	70		-1.21	Lebanon	111
.69	Morocco	30		.13	Brazil	71		-1.22	Austria	112
.68	Mali	31		.12	Poland	72		-1.34	Israel	113
.66	Togo	32		.09	Belarus	73		-1.47	France	114
.64	Albania	33		.07	Netherlands	74		-1.64	Latvia	115
.64	Sri Lanka	34		.05	Chile	75		-1.68	United States	116
.64	Vietnam	35		.05	Czech Republic	76		-1.70	Ukraine	117
.64	Cuba	36		.00	Zambia	77		-1.99	United Kingdom	118
.63	El Salvador	37		-.01	China	78		-2.14	Iceland	119
.62	Venezuela	38		-.03	Libya	79		-2.39	Kuwait	120
.62	India	39		-.08	Slovenia	80		-2.42	Japan	121
.60	Iran	40		-.12	Saudi Arabia	81		-2.63	Singapore	122
.58	Kazakhstan	41		-.12	Thailand	82				

Reducing Population Pressure

This indicator includes the following variables:

Total Fertility Rate

Percentage Change in Projected Population Between 2000 and 2050

High numbers represent higher sustainability; zero represents the mean.

		RANK			RANK			RANK
1.08	Bulgaria	1	.69	Ireland	42	-.39	Mozambique	83
1.08	Czech Republic	2	.66	Australia	43	-.40	Kuwait	84
1.08	Spain	3	.65	Iceland	44	-.42	Singapore	85
1.08	Latvia	4	.62	Mauritius	45	-.43	Algeria	86
1.07	Russian Federation	5	.58	South Africa	46	-.44	Ghana	87
1.07	Italy	6	.52	Uruguay	47	-.46	Malaysia	88
1.07	Estonia	7	.51	Sri Lanka	48	-.56	Honduras	89
1.06	Slovenia	8	.44	Azerbaijan	49	-.57	Bolivia	90
1.05	Belarus	9	.40	United States	50	-.66	El Salvador	91
1.05	Ukraine	10	.38	Kyrgyz Republic	51	-.66	Malawi	92
1.04	Greece	11	.33	Albania	52	-.67	Rwanda	93
1.04	Armenia	12	.32	Brazil	53	-.70	Haiti	94
1.04	Hungary	13	.30	Chile	54	-.74	Libya	95
1.03	Germany	14	.27	Uzbekistan	55	-.77	Central African Republic	96
1.03	Romania	15	.26	Zimbabwe	56	-.80	Sudan	97
1.03	Austria	16	.25	Lebanon	57	-.85	Papua New Guinea	98
1.03	Lithuania	17	.25	Jamaica	58	-.86	Nepal	99
1.03	Japan	18	.24	Argentina	59	-.95	Syria	100
1.02	Slovak Republic	19	.23	Botswana	60	-.96	Paraguay	101
1.01	Poland	20	.23	Panama	61	-1.00	Pakistan	102
1.00	Croatia	21	.22	Turkey	62	-1.00	Nicaragua	103
.99	Portugal	22	.20	Indonesia	63	-1.06	Jordan	104
.98	Moldova	23	.19	Vietnam	64	-1.19	Togo	105
.97	Cuba	24	.18	Mexico	65	-1.19	Cameroon	106
.97	Switzerland	25	.11	Israel	66	-1.20	Gabon	107
.96	Belgium	26	.10	Iran	67	-1.33	Zambia	108
.95	Sweden	27	.10	Tunisia	68	-1.37	Bhutan	109
.92	Kazakhstan	28	.06	Mongolia	69	-1.40	Guatemala	110
.92	Finland	29	-.01	Morocco	70	-1.51	Senegal	111
.92	South Korea	30	-.08	Venezuela	71	-1.54	Tanzania	112
.88	Netherlands	31	-.08	Fiji	72	-1.61	Nigeria	113
.86	United Kingdom	32	-.10	Bangladesh	73	-1.77	Saudi Arabia	114
.83	Macedonia	33	-.10	India	74	-1.93	Burundi	115
.82	China	34	-.14	Ecuador	75	-2.02	Madagascar	116
.82	France	35	-.16	Dominican Republic	76	-2.03	Benin	117
.78	Denmark	36	-.17	Egypt	77	-2.11	Mali	118
.78	Norway	37	-.19	Kenya	78	-2.17	Niger	119
.76	Trinidad and Tobago	38	-.19	Colombia	79	-2.21	Burkina Faso	120
.73	Thailand	39	-.25	Peru	80	-2.23	Ethiopia	121
.72	Canada	40	-.31	Philippines	81	-2.26	Uganda	122
.70	New Zealand	41	-.36	Costa Rica	82			

Basic Human Sustenance

This indicator includes the following variables:

Daily Per Capita Calorie Supply as a Percentage of Total Requirements
Percentage of Population with Access to Improved Drinking-Water Supply

High numbers represent higher sustainability; zero represents the mean.

		RANK
.97	Belgium	1
.97	Iceland	2
.97	Japan	3
.95	France	4
.94	Germany	5
.93	Ireland	6
.92	Australia	7
.92	Austria	8
.92	Belarus	9
.92	Bulgaria	10
.92	Canada	11
.92	Denmark	12
.92	Lebanon	13
.92	Mauritius	14
.92	Norway	15
.92	Singapore	16
.92	Slovak Republic	17
.92	Slovenia	18
.92	Switzerland	19
.92	United Kingdom	20
.92	United States	21
.92	Italy	22
.92	New Zealand	23
.91	Spain	24
.89	Hungary	25
.87	Russian Federation	26
.87	Israel	27
.87	Costa Rica	28
.85	Greece	29
.82	Portugal	30
.78	Cuba	31
.78	Egypt	32
.78	Iran	33
.78	Saudi Arabia	34
.77	Czech Republic	35
.75	Kuwait	36
.75	Algeria	37
.69	Poland	38
.69	South Korea	39
.68	Moldova	40
.68	Netherlands	41

		RANK
.66	Estonia	42
.65	Croatia	43
.64	Finland	44
.63	Lithuania	45
.59	Malaysia	46
.58	Kazakhstan	47
.56	Latvia	48
.56	Sweden	49
.52	Macedonia	50
.51	Mexico	51
.51	South Africa	52
.42	Turkey	53
.41	Ukraine	54
.40	Jordan	55
.39	Morocco	56
.34	Paraguay	57
.33	Syria	58
.33	Tunisia	59
.30	Argentina	60
.30	Brazil	61
.29	Azerbaijan	62
.28	Armenia	63
.27	Trinidad and Tobago	64
.21	Indonesia	65
.20	Uzbekistan	66
.10	Libya	67
.09	Uruguay	68
.09	Colombia	69
.02	Kyrgyz Republic	70
.02	Chile	71
.00	Guatemala	72
-.09	Albania	73
-.11	Philippines	74
-.14	China	75
-.16	Botswana	76
-.18	Jamaica	77
-.20	Bhutan	78
-.20	India	79
-.26	Honduras	80
-.28	Pakistan	81
-.35	Sri Lanka	82

		RANK
-.35	Panama	83
-.36	Thailand	84
-.40	Venezuela	85
-.43	Dominican Republic	86
-.45	Nepal	87
-.46	Bangladesh	88
-.48	Romania	89
-.54	Ecuador	90
-.55	Nicaragua	91
-.57	Zimbabwe	92
-.58	El Salvador	93
-.61	Gabon	94
-.62	Senegal	95
-.68	Fiji	96
-.70	Vietnam	97
-.82	Benin	98
-1.03	Papua New Guinea	99
-1.09	Mali	100
-1.10	Peru	101
-1.15	Sudan	102
-1.16	Bolivia	103
-1.19	Mongolia	104
-1.21	Cameroon	105
-1.24	Ghana	106
-1.29	Togo	107
-1.30	Niger	108
-1.44	Nigeria	109
-1.45	Tanzania	110
-1.48	Zambia	111
-1.52	Burkina Faso	112
-1.57	Burundi	113
-1.65	Malawi	114
-1.65	Uganda	115
-1.66	Madagascar	116
-1.80	Central African Republic	117
-1.80	Mozambique	118
-1.84	Kenya	119
-1.93	Haiti	120
-2.33	Ethiopia	121
-2.33	Rwanda	122

Environmental Health

This indicator includes the following variables:

Child Death Rate from Respiratory Diseases

Death Rate from Intestinal Infectious Diseases

Under-5 Mortality Rate

Note: These values differ somewhat from the ESI released January 2001 because of an error in the Under-5 Mortality Rate which is corrected here.

High numbers represent higher sustainability; zero represents the mean.

		RANK				RANK				RANK
.99	Denmark	1		.62	Latvia	42		-.11	Honduras	83
.99	Switzerland	2		.62	Costa Rica	43		-.12	Indonesia	84
.98	Austria	3		.58	Colombia	44		-.27	South Africa	85
.97	Germany	4		.55	Moldova	45		-.28	Guatemala	86
.97	Italy	5		.53	Russian Federation	46		-.29	El Salvador	87
.97	Netherlands	6		.53	Belarus	47		-.33	Zimbabwe	88
.96	Slovenia	7		.52	Ukraine	48		-.35	Botswana	89
.96	Sweden	8		.51	Malaysia	49		-.38	Bolivia	90
.96	Canada	9		.50	Albania	50		-.49	Morocco	91
.96	Greece	10		.50	Romania	51		-.77	Egypt	92
.96	Spain	11		.42	Armenia	52		-.78	India	93
.96	Australia	12		.41	Fiji	53		-.81	Ghana	94
.95	Israel	13		.37	Jamaica	54		-.89	Gabon	95
.95	Finland	14		.35	Kazakhstan	55		-.90	Papua New Guinea	96
.95	France	15		.35	Panama	56		-.97	Mongolia	97
.95	New Zealand	16		.32	Sri Lanka	57		-1.05	Kenya	98
.95	Czech Republic	17		.30	Macedonia	58		-1.08	Nepal	99
.95	Japan	18		.28	Saudi Arabia	59		-1.10	Pakistan	100
.95	Belgium	19		.26	Thailand	60		-1.13	Bhutan	101
.95	Ireland	20		.25	Paraguay	61		-1.15	Sudan	102
.94	Norway	21		.25	Azerbaijan	62		-1.16	Bangladesh	103
.94	United Kingdom	22		.24	Brazil	63		-1.18	Cameroon	104
.94	Portugal	23		.24	Lebanon	64		-1.33	Togo	105
.94	United States	24		.21	Libya	65		-1.36	Madagascar	106
.93	Iceland	25		.19	Venezuela	66		-1.37	Haiti	107
.93	Croatia	26		.19	Turkey	67		-1.41	Senegal	108
.93	Poland	27		.18	Jordan	68		-1.50	Uganda	109
.93	Singapore	28		.17	Ecuador	69		-1.53	Tanzania	110
.91	Hungary	29		.15	Iran	70		-1.67	Benin	111
.91	Kuwait	30		.14	Peru	71		-1.69	Nigeria	112
.88	South Korea	31		.14	Tunisia	72		-1.80	Rwanda	113
.87	Slovak Republic	32		.12	Mexico	73		-1.84	Zambia	114
.85	Lithuania	33		.09	Dominican Republic	74		-1.86	Central African Republic	115
.84	Estonia	34		.08	Kyrgyz Republic	75		-2.00	Malawi	116
.79	Mauritius	35		.06	China	76		-2.02	Mali	117
.77	Chile	36		.06	Philippines	77		-2.05	Burkina Faso	118
.76	Bulgaria	37		.05	Uzbekistan	78		-2.06	Burundi	119
.76	Argentina	38		.02	Algeria	79		-2.08	Ethiopia	120
.73	Trinidad and Tobago	39		-.01	Syria	80		-2.15	Mozambique	121
.71	Uruguay	40		-.04	Vietnam	81		-2.28	Niger	122
.67	Cuba	41		-.10	Nicaragua	82				

Science/Technology

This indicator includes the following variables:

Research & Development Scientists and Engineers Per Million Population
Expenditure for Research and Development as a Percentage of GNP
Scientific and Technical Articles Per Million Population

High numbers represent higher sustainability; zero represents the mean.

Score	Country	RANK
2.61	Sweden	1
2.50	Israel	2
2.30	Switzerland	3
2.27	United States	4
2.17	Japan	5
2.04	Finland	6
1.92	Denmark	7
1.92	Australia	8
1.80	Canada	9
1.79	Germany	10
1.72	France	11
1.72	Iceland	12
1.68	Norway	13
1.63	Netherlands	14
1.62	United Kingdom	15
1.20	South Korea	16
1.18	Belgium	17
1.09	New Zealand	18
.97	Singapore	19
.97	Italy	20
.92	Ireland	21
.91	Russian Federation	22
.90	Slovenia	23
.85	Austria	24
.85	Belarus	25
.76	Azerbaijan	26
.64	Uzbekistan	27
.58	Lithuania	28
.56	Slovak Republic	29
.55	Ukraine	30
.51	Estonia	31
.45	Czech Republic	32
.43	Armenia	33
.42	Spain	34
.42	Cuba	35
.41	Croatia	36
.25	Romania	37
.20	Poland	38
.19	Bulgaria	39
.18	El Salvador	40
.14	Hungary	41

Score	Country	RANK
.05	Trinidad and Tobago	42
.05	Macedonia	43
.03	Bolivia	44
.01	Mongolia	45
-.03	Portugal	46
-.03	Greece	47
-.05	Latvia	48
-.12	South Africa	49
-.17	Uruguay	50
-.22	Moldova	51
-.25	Lebanon	52
-.26	Fiji	53
-.28	Iran	54
-.30	Chile	55
-.32	Saudi Arabia	56
-.34	Pakistan	57
-.36	Argentina	58
-.36	Kazakhstan	59
-.37	Brazil	60
-.38	China	61
-.39	Benin	62
-.39	Kyrgyz Republic	63
-.41	Costa Rica	64
-.42	Vietnam	65
-.42	India	66
-.43	Botswana	67
-.45	Morocco	68
-.45	Dominican Republic	69
-.45	Mauritius	70
-.46	Venezuela	71
-.49	Turkey	72
-.50	Libya	73
-.52	Nicaragua	74
-.52	Togo	75
-.52	Uganda	76
-.53	Algeria	77
-.53	Egypt	78
-.55	Peru	79
-.57	Mexico	80
-.57	Honduras	81
-.58	Paraguay	82

Score	Country	RANK
-.59	Tunisia	83
-.59	Ghana	84
-.59	Zimbabwe	85
-.59	Kuwait	86
-.60	Sri Lanka	87
-.60	Albania	88
-.60	Philippines	89
-.62	Jordan	90
-.63	Malaysia	91
-.63	Burundi	92
-.64	Central African Republic	93
-.64	Papua New Guinea	94
-.64	Panama	95
-.65	Gabon	96
-.66	Indonesia	97
-.66	Guatemala	98
-.67	Thailand	99
-.68	Syria	100
-.69	Burkina Faso	101
-.70	Ecuador	102
-.70	Madagascar	103
-.71	Colombia	104
-.73	Cameroon	105
-.74	Bangladesh	106
-.74	Rwanda	107
-.75	Kenya	108
-.75	Nigeria	109
-.76	Jamaica	110
-.77	Senegal	111
-.89	Bhutan	112
-.91	Sudan	113
-.92	Nepal	114
-1.06	Haiti	115
-1.15	Zambia	116
-1.17	Tanzania	117
-1.29	Malawi	118
-1.31	Mali	119
-1.46	Ethiopia	120
-1.46	Mozambique	121
-1.46	Niger	122

Capacity for Debate

This indicator includes the following variables:

IUCN Member Organizations Per Million Population
Civil and Political Liberties

High numbers represent higher sustainability; zero represents the mean.

Score	Country	RANK
2.41	Iceland	1
2.27	Panama	2
2.14	Botswana	3
1.82	Costa Rica	4
1.66	Australia	5
1.63	Fiji	6
1.59	New Zealand	7
1.32	Mauritius	8
1.13	Norway	9
1.13	Uruguay	10
1.10	Denmark	11
1.08	Netherlands	12
.97	Jordan	13
.97	Switzerland	14
.90	Canada	15
.90	Estonia	16
.85	Finland	17
.82	Trinidad and Tobago	18
.80	Lebanon	19
.76	Jamaica	20
.75	Ecuador	21
.75	Ireland	22
.73	Sweden	23
.72	Bolivia	24
.70	Austria	25
.69	El Salvador	26
.62	Israel	27
.55	United Kingdom	28
.52	Spain	29
.51	Belgium	30
.47	South Africa	31
.42	Slovak Republic	32
.42	France	33
.40	Lithuania	34
.39	Slovenia	35
.38	Portugal	36
.36	Czech Republic	37
.34	Guatemala	38
.32	Honduras	39
.32	Zimbabwe	40
.31	United States	41
.30	Hungary	42
.30	Greece	43
.29	Latvia	44
.26	Italy	45
.22	Argentina	46
.18	Germany	47
.18	Poland	48
.16	Kuwait	49
.15	Japan	50
.13	Paraguay	51
.10	Benin	52
.08	Mali	53
.07	Mongolia	54
.05	Chile	55
.05	Dominican Republic	56
-.01	South Korea	57
-.02	Romania	58
-.02	Macedonia	59
-.02	Nicaragua	60
-.02	Singapore	61
-.03	Sri Lanka	62
-.06	Papua New Guinea	63
-.07	Moldova	64
-.08	Bulgaria	65
-.13	Zambia	66
-.16	Malawi	67
-.19	Nepal	68
-.20	Croatia	69
-.20	Thailand	70
-.20	Central African Republic	71
-.21	Philippines	72
-.23	India	73
-.24	Ghana	74
-.28	Senegal	75
-.32	Madagascar	76
-.32	Armenia	77
-.37	Mozambique	78
-.41	Venezuela	79
-.42	Colombia	80
-.43	Burkina Faso	81
-.43	Bangladesh	82
-.44	Mexico	83
-.44	Brazil	84
-.47	Ukraine	85
-.49	Nigeria	86
-.50	Haiti	87
-.52	Albania	88
-.53	Gabon	89
-.54	Peru	90
-.57	Tanzania	91
-.62	Morocco	92
-.64	Indonesia	93
-.64	Tunisia	94
-.66	Azerbaijan	95
-.70	Malaysia	96
-.72	Uganda	97
-.73	Togo	98
-.74	Turkey	99
-.75	Niger	100
-.75	Russian Federation	101
-.76	Cuba	102
-.77	Kyrgyz Republic	103
-.81	Bhutan	104
-.81	Belarus	105
-.86	Kenya	106
-.90	Kazakhstan	107
-.91	Ethiopia	108
-.98	Algeria	109
-.98	Burundi	110
-1.00	Rwanda	111
-1.02	Egypt	112
-1.07	Pakistan	113
-1.19	Iran	114
-1.28	Cameroon	115
-1.30	Uzbekistan	116
-1.32	Libya	117
-1.32	China	118
-1.34	Saudi Arabia	119
-1.42	Syria	120
-1.44	Vietnam	121
-1.44	Sudan	122

Regulation and Management

This indicator includes the following variables:

Stringency and Consistency of Environmental Regulations
Degree to which Environmental Regulations Promote Innovation
Percentage of Land Area Under Protected Status
Number of Sectoral Environmental Impact Assessment Guidelines

High numbers represent higher sustainability; zero represents the mean.

Score	Country	RANK
1.54	United Kingdom	1
1.54	Denmark	2
1.46	Switzerland	3
1.34	Germany	4
1.30	United States	5
1.26	Austria	6
1.24	Dominican Republic	7
1.21	Finland	8
1.21	Canada	9
1.10	New Zealand	10
.89	Slovak Republic	11
.84	Sweden	12
.82	France	13
.78	Pakistan	14
.75	Netherlands	15
.72	Belgium	16
.68	Singapore	17
.66	Chile	18
.65	Nepal	19
.55	Bhutan	20
.53	Costa Rica	21
.43	Malaysia	22
.43	Spain	23
.42	Norway	24
.40	Australia	25
.40	Panama	26
.36	Botswana	27
.36	South Africa	28
.35	Portugal	29
.35	Japan	30
.35	Sri Lanka	31
.34	Tanzania	32
.29	Cuba	33
.28	Venezuela	34
.25	Guatemala	35
.21	Malawi	36
.19	Iceland	37
.17	Czech Republic	38
.12	Thailand	39
.12	Zimbabwe	40
.11	Egypt	41

Score	Country	RANK
.10	Rwanda	42
.08	Italy	43
.07	Ireland	44
.05	Israel	45
.03	Paraguay	46
.01	India	47
-.04	Latvia	48
-.05	Bolivia	49
-.06	Ecuador	50
-.08	Estonia	51
-.12	Senegal	52
-.18	Burkina Faso	53
-.19	Mongolia	54
-.20	Niger	55
-.21	Lithuania	56
-.21	Honduras	57
-.23	Indonesia	58
-.24	Uganda	59
-.28	South Korea	60
-.29	Brazil	61
-.30	Kenya	62
-.30	Colombia	63
-.31	Mozambique	64
-.31	Peru	65
-.31	Zambia	66
-.33	Bangladesh	67
-.33	Central African Republic	68
-.35	Togo	69
-.36	Hungary	70
-.36	Argentina	71
-.37	Armenia	72
-.38	Nicaragua	73
-.39	Ghana	74
-.39	Russian Federation	75
-.40	Jordan	76
-.41	Macedonia	77
-.41	Benin	78
-.43	Croatia	79
-.44	Poland	80
-.45	Kuwait	81
-.45	Mexico	82

Score	Country	RANK
-.45	Greece	83
-.49	Nigeria	84
-.50	Slovenia	85
-.51	Burundi	86
-.51	Turkey	87
-.51	China	88
-.51	Ethiopia	89
-.51	Azerbaijan	90
-.54	Iran	91
-.57	Romania	92
-.58	Cameroon	93
-.60	Belarus	94
-.63	Mali	95
-.64	Sudan	96
-.64	Mauritius	97
-.64	Kyrgyz Republic	98
-.68	Trinidad and Tobago	99
-.69	Gabon	100
-.70	Albania	101
-.70	Kazakhstan	102
-.71	Bulgaria	103
-.72	Algeria	104
-.72	Philippines	105
-.73	Saudi Arabia	106
-.75	Uzbekistan	107
-.75	Madagascar	108
-.80	Moldova	109
-.81	Vietnam	110
-.81	Fiji	111
-.84	Morocco	112
-.86	Haiti	113
-.86	Lebanon	114
-.86	Tunisia	115
-.87	Uruguay	116
-.87	Jamaica	117
-.88	Libya	118
-.88	Papua New Guinea	119
-.88	Syria	120
-.93	Ukraine	121
-1.32	El Salvador	122

Private Sector Responsiveness

This indicator includes the following variables:

Number of ISO 14001 Certified Companies Per Million Dollars GDP
Dow Jones Sustainbility Group Index
Average Innovest EcoValue 21 Rating of Firms
World Business Council for Sustainable Development Members
Levels of Environmental Competitiveness

High numbers represent higher sustainability; zero represents the mean.

Score	Country	RANK
2.12	Switzerland	1
1.83	Japan	2
1.09	Germany	3
1.02	United Kingdom	4
.93	New Zealand	5
.89	Finland	6
.86	Czech Republic	7
.84	United States	8
.83	Hungary	9
.82	Costa Rica	10
.78	Australia	11
.72	Denmark	12
.71	Canada	13
.68	Brazil	14
.68	Slovenia	15
.67	Sweden	16
.62	South Korea	17
.61	Russian Federation	18
.58	Singapore	19
.46	Thailand	20
.42	Austria	21
.40	China	22
.38	Croatia	23
.38	Paraguay	24
.37	Netherlands	25
.25	Slovak Republic	26
.22	Algeria	27
.20	Mexico	28
.15	Lebanon	29
.14	France	30
.05	Israel	31
.05	Zambia	32
.04	Uruguay	33
.03	Norway	34
.03	Jordan	35
.02	Belgium	36
.00	Ireland	37
-.02	Estonia	38
-.03	Egypt	39
-.03	Fiji	40
-.03	Iceland	41

Score	Country	RANK
-.07	South Africa	42
-.10	Malaysia	43
-.12	Spain	44
-.18	Portugal	45
-.20	Chile	46
-.24	Honduras	47
-.26	Argentina	48
-.28	Turkey	49
-.33	Trinidad and Tobago	50
-.35	Italy	51
-.37	Mauritius	52
-.39	Latvia	53
-.39	Morocco	54
-.39	Zimbabwe	55
-.40	Syria	56
-.41	Iran	57
-.42	Ecuador	58
-.42	Lithuania	59
-.42	Sri Lanka	60
-.43	Cuba	61
-.43	Kuwait	62
-.43	Libya	63
-.43	Rwanda	64
-.43	Guatemala	65
-.43	Saudi Arabia	66
-.43	Dominican Republic	67
-.44	Peru	68
-.44	India	69
-.45	Tunisia	70
-.46	Nigeria	71
-.46	Poland	72
-.46	Pakistan	73
-.47	Romania	74
-.48	Albania	75
-.48	Armenia	76
-.48	Azerbaijan	77
-.48	Bangladesh	78
-.48	Belarus	79
-.48	Benin	80
-.48	Bhutan	81
-.48	Botswana	82

Score	Country	RANK
-.48	Burkina Faso	83
-.48	Burundi	84
-.48	Cameroon	85
-.48	Central African Republic	86
-.48	Ethiopia	87
-.48	Gabon	88
-.48	Ghana	89
-.48	Haiti	90
-.48	Jamaica	91
-.48	Kazakhstan	92
-.48	Kenya	93
-.48	Kyrgyz Republic	94
-.48	Macedonia	95
-.48	Madagascar	96
-.48	Malawi	97
-.48	Mali	98
-.48	Moldova	99
-.48	Mongolia	100
-.48	Mozambique	101
-.48	Nepal	102
-.48	Nicaragua	103
-.48	Niger	104
-.48	Panama	105
-.48	Papua New Guinea	106
-.48	Senegal	107
-.48	Sudan	108
-.48	Tanzania	109
-.48	Togo	110
-.48	Uganda	111
-.48	Uzbekistan	112
-.49	Philippines	113
-.50	Colombia	114
-.50	Greece	115
-.70	Vietnam	116
-.73	Indonesia	117
-.73	Bulgaria	118
-.77	Bolivia	119
-.79	Venezuela	120
-.87	El Salvador	121
-.89	Ukraine	122

Component and Indicator Scores | **Annex 2**

129

Environmental Information

This indicator includes the following variables:

Availability of Sustainable Development Information at the National Level
Environmental Strategies and Action Plans
Number of ESI Variables Missing from Selected Data Sets

High numbers represent higher sustainability; zero represents the mean.

Score	Country	RANK
2.25	Netherlands	1
1.88	Norway	2
1.57	United States	3
1.54	Finland	4
1.31	United Kingdom	5
1.30	Austria	6
1.12	France	7
1.12	China	8
1.04	Slovak Republic	9
1.01	Indonesia	10
.97	Portugal	11
.89	Switzerland	12
.84	Malaysia	13
.83	Hungary	14
.83	India	15
.81	Ecuador	16
.79	Japan	17
.79	Germany	18
.79	Poland	19
.79	Australia	20
.77	Egypt	21
.73	Ireland	22
.70	Czech Republic	23
.69	Argentina	24
.68	Canada	25
.66	Chile	26
.64	Mexico	27
.63	Denmark	28
.62	Colombia	29
.61	Spain	30
.58	Italy	31
.56	Thailand	32
.56	Israel	33
.56	Lithuania	34
.51	Sweden	35
.46	Nepal	36
.39	Singapore	37
.38	Sri Lanka	38
.36	Ukraine	39
.30	Pakistan	40
.30	Iceland	41

Score	Country	RANK
.29	Nicaragua	42
.27	Estonia	43
.26	Slovenia	44
.24	Cuba	45
.23	Russian Federation	46
.23	Latvia	47
.23	South Korea	48
.21	Moldova	49
.20	South Africa	50
.20	Mongolia	51
.19	El Salvador	52
.16	Albania	53
.12	Costa Rica	54
.12	Jamaica	55
.08	Uganda	56
-.01	Turkey	57
-.02	Uruguay	58
-.04	Tanzania	59
-.05	Philippines	60
-.07	Belgium	61
-.08	Brazil	62
-.12	Bulgaria	63
-.18	New Zealand	64
-.18	Macedonia	65
-.18	Benin	66
-.18	Vietnam	67
-.19	Bolivia	68
-.21	Fiji	69
-.22	Guatemala	70
-.23	Romania	71
-.23	Iran	72
-.26	Kenya	73
-.27	Ghana	74
-.29	Croatia	75
-.30	Zimbabwe	76
-.30	Tunisia	77
-.34	Peru	78
-.35	Venezuela	79
-.35	Greece	80
-.46	Senegal	81
-.51	Jordan	82

Score	Country	RANK
-.52	Trinidad and Tobago	83
-.56	Nigeria	84
-.58	Honduras	85
-.59	Gabon	86
-.60	Belarus	87
-.62	Botswana	88
-.62	Mozambique	89
-.63	Morocco	90
-.64	Mauritius	91
-.65	Togo	92
-.67	Malawi	93
-.67	Ethiopia	94
-.68	Kuwait	95
-.70	Kazakhstan	96
-.71	Cameroon	97
-.72	Zambia	98
-.75	Algeria	99
-.75	Dominican Republic	100
-.76	Niger	101
-.76	Papua New Guinea	102
-.77	Bangladesh	103
-.78	Mali	104
-.80	Azerbaijan	105
-.81	Panama	106
-.82	Burkina Faso	107
-.85	Sudan	108
-.86	Central African Republic	109
-.87	Uzbekistan	110
-.89	Kyrgyz Republic	111
-.90	Rwanda	112
-.91	Bhutan	113
-.91	Libya	114
-.91	Madagascar	115
-.92	Burundi	116
-.95	Syria	117
-.95	Lebanon	118
-.96	Armenia	119
-.98	Saudi Arabia	120
-1.15	Paraguay	121
-1.44	Haiti	122

Eco-Efficiency

This indicator includes the following variables:

Energy Efficiency (Total Energy Consumption Per Unit GDP)
Renewable Energy Production as a Percentage of Total Energy Consumption

High numbers represent higher sustainability; zero represents the mean.

		RANK			RANK			RANK
.95	Uganda	1	.50	Turkey	42	-.18	Venezuela	83
.93	Ethiopia	2	.50	Zambia	43	-.27	South Korea	84
.88	Cameroon	3	.49	Spain	44	-.29	Iran	85
.86	Switzerland	4	.49	France	45	-.31	China	86
.85	Norway	5	.48	Denmark	46	-.37	Jamaica	87
.84	Malawi	6	.48	Panama	47	-.41	Cuba	88
.83	Uruguay	7	.48	Japan	48	-.45	Lebanon	89
.83	Bhutan	8	.47	Chile	49	-.54	Kyrgyz Republic	90
.83	Paraguay	9	.40	Philippines	50	-.56	Tunisia	91
.82	Nepal	10	.40	Armenia	51	-.61	Israel	92
.82	Austria	11	.36	Colombia	52	-.65	Vietnam	93
.80	Sweden	12	.35	Bolivia	53	-.69	Romania	94
.80	Mali	13	.34	Dominican Republic	54	-.70	Moldova	95
.80	Iceland	14	.32	Morocco	55	-.77	Niger	96
.80	Ghana	15	.31	Australia	56	-.77	Poland	97
.77	Madagascar	16	.31	Albania	57	-.81	Senegal	98
.77	Tanzania	17	.29	Latvia	58	-.84	Botswana	99
.74	Honduras	18	.28	Ireland	59	-.88	Algeria	100
.72	Burundi	19	.27	Mauritius	60	-.88	South Africa	101
.71	Sudan	20	.26	United States	61	-.93	Hungary	102
.71	Mozambique	21	.25	Syria	62	-.94	Lithuania	103
.70	Peru	22	.24	Germany	63	-.95	Estonia	104
.70	Brazil	23	.22	Mexico	64	-.96	Macedonia	105
.69	Costa Rica	24	.22	Ecuador	65	-.97	Slovak Republic	106
.69	Finland	25	.20	Central African Republic	66	-.98	Bulgaria	107
.67	Rwanda	26	.19	Greece	67	-.99	Jordan	108
.66	New Zealand	27	.13	Zimbabwe	68	-1.04	Czech Republic	109
.64	Fiji	28	.10	Croatia	69	-1.12	Singapore	110
.63	Haiti	29	.10	United Kingdom	70	-1.19	Russian Federation	111
.61	Guatemala	30	.07	Nigeria	71	-1.20	Libya	112
.60	Sri Lanka	31	.07	Nicaragua	72	-1.38	Kuwait	113
.60	El Salvador	32	.06	Bangladesh	73	-1.39	Kazakhstan	114
.60	Kenya	33	.05	Pakistan	74	-1.42	Benin	115
.60	Burkina Faso	34	.04	Thailand	75	-1.48	Saudi Arabia	116
.56	Portugal	35	.02	Netherlands	76	-1.51	Belarus	117
.55	Canada	36	-.03	India	77	-1.64	Uzbekistan	118
.55	Papua New Guinea	37	-.05	Indonesia	78	-1.67	Azerbaijan	119
.54	Gabon	38	-.09	Togo	79	-1.71	Mongolia	120
.52	Argentina	39	-.09	Egypt	80	-1.77	Ukraine	121
.51	Italy	40	-.15	Belgium	81	-2.16	Trinidad and Tobago	122
.51	Slovenia	41	-.17	Malaysia	82			

Reducing Public Choice Failures

This indicator includes the following variables:

Price of Premium Gasoline
Subsidies for Energy or Materials Usage
Reducing Corruption

High numbers represent higher sustainability; zero represents the mean.

Score	Country	RANK
2.25	Finland	1
1.80	Netherlands	2
1.62	Denmark	3
1.61	Iceland	4
1.60	United Kingdom	5
1.57	New Zealand	6
1.48	Sweden	7
1.44	Ireland	8
1.38	Austria	9
1.38	Switzerland	10
1.36	France	11
1.35	Norway	12
1.25	Singapore	13
1.11	Germany	14
1.10	Belgium	15
1.09	Israel	16
.98	Italy	17
.95	Portugal	18
.86	Japan	19
.82	Uruguay	20
.76	Australia	21
.72	Spain	22
.70	Slovenia	23
.68	Canada	24
.63	Argentina	25
.54	Chile	26
.45	Morocco	27
.41	Sri Lanka	28
.36	Peru	29
.36	Hungary	30
.35	Brazil	31
.31	South Korea	32
.28	Fiji	33
.27	Czech Republic	34
.26	Uganda	35
.25	Greece	36
.24	United States	37
.20	Bolivia	38
.17	Turkey	39
.15	Mauritius	40
.11	Senegal	41

Score	Country	RANK
.09	Mali	42
.07	Estonia	43
.04	Tunisia	44
-.01	Cuba	45
-.02	Burkina Faso	46
-.02	Albania	47
-.02	South Africa	48
-.06	Macedonia	49
-.08	Trinidad and Tobago	50
-.09	Croatia	51
-.12	Lithuania	52
-.13	Costa Rica	53
-.14	Kenya	54
-.17	Central African Republic	55
-.18	Malaysia	56
-.19	Jordan	57
-.20	Latvia	58
-.22	Botswana	59
-.23	Poland	60
-.24	Malawi	61
-.27	Niger	62
-.28	Haiti	63
-.32	Rwanda	64
-.34	Burundi	65
-.35	Romania	66
-.35	Mozambique	67
-.37	Bangladesh	68
-.37	Bhutan	69
-.39	Nepal	70
-.40	Slovak Republic	71
-.41	Tanzania	72
-.43	Zambia	73
-.43	Kuwait	74
-.43	Togo	75
-.45	El Salvador	76
-.46	Moldova	77
-.46	Jamaica	78
-.46	Gabon	79
-.46	Madagascar	80
-.49	Cameroon	81
-.51	Philippines	82

Score	Country	RANK
-.55	Mexico	83
-.55	Bulgaria	84
-.57	Panama	85
-.61	Armenia	86
-.62	Kyrgyz Republic	87
-.64	Colombia	88
-.65	Pakistan	89
-.65	Lebanon	90
-.65	Ethiopia	91
-.65	Ghana	92
-.66	Honduras	93
-.66	Nicaragua	94
-.67	Syria	95
-.68	Egypt	96
-.70	China	97
-.72	Thailand	98
-.73	Russian Federation	99
-.73	Paraguay	100
-.73	Mongolia	101
-.76	Dominican Republic	102
-.77	Vietnam	103
-.77	Guatemala	104
-.77	Azerbaijan	105
-.78	Benin	106
-.78	Papua New Guinea	107
-.81	Belarus	108
-.82	India	109
-.89	Ukraine	110
-.92	Zimbabwe	111
-.98	Algeria	112
-1.00	Kazakhstan	113
-1.02	Sudan	114
-1.10	Saudi Arabia	115
-1.15	Libya	116
-1.21	Venezuela	117
-1.30	Iran	118
-1.35	Ecuador	119
-1.36	Nigeria	120
-1.36	Uzbekistan	121
-1.54	Indonesia	122

International Commitment

This indicator includes the following variables:

Number of Memberships in Environmental Intergovernmental Organizations
Percentage of Convention on International Trade in Endangered Species (CITES) Reporting Requirements Met
Levels of Participation in the Vienna Convention and the Montreal Protocol on Ozone Depleting Substances
Compliance with Environmental Agreements

High numbers represent higher sustainability; zero represents the mean.

Score	Country	RANK	Score	Country	RANK	Score	Country	RANK
1.58	Netherlands	1	.17	Uganda	42	-.25	Philippines	83
1.53	Germany	2	.16	Mexico	43	-.28	Zambia	84
1.50	Sweden	3	.16	Togo	44	-.37	Paraguay	85
1.29	Norway	4	.15	Tanzania	45	-.40	Slovenia	86
1.28	Denmark	5	.15	Brazil	46	-.42	Vietnam	87
1.20	Austria	6	.14	Costa Rica	47	-.43	Romania	88
1.19	France	7	.13	Latvia	48	-.45	Bangladesh	89
1.12	United Kingdom	8	.13	Malawi	49	-.46	Benin	90
1.11	Finland	9	.12	China	50	-.46	Lebanon	91
1.09	Switzerland	10	.12	Chile	51	-.46	Nepal	92
1.01	Spain	11	.10	Egypt	52	-.46	Guatemala	93
.91	Japan	12	.09	Mali	53	-.50	Uzbekistan	94
.91	Belgium	13	.09	Indonesia	54	-.51	Croatia	95
.91	Canada	14	.08	Russian Federation	55	-.52	Dominican Republic	96
.88	Australia	15	.07	Algeria	56	-.65	Gabon	97
.81	Italy	16	.04	Bulgaria	57	-.66	Macedonia	98
.80	Tunisia	17	.02	Ghana	58	-.67	Ethiopia	99
.78	United States	18	.02	Niger	59	-.67	Nigeria	100
.68	New Zealand	19	.00	Iran	60	-.67	Sudan	101
.62	Greece	20	.00	Estonia	61	-.69	Madagascar	102
.58	Singapore	21	-.01	Argentina	62	-.71	Kuwait	103
.56	South Korea	22	-.02	South Africa	63	-.72	Azerbaijan	104
.53	Czech Republic	23	-.02	Mongolia	64	-.75	Belarus	105
.53	Senegal	24	-.02	Turkey	65	-.81	Fiji	106
.52	Panama	25	-.03	Mauritius	66	-.81	Lithuania	107
.47	Hungary	26	-.06	Ecuador	67	-.81	Saudi Arabia	108
.44	Portugal	27	-.06	Thailand	68	-1.03	Ukraine	109
.42	Sri Lanka	28	-.06	Botswana	69	-1.07	Haiti	110
.39	Cameroon	29	-.06	Iceland	70	-1.17	Central African Republic	111
.30	India	30	-.10	Ireland	71	-1.27	El Salvador	112
.30	Slovak Republic	31	-.10	Bolivia	72	-1.32	Honduras	113
.29	Cuba	32	-.13	Israel	73	-1.47	Libya	114
.29	Morocco	33	-.15	Jordan	74	-1.47	Burundi	115
.28	Trinidad and Tobago	34	-.16	Venezuela	75	-1.57	Rwanda	116
.27	Colombia	35	-.18	Mozambique	76	-1.68	Albania	117
.27	Poland	36	-.18	Peru	77	-1.73	Kazakhstan	118
.27	Kenya	37	-.19	Zimbabwe	78	-1.73	Moldova	119
.26	Nicaragua	38	-.20	Syria	79	-1.78	Armenia	120
.24	Pakistan	39	-.21	Papua New Guinea	80	-1.78	Bhutan	121
.24	Malaysia	40	-.23	Burkina Faso	81	-1.78	Kyrgyz Republic	122
.19	Uruguay	41	-.24	Jamaica	82			

Global-Scale Funding/Participation

This indicator includes the following variables:

Montreal Protocol Multilateral Fund Participation
Global Environmental Facility Participation

High numbers represent higher sustainability; zero represents the mean.

Score	Country	RANK
2.34	Lithuania	1
2.28	Bulgaria	2
2.12	Azerbaijan	3
1.97	Slovak Republic	4
1.85	Czech Republic	5
1.56	Mauritius	6
1.27	Uruguay	7
1.11	Malaysia	8
1.05	Costa Rica	9
.84	Jamaica	10
.79	Hungary	11
.75	Sweden	12
.73	Canada	13
.71	Finland	14
.69	New Zealand	15
.68	Denmark	16
.67	Panama	17
.63	Switzerland	18
.63	Norway	19
.63	Greece	20
.62	Australia	21
.62	Egypt	22
.61	Belgium	23
.56	Netherlands	24
.53	Austria	25
.49	United States	26
.49	Ireland	27
.49	Poland	28
.45	Dominican Republic	29
.44	Germany	30
.44	Spain	31
.44	Argentina	32
.43	United Kingdom	33
.40	France	34
.40	Mexico	35
.39	Japan	36
.39	Venezuela	37
.34	South Africa	38
.31	Thailand	39
.29	Lebanon	40
.28	Peru	41

Score	Country	RANK
.28	Italy	42
.25	Portugal	43
.21	Philippines	44
.20	El Salvador	45
.16	Guatemala	46
.14	Bangladesh	47
.14	Brazil	48
.13	Bhutan	49
.12	Sri Lanka	50
.12	Slovenia	51
.11	Ghana	52
.11	Pakistan	53
.07	Jordan	54
.06	Bolivia	55
.06	India	56
.04	Latvia	57
.04	Uzbekistan	58
.04	Mongolia	59
.02	Papua New Guinea	60
.02	China	61
.02	Cuba	62
.01	Central African Republic	63
.01	Ecuador	64
-.01	Zimbabwe	65
-.03	Belarus	66
-.03	Turkey	67
-.04	Ukraine	68
-.04	Benin	69
-.06	Nicaragua	70
-.07	Uganda	71
-.08	Madagascar	72
-.08	Honduras	73
-.10	Cameroon	74
-.10	Mozambique	75
-.12	Armenia	76
-.12	Senegal	77
-.13	Tunisia	78
-.14	Mali	79
-.14	Romania	80
-.14	Nepal	81
-.15	Algeria	82

Score	Country	RANK
-.15	Trinidad and Tobago	83
-.17	Syria	84
-.17	Vietnam	85
-.17	Russian Federation	86
-.18	Chile	87
-.20	Niger	88
-.23	Burkina Faso	89
-.25	Indonesia	90
-.27	Kenya	91
-.31	Sudan	92
-.52	Gabon	93
-.58	Colombia	94
-.59	Iceland	95
-.77	Israel	96
-.87	Morocco	97
-.93	Paraguay	98
-1.00	Moldova	99
-1.03	Singapore	100
-1.06	Nigeria	101
-1.07	Kuwait	102
-1.07	Estonia	103
-1.09	Malawi	104
-1.12	Burundi	105
-1.13	Zambia	106
-1.13	Botswana	107
-1.15	Tanzania	108
-1.15	Iran	109
-1.17	Albania	110
-1.17	Croatia	111
-1.17	Ethiopia	112
-1.17	Fiji	113
-1.17	Haiti	114
-1.17	Kazakhstan	115
-1.17	Kyrgyz Republic	116
-1.17	Libya	117
-1.17	Macedonia	118
-1.17	Rwanda	119
-1.17	Saudi Arabia	120
-1.17	South Korea	121
-1.17	Togo	122

Protecting International Commons

This indicator includes the following variables:

Forest Stewardship Council (FSC) Accredited Forest Area as a Percentage of Total Forest Area
Ecological Footprint "Deficit"
Carbon-Dioxide (CO2) Emissions (Total times Per Capita)
Historic Cumulative Carbon-Dioxide (CO2) Emissions
Cluorofluorocarbon (CFC) Consumption (Total times Per Capita)
Sulfur Dioxide (SO2) Exports

High numbers represent higher sustainability; zero represents the mean.

Score	Country	RANK		Score	Country	RANK		Score	Country	RANK
1.74	Central African Republic	1		.30	Zimbabwe	42		-.30	Cuba	83
1.43	Papua New Guinea	2		.28	Haiti	43		-.33	Israel	84
1.38	Bolivia	3		.26	Slovak Republic	44		-.36	Poland	85
1.17	Gabon	4		.24	Albania	45		-.36	Austria	86
.99	Uganda	5		.22	Kenya	46		-.37	Bulgaria	87
.87	Mozambique	6		.22	Czech Republic	47		-.38	Singapore	88
.87	Benin	7		.22	Moldova	48		-.38	Trinidad and Tobago	89
.83	Burundi	8		.19	Armenia	49		-.39	Lebanon	90
.82	Mali	9		.11	Canada	50		-.39	Mexico	91
.80	Madagascar	10		.11	Bangladesh	51		-.41	Ireland	92
.79	Ethiopia	11		.11	El Salvador	52		-.43	Argentina	93
.78	Burkina Faso	12		.10	Brazil	53		-.46	Chile	94
.76	Honduras	13		.10	Uruguay	54		-.48	Portugal	95
.76	Malawi	14		.08	Hungary	55		-.49	Ukraine	96
.75	Zambia	15		.08	Slovenia	56		-.49	Egypt	97
.74	Guatemala	16		.05	Panama	57		-.51	Romania	98
.72	Nepal	17		.04	Macedonia	58		-.52	Nigeria	99
.67	Fiji	18		.03	Australia	59		-.53	Denmark	100
.65	New Zealand	19		.03	South Africa	60		-.54	France	101
.65	Niger	20		.02	Lithuania	61		-.55	Syria	102
.62	Bhutan	21		.02	Vietnam	62		-.57	Algeria	103
.62	Costa Rica	22		.01	Ecuador	63		-.59	Venezuela	104
.61	Nicaragua	23		.00	Norway	64		-.64	Libya	105
.58	Cameroon	24		-.07	Netherlands	65		-.65	United Kingdom	106
.57	Rwanda	25		-.07	Dominican Republic	66		-.67	Greece	107
.53	Sri Lanka	26		-.07	Kyrgyz Republic	67		-.68	Japan	108
.53	Iceland	27		-.09	Malaysia	68		-.73	Italy	109
.52	Tanzania	28		-.11	Indonesia	69		-.73	Kazakhstan	110
.51	Botswana	29		-.15	Colombia	70		-.74	Germany	111
.50	Ghana	30		-.17	Pakistan	71		-.76	Thailand	112
.47	Croatia	31		-.17	Belgium	72		-.76	Turkey	113
.43	Paraguay	32		-.18	Estonia	73		-.79	India	114
.40	Mongolia	33		-.19	Jamaica	74		-.79	United States	115
.39	Sudan	34		-.24	Uzbekistan	75		-.88	Iran	116
.38	Mauritius	35		-.25	Finland	76		-.90	South Korea	117
.37	Peru	36		-.25	Morocco	77		-.93	Kuwait	118
.37	Togo	37		-.26	Tunisia	78		-1.00	Spain	119
.35	Senegal	38		-.27	Azerbaijan	79		-1.03	Saudi Arabia	120
.34	Sweden	39		-.27	Belarus	80		-1.16	Russian Federation	121
.33	Switzerland	40		-.30	Philippines	81		-1.63	China	122
.31	Latvia	41		-.30	Jordan	82				

Reference Variable

Percent land with population density > 5/square km

		RANK			RANK			RANK
1.00	Albania	1	1.00	Greece	42	1.00	Pakistan	83
.15	Algeria	2	.87	Guatemala	43	.77	Panama	84
.32	Argentina	3	.99	Haiti	44	.60	Papua New Guinea	85
1.00	Armenia	4	.83	Honduras	45	.35	Paraguay	86
.03	Australia	5	1.00	Hungary	46	.45	Peru	87
1.00	Austria	6	.03	Iceland	47	.97	Philippines	88
.99	Azerbaijan	7	1.00	India	48	1.00	Poland	89
1.00	Bangladesh	8	.86	Indonesia	49	.98	Portugal	90
1.00	Belarus	9	.99	Iran	50	1.00	Romania	91
.93	Belgium	10	1.00	Ireland	51	.19	Russian Federation	92
1.00	Benin	11	1.00	Israel	52	1.00	Rwanda	93
.84	Bhutan	12	1.00	Italy	53	.42	Saudi Arabia	94
.24	Bolivia	13	1.00	Jamaica	54	.86	Senegal	95
.17	Botswana	14	.98	Japan	55	.91	Singapore	96
.40	Brazil	15	.57	Jordan	56	1.00	Slovak Republic	97
1.00	Bulgaria	16	.22	Kazakhstan	57	1.00	Slovenia	98
.95	Burkina Faso	17	.48	Kenya	58	.50	South Africa	99
1.00	Burundi	18	.94	Kuwait	59	.98	South Korea	100
.83	Cameroon	19	.85	Kyrgyz Republic	60	.86	Spain	101
.04	Canada	20	1.00	Latvia	61	1.00	Sri Lanka	102
.37	Central African Republic	21	1.00	Lebanon	62	.53	Sudan	103
.39	Chile	22	.06	Libya	63	.53	Sweden	104
.65	China	23	1.00	Lithuania	64	.98	Switzerland	105
.50	Colombia	24	.99	Macedonia	65	1.00	Syria	106
1.00	Costa Rica	25	.78	Madagascar	66	.98	Tanzania	107
1.00	Croatia	26	1.00	Malawi	67	.99	Thailand	108
.96	Cuba	27	.67	Malaysia	68	1.00	Togo	109
1.00	Czech Republic	28	.31	Mali	69	.95	Trinidad and Tobago	110
1.00	Denmark	29	.99	Mauritius	70	.72	Tunisia	111
1.00	Dominican Republic	30	.70	Mexico	71	1.00	Turkey	112
.60	Ecuador	31	1.00	Moldova	72	1.00	Uganda	113
.18	Egypt	32	.06	Mongolia	73	1.00	Ukraine	114
.99	El Salvador	33	.76	Morocco	74	.94	United Kingdom	115
1.00	Estonia	34	.71	Mozambique	75	.38	United States	116
.90	Ethiopia	35	.93	Nepal	76	1.00	Uruguay	117
.99	Fiji	36	1.00	Netherlands	77	.53	Uzbekistan	118
.54	Finland	37	.22	New Zealand	78	.41	Venezuela	119
.98	France	38	.78	Nicaragua	79	1.00	Vietnam	120
.09	Gabon	39	.21	Niger	80	.54	Zambia	121
1.00	Germany	40	1.00	Nigeria	81	.93	Zimbabwe	122
1.00	Ghana	41	.40	Norway	82			

Annex 3
Country Profiles

The following pages provide detailed information about the environmental performance of the 122 countries in the Environmental Sustainability Index.

At the top of each country's profile, we report the Environmental Sustainability Index score and the average Index score for the country's peer group as defined by GDP per capita (Purchasing Power Parity).[1] Peer groups were assigned by dividing the countries of the index into five equal groups, sorted by GDP per capita, as follows:

Quintile	GDP per Capita	Average ESI Score
1	$14,375 - $29,605	65.2
2	$6,190 - $14,375	52.2
3	$3,330 - $6,190	45.7
4	$1,540 - $3,300	45.2
5	$480 - $1,540	39.3

We use income to assign peer groups not because we wish to reinforce the view that income determines environmental performance. To the contrary, one of our conclusions is that, within similar levels of economic performance, countries exhibit significant variation in their levels of environmental results and sustainability. By comparing a country's Index score with that of others in its peer group, one can get a useful measure of how effective its environmental efforts are.

Below this highly aggregated summary, we present the country's scores on each of the 22 core indicators in bar graph form. The narrow, dark green bars represent the scores of the country (with values appearing in bold type), and the wider, more lightly shaded bars show the average scores for the peer group (with values appearing in normal type). These scores represent the average of the standardized ("z") scores of the variables that comprise the indicators, as explained in more detail in Annex 1. Higher numbers represent better performance.

[1] For four countries that lacked GDP per capita data we assigned peer groups based on Human Development Index quintiles.

Albania

45.1   ESI
76   Ranking
$2,804   GDP per Capita
45.3   Peer Group ESI
46   Variable Coverage (out of 67)
14   Missing Variables Imputed

0.00 — Indicator value
0.00 ▬ Reference (average value for peer group)

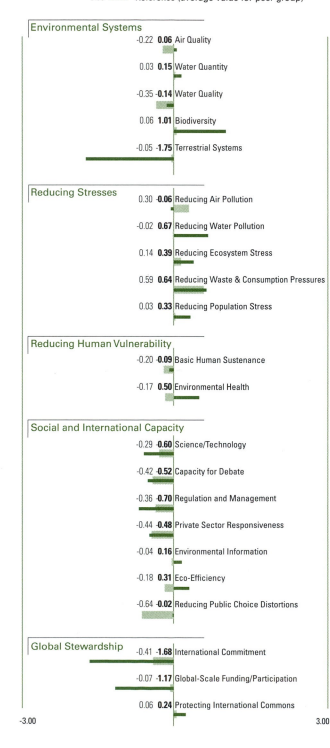

Environmental Systems

-0.22	**0.06**	Air Quality
0.03	**0.15**	Water Quantity
-0.35	**-0.14**	Water Quality
0.06	**1.01**	Biodiversity
-0.05	**-1.75**	Terrestrial Systems

Reducing Stresses

0.30	**-0.06**	Reducing Air Pollution
-0.02	**0.67**	Reducing Water Pollution
0.14	**0.39**	Reducing Ecosystem Stress
0.59	**0.64**	Reducing Waste & Consumption Pressures
0.03	**0.33**	Reducing Population Stress

Reducing Human Vulnerability

-0.20	**-0.09**	Basic Human Sustenance
-0.17	**0.50**	Environmental Health

Social and International Capacity

-0.29	**-0.60**	Science/Technology
-0.42	**-0.52**	Capacity for Debate
-0.36	**-0.70**	Regulation and Management
-0.44	**-0.48**	Private Sector Responsiveness
-0.04	**0.16**	Environmental Information
-0.18	**0.31**	Eco-Efficiency
-0.64	**-0.02**	Reducing Public Choice Distortions

Global Stewardship

-0.41	**-1.68**	International Commitment
-0.07	**-1.17**	Global-Scale Funding/Participation
0.06	**0.24**	Protecting International Commons

-3.00   3.00

Algeria

40.6   ESI
94   Ranking
$4,792   GDP per Capita
45.8   Peer Group ESI
46   Variable Coverage (out of 67)
12   Missing Variables Imputed

0.00 — Indicator value
0.00 ▬ Reference (average value for peer group)

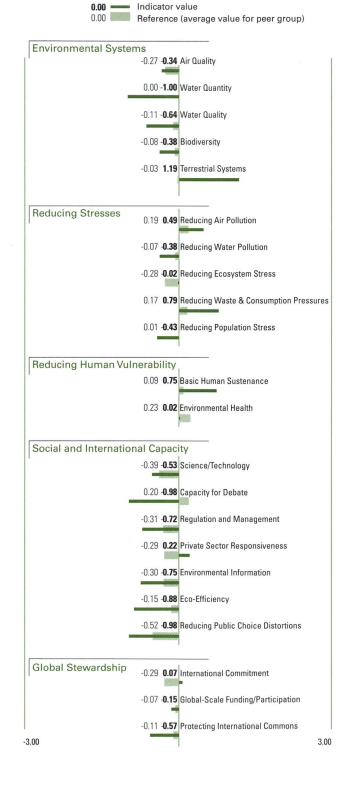

Environmental Systems

-0.27	**-0.34**	Air Quality
0.00	**-1.00**	Water Quantity
-0.11	**-0.64**	Water Quality
-0.08	**-0.38**	Biodiversity
-0.03	**1.19**	Terrestrial Systems

Reducing Stresses

0.19	**0.49**	Reducing Air Pollution
-0.07	**-0.38**	Reducing Water Pollution
-0.28	**-0.02**	Reducing Ecosystem Stress
0.17	**0.79**	Reducing Waste & Consumption Pressures
0.01	**-0.43**	Reducing Population Stress

Reducing Human Vulnerability

0.09	**0.75**	Basic Human Sustenance
0.23	**0.02**	Environmental Health

Social and International Capacity

-0.39	**-0.53**	Science/Technology
0.20	**-0.98**	Capacity for Debate
-0.31	**-0.72**	Regulation and Management
-0.29	**0.22**	Private Sector Responsiveness
-0.30	**-0.75**	Environmental Information
-0.15	**-0.88**	Eco-Efficiency
-0.52	**-0.98**	Reducing Public Choice Distortions

Global Stewardship

-0.29	**0.07**	International Commitment
-0.07	**-0.15**	Global-Scale Funding/Participation
-0.11	**-0.57**	Protecting International Commons

-3.00   3.00

Argentina

62.9   ESI
19   Ranking
$12,013 GDP per Capita
52.2   Peer Group ESI
62   Variable Coverage (out of 67)
2   Missing Variables Imputed

0.00 ── Indicator value
0.00 ▬ Reference (average value for peer group)

Armenia

50.7   ESI
48   Ranking
$2,072 GDP per Capita
45.3   Peer Group ESI
42   Variable Coverage (out of 67)
18   Missing Variables Imputed

0.00 ── Indicator value
0.00 ▬ Reference (average value for peer group)

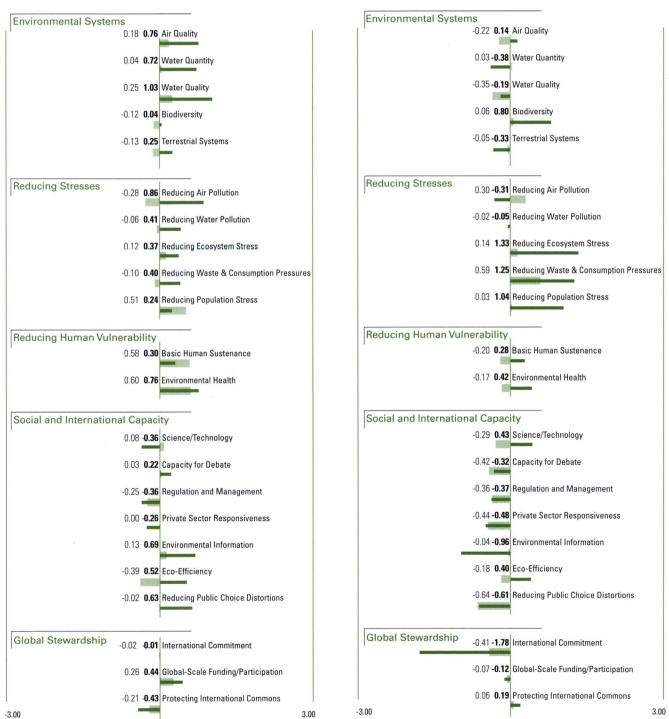

Environmental Systems

Argentina:
- 0.18 **0.76** Air Quality
- 0.04 **0.72** Water Quantity
- 0.25 **1.03** Water Quality
- -0.12 **0.04** Biodiversity
- -0.13 **0.25** Terrestrial Systems

Armenia:
- -0.22 **0.14** Air Quality
- 0.03 **-0.38** Water Quantity
- -0.35 **-0.19** Water Quality
- 0.06 **0.80** Biodiversity
- -0.05 **-0.33** Terrestrial Systems

Reducing Stresses

Argentina:
- -0.28 **0.86** Reducing Air Pollution
- -0.06 **0.41** Reducing Water Pollution
- 0.12 **0.37** Reducing Ecosystem Stress
- -0.10 **0.40** Reducing Waste & Consumption Pressures
- 0.51 **0.24** Reducing Population Stress

Armenia:
- 0.30 **-0.31** Reducing Air Pollution
- -0.02 **-0.05** Reducing Water Pollution
- 0.14 **1.33** Reducing Ecosystem Stress
- 0.59 **1.25** Reducing Waste & Consumption Pressures
- 0.03 **1.04** Reducing Population Stress

Reducing Human Vulnerability

Argentina:
- 0.58 **0.30** Basic Human Sustenance
- 0.60 **0.76** Environmental Health

Armenia:
- -0.20 **0.28** Basic Human Sustenance
- -0.17 **0.42** Environmental Health

Social and International Capacity

Argentina:
- 0.08 **-0.36** Science/Technology
- 0.03 **0.22** Capacity for Debate
- -0.25 **-0.36** Regulation and Management
- 0.00 **-0.26** Private Sector Responsiveness
- 0.13 **0.69** Environmental Information
- -0.39 **0.52** Eco-Efficiency
- -0.02 **0.63** Reducing Public Choice Distortions

Armenia:
- -0.29 **0.43** Science/Technology
- -0.42 **-0.32** Capacity for Debate
- -0.36 **-0.37** Regulation and Management
- -0.44 **-0.48** Private Sector Responsiveness
- -0.04 **-0.96** Environmental Information
- -0.18 **0.40** Eco-Efficiency
- -0.64 **-0.61** Reducing Public Choice Distortions

Global Stewardship

Argentina:
- -0.02 **-0.01** International Commitment
- 0.26 **0.44** Global-Scale Funding/Participation
- -0.21 **-0.43** Protecting International Commons

Armenia:
- -0.41 **-1.78** International Commitment
- -0.07 **-0.12** Global-Scale Funding/Participation
- 0.06 **0.19** Protecting International Commons

-3.00   3.00
-3.00   3.00

Australia

70.9   ESI
7   Ranking
$22,452   GDP per Capita
65.2   Peer Group ESI
60   Variable Coverage (out of 67)
6   Missing Variables Imputed

0.00 ▬▬ Indicator value
0.00 ▬▬ Reference (average value for peer group)

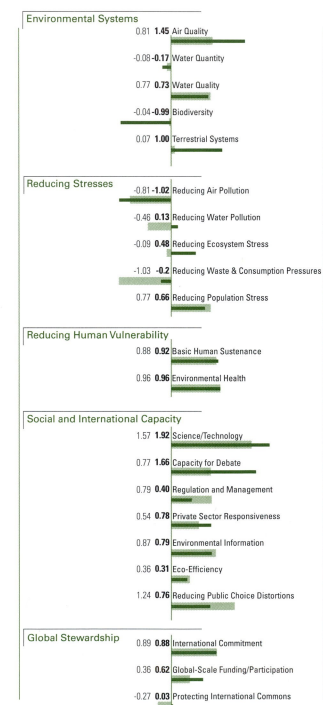

Environmental Systems

0.81	**1.45**	Air Quality
-0.08	**-0.17**	Water Quantity
0.77	**0.73**	Water Quality
-0.04	**-0.99**	Biodiversity
0.07	**1.00**	Terrestrial Systems

Reducing Stresses

-0.81	**-1.02**	Reducing Air Pollution
-0.46	**0.13**	Reducing Water Pollution
-0.09	**0.48**	Reducing Ecosystem Stress
-1.03	**-0.2**	Reducing Waste & Consumption Pressures
0.77	**0.66**	Reducing Population Stress

Reducing Human Vulnerability

0.88	**0.92**	Basic Human Sustenance
0.96	**0.96**	Environmental Health

Social and International Capacity

1.57	**1.92**	Science/Technology
0.77	**1.66**	Capacity for Debate
0.79	**0.40**	Regulation and Management
0.54	**0.78**	Private Sector Responsiveness
0.87	**0.79**	Environmental Information
0.36	**0.31**	Eco-Efficiency
1.24	**0.76**	Reducing Public Choice Distortions

Global Stewardship

0.89	**0.88**	International Commitment
0.36	**0.62**	Global-Scale Funding/Participation
-0.27	**0.03**	Protecting International Commons

-3.00   3.00

Austria

68.2   ESI
8   Ranking
$23,166   GDP per Capita
65.2   Peer Group ESI
60   Variable Coverage (out of 67)
5   Missing Variables Imputed

0.00 ▬▬ Indicator value
0.00 ▬▬ Reference (average value for peer group)

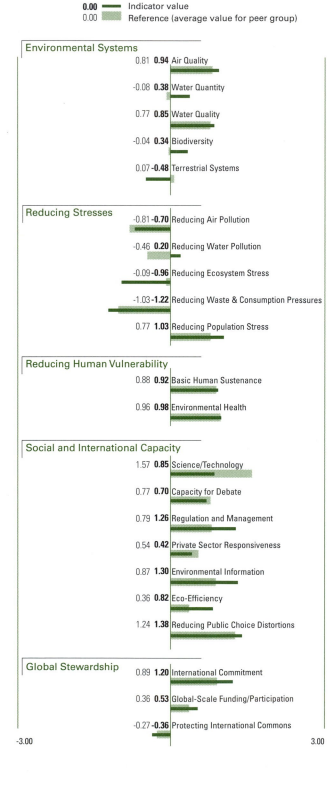

Environmental Systems

0.81	**0.94**	Air Quality
-0.08	**0.38**	Water Quantity
0.77	**0.85**	Water Quality
-0.04	**0.34**	Biodiversity
0.07	**-0.48**	Terrestrial Systems

Reducing Stresses

-0.81	**-0.70**	Reducing Air Pollution
-0.46	**0.20**	Reducing Water Pollution
-0.09	**-0.96**	Reducing Ecosystem Stress
-1.03	**-1.22**	Reducing Waste & Consumption Pressures
0.77	**1.03**	Reducing Population Stress

Reducing Human Vulnerability

0.88	**0.92**	Basic Human Sustenance
0.96	**0.98**	Environmental Health

Social and International Capacity

1.57	**0.85**	Science/Technology
0.77	**0.70**	Capacity for Debate
0.79	**1.26**	Regulation and Management
0.54	**0.42**	Private Sector Responsiveness
0.87	**1.30**	Environmental Information
0.36	**0.82**	Eco-Efficiency
1.24	**1.38**	Reducing Public Choice Distortions

Global Stewardship

0.89	**1.20**	International Commitment
0.36	**0.53**	Global-Scale Funding/Participation
-0.27	**-0.36**	Protecting International Commons

-3.00   3.00

Azerbaijan

46.5	ESI
66	Ranking
$2,175	GDP per Capita
45.3	Peer Group ESI
42	Variable Coverage (out of 67)
17	Missing Variables Imputed

Bangladesh

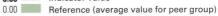

40.4	ESI
95	Ranking
$1,361	GDP per Capita
39.2	Peer Group ESI
49	Variable Coverage (out of 67)
10	Missing Variables Imputed

0.00 ━━ Indicator value
0.00 ▬ Reference (average value for peer group)

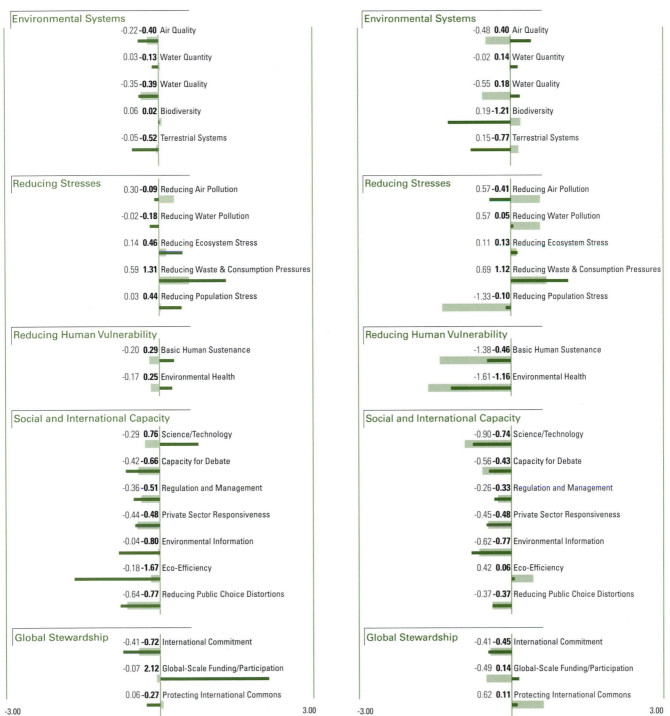

Azerbaijan

Environmental Systems
- -0.22 **-0.40** Air Quality
- 0.03 **-0.13** Water Quantity
- -0.35 **-0.39** Water Quality
- 0.06 **0.02** Biodiversity
- -0.05 **-0.52** Terrestrial Systems

Reducing Stresses
- 0.30 **-0.09** Reducing Air Pollution
- -0.02 **-0.18** Reducing Water Pollution
- 0.14 **0.46** Reducing Ecosystem Stress
- 0.59 **1.31** Reducing Waste & Consumption Pressures
- 0.03 **0.44** Reducing Population Stress

Reducing Human Vulnerability
- -0.20 **0.29** Basic Human Sustenance
- -0.17 **0.25** Environmental Health

Social and International Capacity
- -0.29 **0.76** Science/Technology
- -0.42 **-0.66** Capacity for Debate
- -0.36 **-0.51** Regulation and Management
- -0.44 **-0.48** Private Sector Responsiveness
- -0.04 **-0.80** Environmental Information
- -0.18 **-1.67** Eco-Efficiency
- -0.64 **-0.77** Reducing Public Choice Distortions

Global Stewardship
- -0.41 **-0.72** International Commitment
- -0.07 **2.12** Global-Scale Funding/Participation
- 0.06 **-0.27** Protecting International Commons

-3.00   3.00

Bangladesh

Environmental Systems
- -0.48 **0.40** Air Quality
- -0.02 **0.14** Water Quantity
- -0.55 **0.18** Water Quality
- 0.19 **-1.21** Biodiversity
- 0.15 **-0.77** Terrestrial Systems

Reducing Stresses
- 0.57 **-0.41** Reducing Air Pollution
- 0.57 **0.05** Reducing Water Pollution
- 0.11 **0.13** Reducing Ecosystem Stress
- 0.69 **1.12** Reducing Waste & Consumption Pressures
- -1.33 **-0.10** Reducing Population Stress

Reducing Human Vulnerability
- -1.38 **-0.46** Basic Human Sustenance
- -1.61 **-1.16** Environmental Health

Social and International Capacity
- -0.90 **-0.74** Science/Technology
- -0.56 **-0.43** Capacity for Debate
- -0.26 **-0.33** Regulation and Management
- -0.45 **-0.48** Private Sector Responsiveness
- -0.62 **-0.77** Environmental Information
- 0.42 **0.06** Eco-Efficiency
- -0.37 **-0.37** Reducing Public Choice Distortions

Global Stewardship
- -0.41 **-0.45** International Commitment
- -0.49 **0.14** Global-Scale Funding/Participation
- 0.62 **0.11** Protecting International Commons

-3.00   3.00

Belarus

Country Profiles | Annex 3

142

48.1	ESI
55	Ranking
$6,319	GDP per Capita
52.2	Peer Group ESI
47	Variable Coverage (out of 67)
13	Missing Variables Imputed

0.00 ━━ Indicator value
0.00 � Reference (average value for peer group)

Environmental Systems
0.18 **0.93** Air Quality
0.04 **0.00** Water Quantity
0.25 **-0.22** Water Quality
-0.12 **0.74** Biodiversity
-0.13 **-0.99** Terrestrial Systems

Reducing Stresses
-0.28 **0.19** Reducing Air Pollution
-0.06 **0.07** Reducing Water Pollution
0.12 **0.67** Reducing Ecosystem Stress
-0.10 **0.09** Reducing Waste & Consumption Pressures
0.51 **1.05** Reducing Population Stress

Reducing Human Vulnerability
0.58 **0.92** Basic Human Sustenance
0.60 **0.53** Environmental Health

Social and International Capacity
0.08 **0.85** Science/Technology
0.03 **-0.81** Capacity for Debate
-0.25 **-0.60** Regulation and Management
0.00 **-0.48** Private Sector Responsiveness
0.13 **-0.60** Environmental Information
-0.39 **-1.51** Eco-Efficiency
-0.02 **-0.81** Reducing Public Choice Distortions

Global Stewardship
-0.02 **-0.75** International Commitment
0.26 **-0.03** Global-Scale Funding/Participation
-0.21 **-0.27** Protecting International Commons

-3.00   3.00

Belgium

44.4	ESI
78	Ranking
$23,223	GDP per Capita
65.2	Peer Group ESI
59	Variable Coverage (out of 67)
8	Missing Variables Imputed

0.00 ━━ Indicator value
0.00 ▮ Reference (average value for peer group)

Environmental Systems
0.81 **0.52** Air Quality
-0.08 **-0.36** Water Quantity
0.77 **-2.25** Water Quality
-0.04 **0.29** Biodiversity
0.07 **-1.48** Terrestrial Systems

Reducing Stresses
-0.81 **-2.88** Reducing Air Pollution
-0.46 **-2.20** Reducing Water Pollution
-0.09 **-1.52** Reducing Ecosystem Stress
-1.03 **-0.79** Reducing Waste & Consumption Pressures
0.77 **0.96** Reducing Population Stress

Reducing Human Vulnerability
0.88 **0.97** Basic Human Sustenance
0.96 **0.95** Environmental Health

Social and International Capacity
1.57 **1.18** Science/Technology
0.77 **0.51** Capacity for Debate
0.79 **0.72** Regulation and Management
0.54 **0.02** Private Sector Responsiveness
0.87 **-0.07** Environmental Information
0.36 **-0.15** Eco-Efficiency
1.24 **1.10** Reducing Public Choice Distortions

Global Stewardship
0.89 **0.91** International Commitment
0.36 **0.61** Global-Scale Funding/Participation
-0.27 **-0.17** Protecting International Commons

-3.00   3.00

Benin

39.2	ESI
102	Ranking
$867	GDP per Capita
39.2	Peer Group ESI
43	Variable Coverage (out of 67)
15	Missing Variables Imputed

0.00 ——— Indicator value
0.00 ▬ Reference (average value for peer group)

Environmental Systems
-0.48 **-0.61** Air Quality
-0.02 **0.21** Water Quantity
-0.55 **-0.70** Water Quality
0.19 **1.03** Biodiversity
0.15 **0.69** Terrestrial Systems

Reducing Stresses
0.57 **0.38** Reducing Air Pollution
0.57 **0.41** Reducing Water Pollution
0.11 **-0.03** Reducing Ecosystem Stress
0.69 **0.31** Reducing Waste & Consumption Pressures
-1.33 **-2.03** Reducing Population Stress

Reducing Human Vulnerability
-1.38 **-0.82** Basic Human Sustenance
-1.61 **-1.67** Environmental Health

Social and International Capacity
-0.90 **-0.39** Science/Technology
-0.56 **0.10** Capacity for Debate
-0.26 **-0.41** Regulation and Management
-0.45 **-0.48** Private Sector Responsiveness
-0.62 **-0.18** Environmental Information
0.42 **-1.42** Eco-Efficiency
-0.37 **-0.78** Reducing Public Choice Distortions

Global Stewardship
-0.41 **-0.46** International Commitment
-0.49 **-0.04** Global-Scale Funding/Participation
0.62 **0.87** Protecting International Commons

-3.00   3.00

Bhutan

46.3	ESI
71	Ranking
$1,536	GDP per Capita
39.2	Peer Group ESI
42	Variable Coverage (out of 67)
18	Missing Variables Imputed

0.00 ——— Indicator value
0.00 ▬ Reference (average value for peer group)

Environmental Systems
-0.48 **0.00** Air Quality
-0.02 **0.73** Water Quantity
-0.55 **-0.49** Water Quality
0.19 **-0.40** Biodiversity
0.15 **0.88** Terrestrial Systems

Reducing Stresses
0.57 **1.36** Reducing Air Pollution
0.57 **0.96** Reducing Water Pollution
0.11 **0.34** Reducing Ecosystem Stress
0.69 **0.36** Reducing Waste & Consumption Pressures
-1.33 **-1.37** Reducing Population Stress

Reducing Human Vulnerability
-1.38 **-0.20** Basic Human Sustenance
-1.61 **-1.13** Environmental Health

Social and International Capacity
-0.90 **-0.89** Science/Technology
-0.56 **-0.81** Capacity for Debate
-0.26 **0.55** Regulation and Management
-0.45 **-0.48** Private Sector Responsiveness
-0.62 **-0.91** Environmental Information
0.42 **0.83** Eco-Efficiency
-0.37 **-0.37** Reducing Public Choice Distortions

Global Stewardship
-0.41 **-1.78** International Commitment
-0.49 **0.13** Global-Scale Funding/Participation
0.62 **0.62** Protecting International Commons

-3.00   3.00

Bolivia

144

58.2	ESI
27	Ranking
$2,269	GDP per Capita
45.3	Peer Group ESI
50	Variable Coverage (out of 67)
13	Missing Variables Imputed

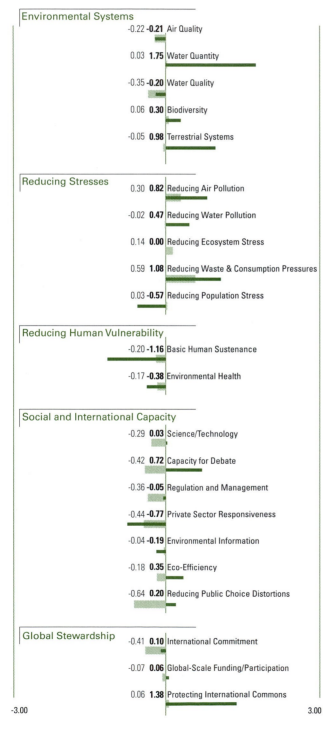

Botswana

53.5	ESI
40	Ranking
$6,103	GDP per Capita
45.8	Peer Group ESI
45	Variable Coverage (out of 67)
13	Missing Variables Imputed

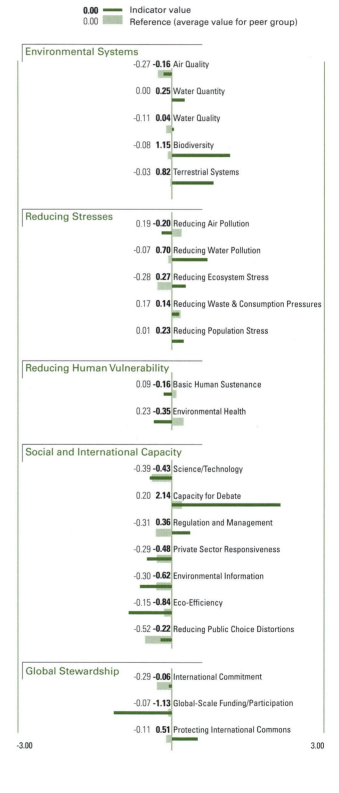

Brazil

57.4	ESI
29	Ranking
$6,625	GDP per Capita
52.2	Peer Group ESI
62	Variable Coverage (out of 67)
3	Missing Variables Imputed

0.00 ▬▬ Indicator value
0.00 ▬ Reference (average value for peer group)

Bulgaria

47.4	ESI
59	Ranking
$4,809	GDP per Capita
45.8	Peer Group ESI
60	Variable Coverage (out of 67)
5	Missing Variables Imputed

0.00 ▬▬ Indicator value
0.00 ▬ Reference (average value for peer group)

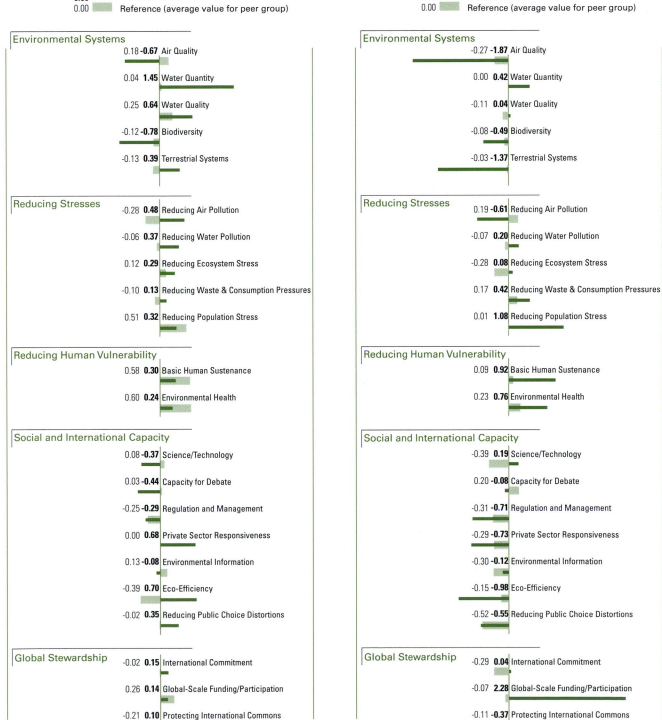

Brazil

Environmental Systems
- 0.18 **-0.67** Air Quality
- 0.04 **1.45** Water Quantity
- 0.25 **0.64** Water Quality
- -0.12 **-0.78** Biodiversity
- -0.13 **0.39** Terrestrial Systems

Reducing Stresses
- -0.28 **0.48** Reducing Air Pollution
- -0.06 **0.37** Reducing Water Pollution
- 0.12 **0.29** Reducing Ecosystem Stress
- -0.10 **0.13** Reducing Waste & Consumption Pressures
- 0.51 **0.32** Reducing Population Stress

Reducing Human Vulnerability
- 0.58 **0.30** Basic Human Sustenance
- 0.60 **0.24** Environmental Health

Social and International Capacity
- 0.08 **-0.37** Science/Technology
- 0.03 **-0.44** Capacity for Debate
- -0.25 **-0.29** Regulation and Management
- 0.00 **0.68** Private Sector Responsiveness
- 0.13 **-0.08** Environmental Information
- -0.39 **0.70** Eco-Efficiency
- -0.02 **0.35** Reducing Public Choice Distortions

Global Stewardship
- -0.02 **0.15** International Commitment
- 0.26 **0.14** Global-Scale Funding/Participation
- -0.21 **0.10** Protecting International Commons

-3.00   3.00

Bulgaria

Environmental Systems
- -0.27 **-1.87** Air Quality
- 0.00 **0.42** Water Quantity
- -0.11 **0.04** Water Quality
- -0.08 **-0.49** Biodiversity
- -0.03 **-1.37** Terrestrial Systems

Reducing Stresses
- 0.19 **-0.61** Reducing Air Pollution
- -0.07 **0.20** Reducing Water Pollution
- -0.28 **0.08** Reducing Ecosystem Stress
- 0.17 **0.42** Reducing Waste & Consumption Pressures
- 0.01 **1.08** Reducing Population Stress

Reducing Human Vulnerability
- 0.09 **0.92** Basic Human Sustenance
- 0.23 **0.76** Environmental Health

Social and International Capacity
- -0.39 **0.19** Science/Technology
- 0.20 **-0.08** Capacity for Debate
- -0.31 **-0.71** Regulation and Management
- -0.29 **-0.73** Private Sector Responsiveness
- -0.30 **-0.12** Environmental Information
- -0.15 **-0.98** Eco-Efficiency
- -0.52 **-0.55** Reducing Public Choice Distortions

Global Stewardship
- -0.29 **0.04** International Commitment
- -0.07 **2.28** Global-Scale Funding/Participation
- -0.11 **-0.37** Protecting International Commons

-3.00   3.00

Burkina Faso

38.3   ESI
105   Ranking
$870   GDP per Capita
39.2   Peer Group ESI
45   Variable Coverage (out of 67)
13   Missing Variables Imputed

Burundi

29.8   ESI
121   Ranking
$570   GDP per Capita
39.2   Peer Group ESI
42   Variable Coverage (out of 67)
16   Missing Variables Imputed

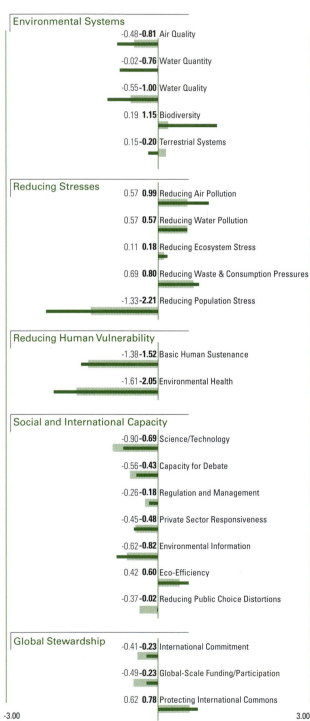

0.00 ━━━ Indicator value
0.00 ▬ Reference (average value for peer group)

Environmental Systems

-0.48 **-0.81** Air Quality
-0.02 **-0.76** Water Quantity
-0.55 **-1.00** Water Quality
0.19 **1.15** Biodiversity
0.15 **-0.20** Terrestrial Systems

Reducing Stresses

0.57 **0.99** Reducing Air Pollution
0.57 **0.57** Reducing Water Pollution
0.11 **0.18** Reducing Ecosystem Stress
0.69 **0.80** Reducing Waste & Consumption Pressures
-1.33 **-2.21** Reducing Population Stress

Reducing Human Vulnerability

-1.38 **-1.52** Basic Human Sustenance
-1.61 **-2.05** Environmental Health

Social and International Capacity

-0.90 **-0.69** Science/Technology
-0.56 **-0.43** Capacity for Debate
-0.26 **-0.18** Regulation and Management
-0.45 **-0.48** Private Sector Responsiveness
-0.62 **-0.82** Environmental Information
0.42 **0.60** Eco-Efficiency
-0.37 **-0.02** Reducing Public Choice Distortions

Global Stewardship

-0.41 **-0.23** International Commitment
-0.49 **-0.23** Global-Scale Funding/Participation
0.62 **0.78** Protecting International Commons

-3.00   3.00

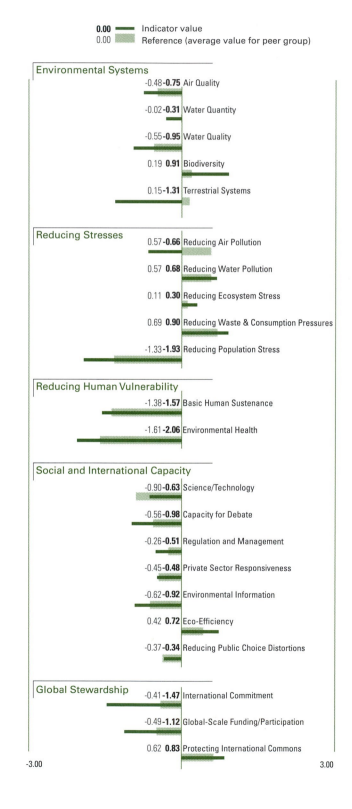

0.00 ━━━ Indicator value
0.00 ▬ Reference (average value for peer group)

Environmental Systems

-0.48 **-0.75** Air Quality
-0.02 **-0.31** Water Quantity
-0.55 **-0.95** Water Quality
0.19 **0.91** Biodiversity
0.15 **-1.31** Terrestrial Systems

Reducing Stresses

0.57 **-0.66** Reducing Air Pollution
0.57 **0.68** Reducing Water Pollution
0.11 **0.30** Reducing Ecosystem Stress
0.69 **0.90** Reducing Waste & Consumption Pressures
-1.33 **-1.93** Reducing Population Stress

Reducing Human Vulnerability

-1.38 **-1.57** Basic Human Sustenance
-1.61 **-2.06** Environmental Health

Social and International Capacity

-0.90 **-0.63** Science/Technology
-0.56 **-0.98** Capacity for Debate
-0.26 **-0.51** Regulation and Management
-0.45 **-0.48** Private Sector Responsiveness
-0.62 **-0.92** Environmental Information
0.42 **0.72** Eco-Efficiency
-0.37 **-0.34** Reducing Public Choice Distortions

Global Stewardship

-0.41 **-1.47** International Commitment
-0.49 **-1.12** Global-Scale Funding/Participation
0.62 **0.83** Protecting International Commons

-3.00   3.00

Cameroon

44.7	ESI
77	Ranking
$1,474	GDP per Capita
39.2	Peer Group ESI
45	Variable Coverage (out of 67)
13	Missing Variables Imputed

0.00 ▬▬ Indicator value
0.00 ▮ Reference (average value for peer group)

Environmental Systems
- -0.48 **-0.50** Air Quality
- -0.02 **0.68** Water Quantity
- -0.55 **-0.20** Water Quality
- 0.19 **0.44** Biodiversity
- 0.15 **0.41** Terrestrial Systems

Reducing Stresses
- 0.57 **0.77** Reducing Air Pollution
- 0.57 **0.90** Reducing Water Pollution
- 0.11 **0.21** Reducing Ecosystem Stress
- 0.69 **0.43** Reducing Waste & Consumption Pressures
- -1.33 **-1.19** Reducing Population Stress

Reducing Human Vulnerability
- -1.38 **-1.21** Basic Human Sustenance
- -1.61 **-1.18** Environmental Health

Social and International Capacity
- -0.90 **-0.73** Science/Technology
- -0.56 **-1.28** Capacity for Debate
- -0.26 **-0.58** Regulation and Management
- -0.45 **-0.48** Private Sector Responsiveness
- -0.62 **-0.71** Environmental Information
- 0.42 **0.88** Eco-Efficiency
- -0.37 **-0.49** Reducing Public Choice Distortions

Global Stewardship
- -0.41 **0.39** International Commitment
- -0.49 **-0.10** Global-Scale Funding/Participation
- 0.62 **0.58** Protecting International Commons

-3.00   3.00

Canada

78.1	ESI
3	Ranking
$23,582	GDP per Capita
65.2	Peer Group ESI
66	Variable Coverage (out of 67)
0	Missing Variables Imputed

0.00 ▬▬ Indicator value
0.00 ▮ Reference (average value for peer group)

Environmental Systems
- 0.81 **0.98** Air Quality
- -0.08 **1.70** Water Quantity
- 0.77 **1.54** Water Quality
- -0.04 **1.12** Biodiversity
- 0.07 **1.45** Terrestrial Systems

Reducing Stresses
- -0.81 **-0.51** Reducing Air Pollution
- -0.46 **0.52** Reducing Water Pollution
- -0.09 **0.34** Reducing Ecosystem Stress
- -1.03 **-0.92** Reducing Waste & Consumption Pressures
- 0.77 **0.72** Reducing Population Stress

Reducing Human Vulnerability
- 0.88 **0.92** Basic Human Sustenance
- 0.96 **0.96** Environmental Health

Social and International Capacity
- 1.57 **1.80** Science/Technology
- 0.77 **0.90** Capacity for Debate
- 0.79 **1.21** Regulation and Management
- 0.54 **0.71** Private Sector Responsiveness
- 0.87 **0.68** Environmental Information
- 0.36 **0.55** Eco-Efficiency
- 1.24 **0.68** Reducing Public Choice Distortions

Global Stewardship
- 0.89 **0.91** International Commitment
- 0.36 **0.73** Global-Scale Funding/Participation
- -0.27 **0.11** Protecting International Commons

-3.00   3.00

Central African Republic   Chile

47.7	ESI
57	Ranking
$1,118	GDP per Capita
39.2	Peer Group ESI
42	Variable Coverage (out of 67)
16	Missing Variables Imputed

56.6	ESI
31	Ranking
$8,787	GDP per Capita
52.2	Peer Group ESI
61	Variable Coverage (out of 67)
4	Missing Variables Imputed

0.00 ── Indicator value
0.00 ▬ Reference (average value for peer group)

0.00 ── Indicator value
0.00 ▬ Reference (average value for peer group)

Country Profiles | **Annex 3**

148

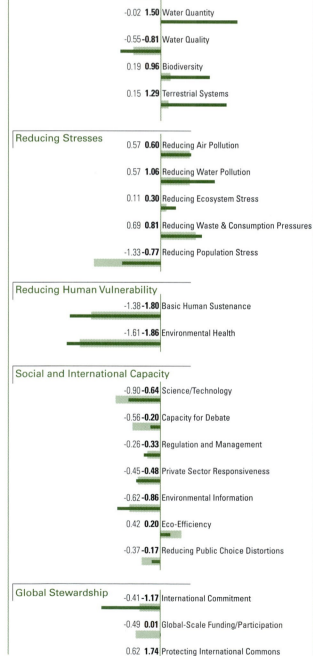

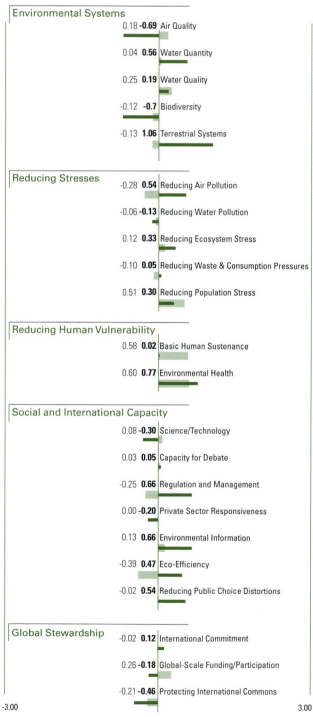

Central African Republic

Environmental Systems
- -0.48 **-0.64** Air Quality
- -0.02 **1.50** Water Quantity
- -0.55 **-0.81** Water Quality
- 0.19 **0.96** Biodiversity
- 0.15 **1.29** Terrestrial Systems

Reducing Stresses
- 0.57 **0.60** Reducing Air Pollution
- 0.57 **1.06** Reducing Water Pollution
- 0.11 **0.30** Reducing Ecosystem Stress
- 0.69 **0.81** Reducing Waste & Consumption Pressures
- -1.33 **-0.77** Reducing Population Stress

Reducing Human Vulnerability
- -1.38 **-1.80** Basic Human Sustenance
- -1.61 **-1.86** Environmental Health

Social and International Capacity
- -0.90 **-0.64** Science/Technology
- -0.56 **-0.20** Capacity for Debate
- -0.26 **-0.33** Regulation and Management
- -0.45 **-0.48** Private Sector Responsiveness
- -0.62 **-0.86** Environmental Information
- 0.42 **0.20** Eco-Efficiency
- -0.37 **-0.17** Reducing Public Choice Distortions

Global Stewardship
- -0.41 **-1.17** International Commitment
- -0.49 **0.01** Global-Scale Funding/Participation
- 0.62 **1.74** Protecting International Commons

-3.00   3.00

Chile

Environmental Systems
- 0.18 **-0.69** Air Quality
- 0.04 **0.56** Water Quantity
- 0.25 **0.19** Water Quality
- -0.12 **-0.7** Biodiversity
- -0.13 **1.06** Terrestrial Systems

Reducing Stresses
- -0.28 **0.54** Reducing Air Pollution
- -0.06 **-0.13** Reducing Water Pollution
- 0.12 **0.33** Reducing Ecosystem Stress
- -0.10 **0.05** Reducing Waste & Consumption Pressures
- 0.51 **0.30** Reducing Population Stress

Reducing Human Vulnerability
- 0.58 **0.02** Basic Human Sustenance
- 0.60 **0.77** Environmental Health

Social and International Capacity
- 0.08 **-0.30** Science/Technology
- 0.03 **0.05** Capacity for Debate
- -0.25 **0.66** Regulation and Management
- 0.00 **-0.20** Private Sector Responsiveness
- 0.13 **0.66** Environmental Information
- -0.39 **0.47** Eco-Efficiency
- -0.02 **0.54** Reducing Public Choice Distortions

Global Stewardship
- -0.02 **0.12** International Commitment
- 0.26 **-0.18** Global-Scale Funding/Participation
- -0.21 **-0.46** Protecting International Commons

-3.00   3.00

China

37.5	ESI
109	Ranking
$3,105	GDP per Capita
45.3	Peer Group ESI
60	Variable Coverage (out of 67)
4	Missing Variables Imputed

Colombia

54.8	ESI
36	Ranking
$6,006	GDP per Capita
45.8	Peer Group ESI
59	Variable Coverage (out of 67)
4	Missing Variables Imputed

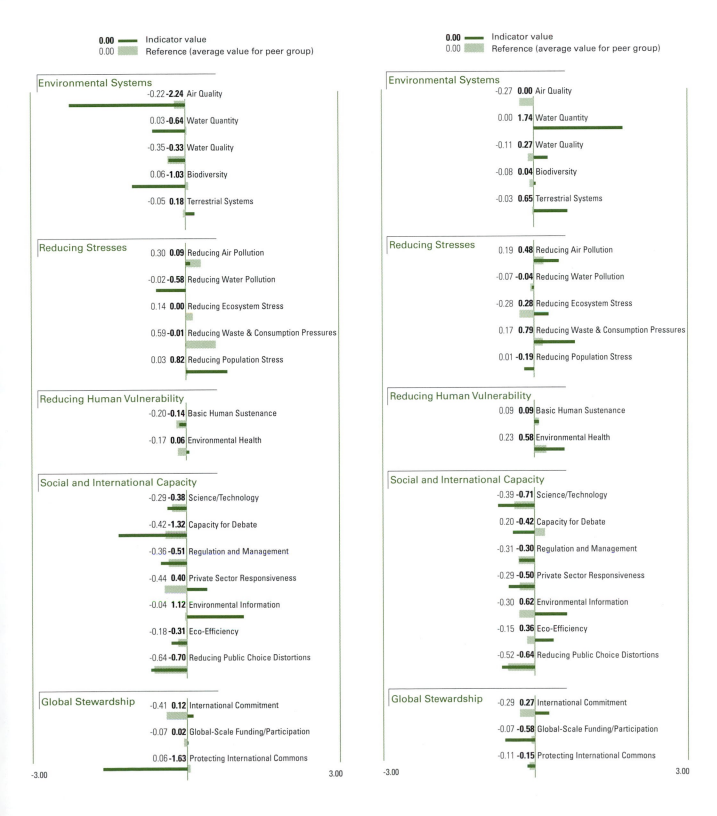

Costa Rica

58.8	ESI
26	Ranking
$5,987	GDP per Capita
45.8	Peer Group ESI
58	Variable Coverage (out of 67)
5	Missing Variables Imputed

0.00 ━━ Indicator value
0.00 ▬ Reference (average value for peer group)

Environmental Systems
-0.27 **-0.47** Air Quality
0.00 **0.79** Water Quantity
-0.11 **0.23** Water Quality
-0.08 **0.52** Biodiversity
-0.03 **-0.92** Terrestrial Systems

Reducing Stresses
0.19 **0.52** Reducing Air Pollution
-0.07 **-1.00** Reducing Water Pollution
-0.28 **-0.92** Reducing Ecosystem Stress
0.17 **-0.24** Reducing Waste & Consumption Pressures
0.01 **-0.36** Reducing Population Stress

Reducing Human Vulnerability
0.09 **0.87** Basic Human Sustenance
0.23 **0.62** Environmental Health

Social and International Capacity
-0.39 **-0.41** Science/Technology
0.20 **1.82** Capacity for Debate
-0.31 **0.53** Regulation and Management
-0.29 **0.82** Private Sector Responsiveness
-0.30 **0.12** Environmental Information
-0.15 **0.69** Eco-Efficiency
-0.52 **-0.13** Reducing Public Choice Distortions

Global Stewardship
-0.29 **0.14** International Commitment
-0.07 **1.05** Global-Scale Funding/Participation
-0.11 **0.62** Protecting International Commons

-3.00   3.00

Croatia

54.1	ESI
39	Ranking
$6,749	GDP per Capita
52.2	Peer Group ESI
48	Variable Coverage (out of 67)
11	Missing Variables Imputed

0.00 ━━ Indicator value
0.00 ▬ Reference (average value for peer group)

Environmental Systems
0.18 **0.28** Air Quality
0.04 **0.72** Water Quantity
0.25 **0.09** Water Quality
-0.12 **0.10** Biodiversity
-0.13 **-0.30** Terrestrial Systems

Reducing Stresses
-0.28 **-0.76** Reducing Air Pollution
-0.06 **0.20** Reducing Water Pollution
0.12 **0.34** Reducing Ecosystem Stress
-0.10 **0.37** Reducing Waste & Consumption Pressures
0.51 **1.00** Reducing Population Stress

Reducing Human Vulnerability
0.58 **0.65** Basic Human Sustenance
0.60 **0.93** Environmental Health

Social and International Capacity
0.08 **0.41** Science/Technology
0.03 **-0.20** Capacity for Debate
-0.25 **-0.43** Regulation and Management
0.00 **0.38** Private Sector Responsiveness
0.13 **-0.29** Environmental Information
-0.39 **0.10** Eco-Efficiency
-0.02 **-0.09** Reducing Public Choice Distortions

Global Stewardship
-0.02 **-0.51** International Commitment
0.26 **-1.17** Global-Scale Funding/Participation
-0.21 **0.47** Protecting International Commons

-3.00   3.00

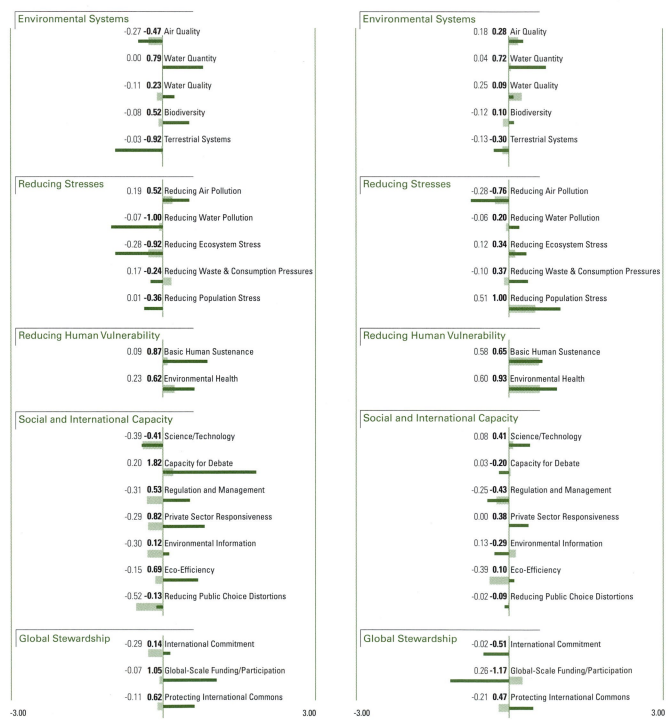

Cuba

54.9	ESI
35	Ranking
N/A*	GDP per Capita
52.2	Peer Group ESI
52	Variable Coverage (out of 67)
6	Missing Variables Imputed

0.00 ━━ Indicator value
0.00 ▭ Reference (average value for peer group)

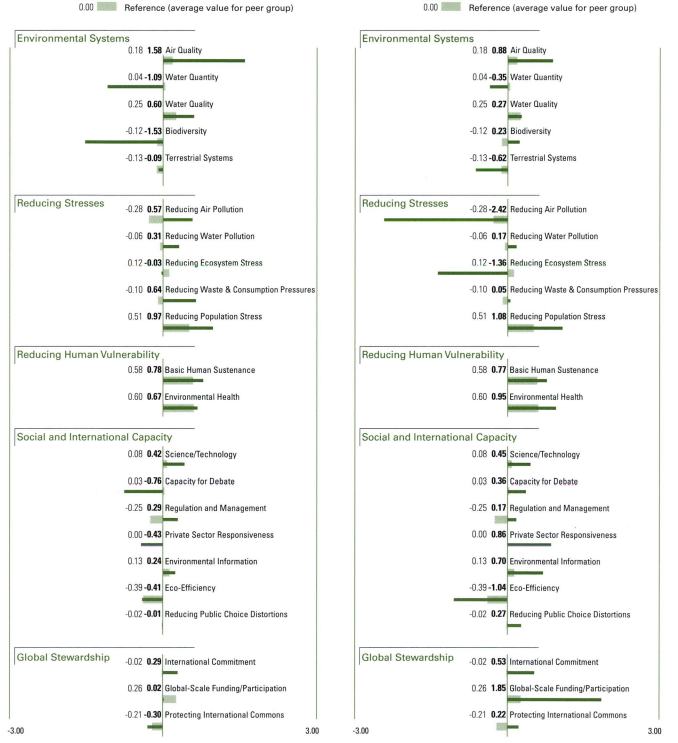

Environmental Systems

0.18	**1.58**	Air Quality
0.04	**-1.09**	Water Quantity
0.25	**0.60**	Water Quality
-0.12	**-1.53**	Biodiversity
-0.13	**-0.09**	Terrestrial Systems

Reducing Stresses

-0.28	**0.57**	Reducing Air Pollution
-0.06	**0.31**	Reducing Water Pollution
0.12	**-0.03**	Reducing Ecosystem Stress
-0.10	**0.64**	Reducing Waste & Consumption Pressures
0.51	**0.97**	Reducing Population Stress

Reducing Human Vulnerability

| 0.58 | **0.78** | Basic Human Sustenance |
| 0.60 | **0.67** | Environmental Health |

Social and International Capacity

0.08	**0.42**	Science/Technology
0.03	**-0.76**	Capacity for Debate
-0.25	**0.29**	Regulation and Management
0.00	**-0.43**	Private Sector Responsiveness
0.13	**0.24**	Environmental Information
-0.39	**-0.41**	Eco-Efficiency
-0.02	**-0.01**	Reducing Public Choice Distortions

Global Stewardship

-0.02	**0.29**	International Commitment
0.26	**0.02**	Global-Scale Funding/Participation
-0.21	**-0.30**	Protecting International Commons

-3.00   3.00

Czech Republic

57.2	ESI
30	Ranking
$12,362	GDP per Capita
52.2	Peer Group ESI
56	Variable Coverage (out of 67)
9	Missing Variables Imputed

0.00 ━━ Indicator value
0.00 ▭ Reference (average value for peer group)

Environmental Systems

0.18	**0.88**	Air Quality
0.04	**-0.35**	Water Quantity
0.25	**0.27**	Water Quality
-0.12	**0.23**	Biodiversity
-0.13	**-0.62**	Terrestrial Systems

Reducing Stresses

-0.28	**-2.42**	Reducing Air Pollution
-0.06	**0.17**	Reducing Water Pollution
0.12	**-1.36**	Reducing Ecosystem Stress
-0.10	**0.05**	Reducing Waste & Consumption Pressures
0.51	**1.08**	Reducing Population Stress

Reducing Human Vulnerability

| 0.58 | **0.77** | Basic Human Sustenance |
| 0.60 | **0.95** | Environmental Health |

Social and International Capacity

0.08	**0.45**	Science/Technology
0.03	**0.36**	Capacity for Debate
-0.25	**0.17**	Regulation and Management
0.00	**0.86**	Private Sector Responsiveness
0.13	**0.70**	Environmental Information
-0.39	**-1.04**	Eco-Efficiency
-0.02	**0.27**	Reducing Public Choice Distortions

Global Stewardship

-0.02	**0.53**	International Commitment
0.26	**1.85**	Global-Scale Funding/Participation
-0.21	**0.22**	Protecting International Commons

-3.00   3.00

* Not available. Peer group assigned based on Human Development Index.

Denmark

67.0   ESI
10   Ranking
$24,218   GDP per Capita
65.2   Peer Group ESI
60   Variable Coverage (out of 67)
7   Missing Variables Imputed

0.00 ▬▬ Indicator value
0.00 ▬▬ Reference (average value for peer group)

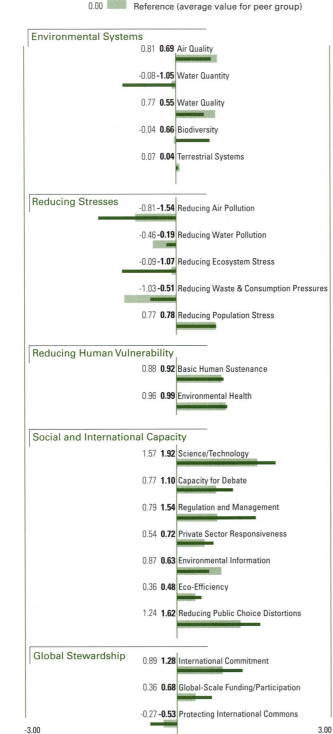

Dominican Republic

45.3   ESI
74   Ranking
$4,598   GDP per Capita
45.8   Peer Group ESI
43   Variable Coverage (out of 67)
15   Missing Variables Imputed

0.00 ▬▬ Indicator value
0.00 ▬▬ Reference (average value for peer group)

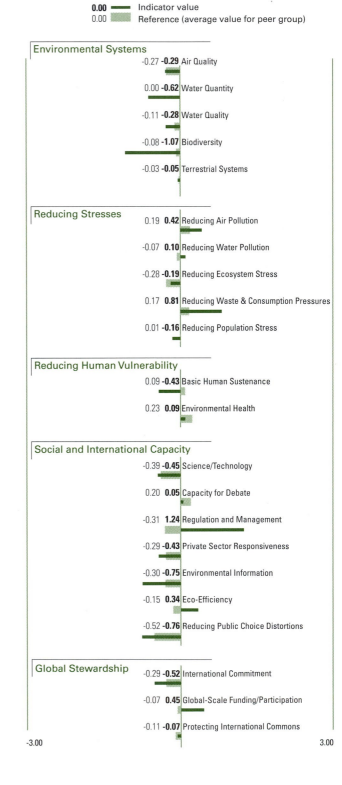

Ecuador

51.8	ESI
44	Ranking
$3,003	GDP per Capita
45.3	Peer Group ESI
56	Variable Coverage (out of 67)
7	Missing Variables Imputed

0.00 ━━━ Indicator value
0.00 ▭ Reference (average value for peer group)

Environmental Systems
-0.22	**0.20**	Air Quality
0.03	**0.81**	Water Quantity
-0.35	**-0.06**	Water Quality
0.06	**0.07**	Biodiversity
-0.05	**0.59**	Terrestrial Systems

Reducing Stresses
0.30	**0.27**	Reducing Air Pollution
-0.02	**0.37**	Reducing Water Pollution
0.14	**-0.20**	Reducing Ecosystem Stress
0.59	**0.23**	Reducing Waste & Consumption Pressures
0.03	**-0.14**	Reducing Population Stress

Reducing Human Vulnerability
-0.20	**-0.54**	Basic Human Sustenance
-0.17	**0.17**	Environmental Health

Social and International Capacity
-0.29	**-0.70**	Science/Technology
-0.42	**0.75**	Capacity for Debate
-0.36	**-0.06**	Regulation and Management
-0.44	**-0.42**	Private Sector Responsiveness
-0.04	**0.81**	Environmental Information
-0.18	**0.22**	Eco-Efficiency
-0.64	**-1.35**	Reducing Public Choice Distortions

Global Stewardship
-0.41	**-0.06**	International Commitment
-0.07	**0.01**	Global-Scale Funding/Participation
0.06	**0.01**	Protecting International Commons

-3.00   3.00

Egypt

46.4	ESI
69	Ranking
$3,041	GDP per Capita
45.3	Peer Group ESI
57	Variable Coverage (out of 67)
7	Missing Variables Imputed

0.00 ━━━ Indicator value
0.00 ▭ Reference (average value for peer group)

Environmental Systems
-0.22	**-0.79**	Air Quality
0.03	**-0.34**	Water Quantity
-0.35	**-0.15**	Water Quality
0.06	**-0.75**	Biodiversity
-0.05	**1.48**	Terrestrial Systems

Reducing Stresses
0.30	**-0.35**	Reducing Air Pollution
-0.02	**-0.64**	Reducing Water Pollution
0.14	**0.46**	Reducing Ecosystem Stress
0.59	**0.49**	Reducing Waste & Consumption Pressures
0.03	**-0.17**	Reducing Population Stress

Reducing Human Vulnerability
-0.20	**0.78**	Basic Human Sustenance
-0.17	**-0.77**	Environmental Health

Social and International Capacity
-0.29	**-0.53**	Science/Technology
-0.42	**-1.02**	Capacity for Debate
-0.36	**0.11**	Regulation and Management
-0.44	**-0.03**	Private Sector Responsiveness
-0.04	**0.77**	Environmental Information
-0.18	**-0.09**	Eco-Efficiency
-0.64	**-0.68**	Reducing Public Choice Distortions

Global Stewardship
-0.41	**0.10**	International Commitment
-0.07	**0.62**	Global-Scale Funding/Participation
0.06	**-0.49**	Protecting International Commons

-3.00   3.00

El Salvador

43.7	ESI
82	Ranking
$4,036	GDP per Capita
45.8	Peer Group ESI
55	Variable Coverage (out of 67)
8	Missing Variables Imputed

0.00 ━━ Indicator value
0.00 ▬ Reference (average value for peer group)

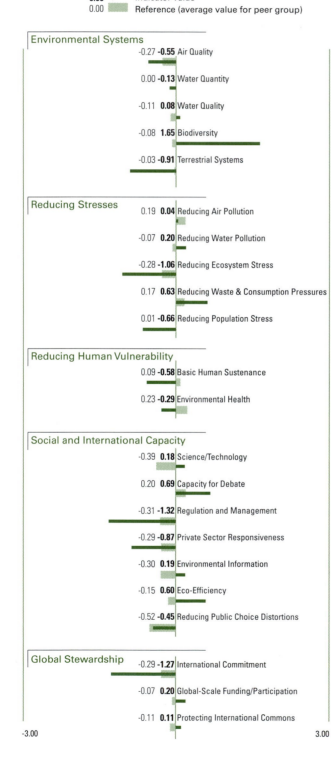

Environmental Systems

-0.27 **-0.55** Air Quality
0.00 **-0.13** Water Quantity
-0.11 **0.08** Water Quality
-0.08 **1.65** Biodiversity
-0.03 **-0.91** Terrestrial Systems

Reducing Stresses

0.19 **0.04** Reducing Air Pollution
-0.07 **0.20** Reducing Water Pollution
-0.28 **-1.06** Reducing Ecosystem Stress
0.17 **0.63** Reducing Waste & Consumption Pressures
0.01 **-0.66** Reducing Population Stress

Reducing Human Vulnerability

0.09 **-0.58** Basic Human Sustenance
0.23 **-0.29** Environmental Health

Social and International Capacity

-0.39 **0.18** Science/Technology
0.20 **0.69** Capacity for Debate
-0.31 **-1.32** Regulation and Management
-0.29 **-0.87** Private Sector Responsiveness
-0.30 **0.19** Environmental Information
-0.15 **0.60** Eco-Efficiency
-0.52 **-0.45** Reducing Public Choice Distortions

Global Stewardship

-0.29 **-1.27** International Commitment
-0.07 **0.20** Global-Scale Funding/Participation
-0.11 **0.11** Protecting International Commons

-3.00   3.00

Estonia

57.7	ESI
28	Ranking
$7,682	GDP per Capita
52.2	Peer Group ESI
48	Variable Coverage (out of 67)
12	Missing Variables Imputed

0.00 ━━ Indicator value
0.00 ▬ Reference (average value for peer group)

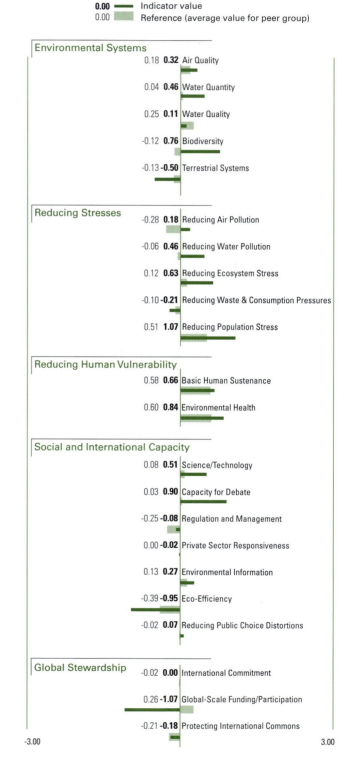

Environmental Systems

0.18 **0.32** Air Quality
0.04 **0.46** Water Quantity
0.25 **0.11** Water Quality
-0.12 **0.76** Biodiversity
-0.13 **-0.50** Terrestrial Systems

Reducing Stresses

-0.28 **0.18** Reducing Air Pollution
-0.06 **0.46** Reducing Water Pollution
0.12 **0.63** Reducing Ecosystem Stress
-0.10 **-0.21** Reducing Waste & Consumption Pressures
0.51 **1.07** Reducing Population Stress

Reducing Human Vulnerability

0.58 **0.66** Basic Human Sustenance
0.60 **0.84** Environmental Health

Social and International Capacity

0.08 **0.51** Science/Technology
0.03 **0.90** Capacity for Debate
-0.25 **-0.08** Regulation and Management
0.00 **-0.02** Private Sector Responsiveness
0.13 **0.27** Environmental Information
-0.39 **-0.95** Eco-Efficiency
-0.02 **0.07** Reducing Public Choice Distortions

Global Stewardship

-0.02 **0.00** International Commitment
0.26 **-1.07** Global-Scale Funding/Participation
-0.21 **-0.18** Protecting International Commons

-3.00   3.00

Ethiopia

31.0	ESI
119	Ranking
$574	GDP per Capita
39.2	Peer Group ESI
44	Variable Coverage (out of 67)
14	Missing Variables Imputed

Fiji

48.0	ESI
56	Ranking
$4,231	GDP per Capita
45.8	Peer Group ESI
43	Variable Coverage (out of 67)
15	Missing Variables Imputed

0.00 ── Indicator value
0.00 ▧ Reference (average value for peer group)

0.00 ── Indicator value
0.00 ▧ Reference (average value for peer group)

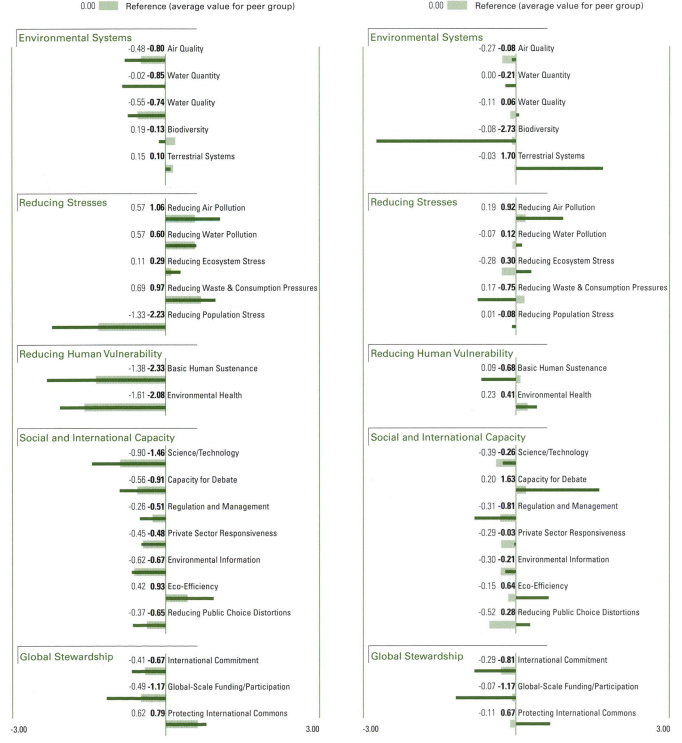

Ethiopia

Environmental Systems
-0.48 **-0.80** Air Quality
-0.02 **-0.85** Water Quantity
-0.55 **-0.74** Water Quality
0.19 **-0.13** Biodiversity
0.15 **0.10** Terrestrial Systems

Reducing Stresses
0.57 **1.06** Reducing Air Pollution
0.57 **0.60** Reducing Water Pollution
0.11 **0.29** Reducing Ecosystem Stress
0.69 **0.97** Reducing Waste & Consumption Pressures
-1.33 **-2.23** Reducing Population Stress

Reducing Human Vulnerability
-1.38 **-2.33** Basic Human Sustenance
-1.61 **-2.08** Environmental Health

Social and International Capacity
-0.90 **-1.46** Science/Technology
-0.56 **-0.91** Capacity for Debate
-0.26 **-0.51** Regulation and Management
-0.45 **-0.48** Private Sector Responsiveness
-0.62 **-0.67** Environmental Information
0.42 **0.93** Eco-Efficiency
-0.37 **-0.65** Reducing Public Choice Distortions

Global Stewardship
-0.41 **-0.67** International Commitment
-0.49 **-1.17** Global-Scale Funding/Participation
0.62 **0.79** Protecting International Commons

-3.00   3.00

Fiji

Environmental Systems
-0.27 **-0.08** Air Quality
0.00 **-0.21** Water Quantity
-0.11 **0.06** Water Quality
-0.08 **-2.73** Biodiversity
-0.03 **1.70** Terrestrial Systems

Reducing Stresses
0.19 **0.92** Reducing Air Pollution
-0.07 **0.12** Reducing Water Pollution
-0.28 **0.30** Reducing Ecosystem Stress
0.17 **-0.75** Reducing Waste & Consumption Pressures
0.01 **-0.08** Reducing Population Stress

Reducing Human Vulnerability
0.09 **-0.68** Basic Human Sustenance
0.23 **0.41** Environmental Health

Social and International Capacity
-0.39 **-0.26** Science/Technology
0.20 **1.63** Capacity for Debate
-0.31 **-0.81** Regulation and Management
-0.29 **-0.03** Private Sector Responsiveness
-0.30 **-0.21** Environmental Information
-0.15 **0.64** Eco-Efficiency
-0.52 **0.28** Reducing Public Choice Distortions

Global Stewardship
-0.29 **-0.81** International Commitment
-0.07 **-1.17** Global-Scale Funding/Participation
-0.11 **0.67** Protecting International Commons

-3.00   3.00

Finland

80.5 ESI
1 Ranking
$20,847 GDP per Capita
65.2 Peer Group ESI
66 Variable Coverage (out of 67)
1 Missing Variables Imputed

0.00 Indicator value
0.00 Reference (average value for peer group)

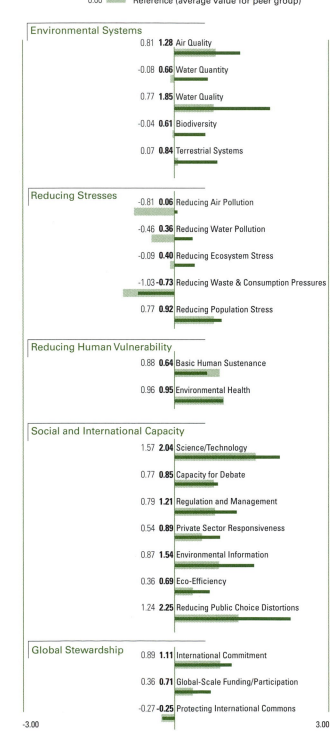

Environmental Systems
0.81 **1.28** Air Quality
-0.08 **0.66** Water Quantity
0.77 **1.85** Water Quality
-0.04 **0.61** Biodiversity
0.07 **0.84** Terrestrial Systems

Reducing Stresses
-0.81 **0.06** Reducing Air Pollution
-0.46 **0.36** Reducing Water Pollution
-0.09 **0.40** Reducing Ecosystem Stress
-1.03 **-0.73** Reducing Waste & Consumption Pressures
0.77 **0.92** Reducing Population Stress

Reducing Human Vulnerability
0.88 **0.64** Basic Human Sustenance
0.96 **0.95** Environmental Health

Social and International Capacity
1.57 **2.04** Science/Technology
0.77 **0.85** Capacity for Debate
0.79 **1.21** Regulation and Management
0.54 **0.89** Private Sector Responsiveness
0.87 **1.54** Environmental Information
0.36 **0.69** Eco-Efficiency
1.24 **2.25** Reducing Public Choice Distortions

Global Stewardship
0.89 **1.11** International Commitment
0.36 **0.71** Global-Scale Funding/Participation
-0.27 **-0.25** Protecting International Commons

-3.00   3.00

France

65.8 ESI
13 Ranking
$21,175 GDP per Capita
65.2 Peer Group ESI
63 Variable Coverage (out of 67)
4 Missing Variables Imputed

0.00 Indicator value
0.00 Reference (average value for peer group)

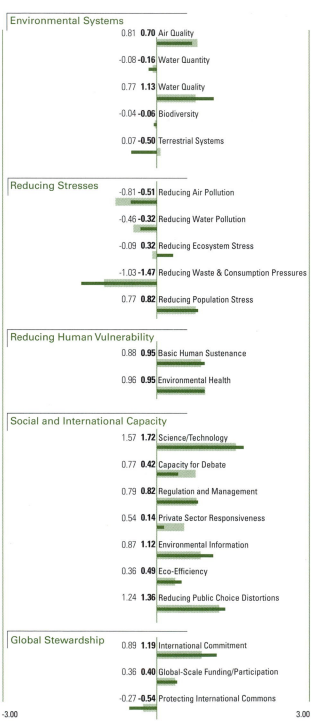

Environmental Systems
0.81 **0.70** Air Quality
-0.08 **-0.16** Water Quantity
0.77 **1.13** Water Quality
-0.04 **-0.06** Biodiversity
0.07 **-0.50** Terrestrial Systems

Reducing Stresses
-0.81 **-0.51** Reducing Air Pollution
-0.46 **-0.32** Reducing Water Pollution
-0.09 **0.32** Reducing Ecosystem Stress
-1.03 **-1.47** Reducing Waste & Consumption Pressures
0.77 **0.82** Reducing Population Stress

Reducing Human Vulnerability
0.88 **0.95** Basic Human Sustenance
0.96 **0.95** Environmental Health

Social and International Capacity
1.57 **1.72** Science/Technology
0.77 **0.42** Capacity for Debate
0.79 **0.82** Regulation and Management
0.54 **0.14** Private Sector Responsiveness
0.87 **1.12** Environmental Information
0.36 **0.49** Eco-Efficiency
1.24 **1.36** Reducing Public Choice Distortions

Global Stewardship
0.89 **1.19** International Commitment
0.36 **0.40** Global-Scale Funding/Participation
-0.27 **-0.54** Protecting International Commons

-3.00   3.00

Gabon

50.2 ESI
49 Ranking
$6,353 GDP per Capita
52.2 Peer Group ESI
43 Variable Coverage (out of 67)
15 Missing Variables Imputed

Germany

64.2 ESI
15 Ranking
$22,169 GDP per Capita
65.2 Peer Group ESI
60 Variable Coverage (out of 67)
7 Missing Variables Imputed

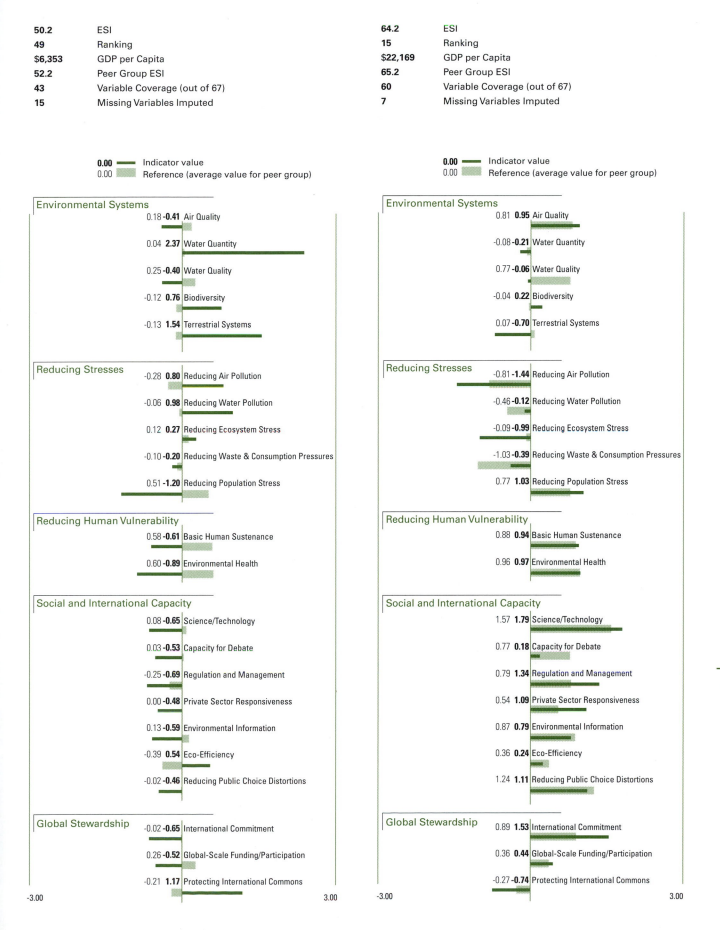

Ghana

46.9 ESI
64 Ranking
$1,735 GDP per Capita
45.3 Peer Group ESI
48 Variable Coverage (out of 67)
10 Missing Variables Imputed

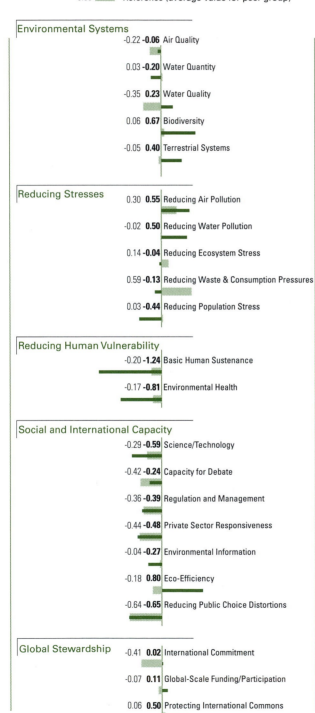

0.00 ── Indicator value
0.00 ▬ Reference (average value for peer group)

Environmental Systems
-0.22 **-0.06** Air Quality
0.03 **-0.20** Water Quantity
-0.35 **0.23** Water Quality
0.06 **0.67** Biodiversity
-0.05 **0.40** Terrestrial Systems

Reducing Stresses
0.30 **0.55** Reducing Air Pollution
-0.02 **0.50** Reducing Water Pollution
0.14 **-0.04** Reducing Ecosystem Stress
0.59 **-0.13** Reducing Waste & Consumption Pressures
0.03 **-0.44** Reducing Population Stress

Reducing Human Vulnerability
-0.20 **-1.24** Basic Human Sustenance
-0.17 **-0.81** Environmental Health

Social and International Capacity
-0.29 **-0.59** Science/Technology
-0.42 **-0.24** Capacity for Debate
-0.36 **-0.39** Regulation and Management
-0.44 **-0.48** Private Sector Responsiveness
-0.04 **-0.27** Environmental Information
-0.18 **0.80** Eco-Efficiency
-0.64 **-0.65** Reducing Public Choice Distortions

Global Stewardship
-0.41 **0.02** International Commitment
-0.07 **0.11** Global-Scale Funding/Participation
0.06 **0.50** Protecting International Commons

-3.00   3.00

Greece

53.1 ESI
41 Ranking
$13,943 GDP per Capita
52.2 Peer Group ESI
58 Variable Coverage (out of 67)
7 Missing Variables Imputed

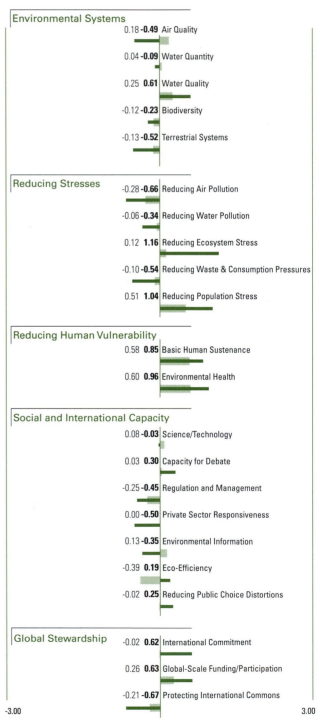

0.00 ── Indicator value
0.00 ▬ Reference (average value for peer group)

Environmental Systems
0.18 **-0.49** Air Quality
0.04 **-0.09** Water Quantity
0.25 **0.61** Water Quality
-0.12 **-0.23** Biodiversity
-0.13 **-0.52** Terrestrial Systems

Reducing Stresses
-0.28 **-0.66** Reducing Air Pollution
-0.06 **-0.34** Reducing Water Pollution
0.12 **1.16** Reducing Ecosystem Stress
-0.10 **-0.54** Reducing Waste & Consumption Pressures
0.51 **1.04** Reducing Population Stress

Reducing Human Vulnerability
0.58 **0.85** Basic Human Sustenance
0.60 **0.96** Environmental Health

Social and International Capacity
0.08 **-0.03** Science/Technology
0.03 **0.30** Capacity for Debate
-0.25 **-0.45** Regulation and Management
0.00 **-0.50** Private Sector Responsiveness
0.13 **-0.35** Environmental Information
-0.39 **0.19** Eco-Efficiency
-0.02 **0.25** Reducing Public Choice Distortions

Global Stewardship
-0.02 **0.62** International Commitment
0.26 **0.63** Global-Scale Funding/Participation
-0.21 **-0.67** Protecting International Commons

-3.00   3.00

Guatemala

47.2	ESI
61	Ranking
$3,505	GDP per Capita
45.8	Peer Group ESI
48	Variable Coverage (out of 67)
11	Missing Variables Imputed

Haiti

24.5	ESI
122	Ranking
$1,383	GDP per Capita
39.2	Peer Group ESI
42	Variable Coverage (out of 67)
16	Missing Variables Imputed

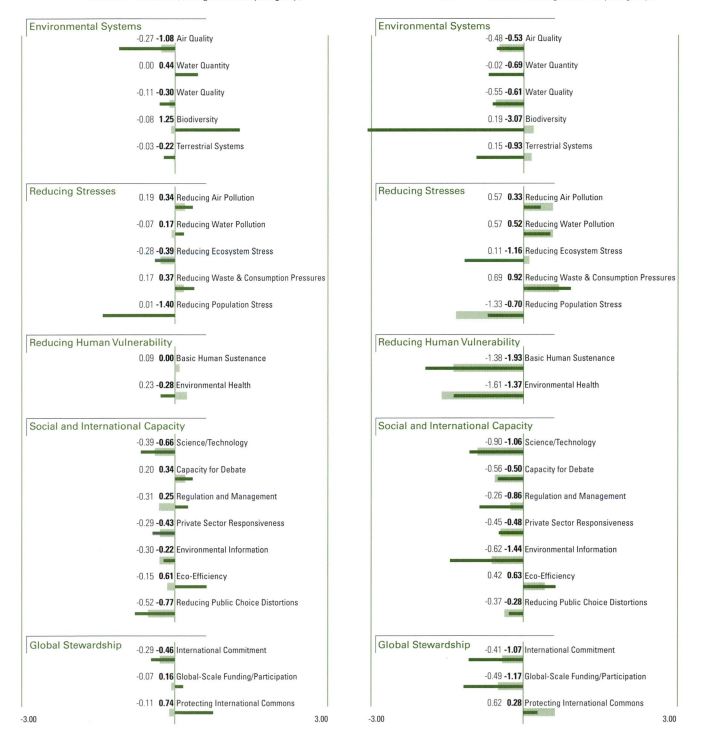

0.00 ▬ Indicator value
0.00 ▬ Reference (average value for peer group)

Environmental Systems

-0.27 **-1.08** Air Quality
0.00 **0.44** Water Quantity
-0.11 **-0.30** Water Quality
-0.08 **1.25** Biodiversity
-0.03 **-0.22** Terrestrial Systems

Reducing Stresses

0.19 **0.34** Reducing Air Pollution
-0.07 **0.17** Reducing Water Pollution
-0.28 **-0.39** Reducing Ecosystem Stress
0.17 **0.37** Reducing Waste & Consumption Pressures
0.01 **-1.40** Reducing Population Stress

Reducing Human Vulnerability

0.09 **0.00** Basic Human Sustenance
0.23 -0.28 Environmental Health

Social and International Capacity

-0.39 **-0.66** Science/Technology
0.20 **0.34** Capacity for Debate
-0.31 **0.25** Regulation and Management
-0.29 **-0.43** Private Sector Responsiveness
-0.30 **-0.22** Environmental Information
-0.15 **0.61** Eco-Efficiency
-0.52 **-0.77** Reducing Public Choice Distortions

Global Stewardship

-0.29 **-0.46** International Commitment
-0.07 **0.16** Global-Scale Funding/Participation
-0.11 **0.74** Protecting International Commons

-3.00   3.00

0.00 ▬ Indicator value
0.00 ▬ Reference (average value for peer group)

Environmental Systems

-0.48 **-0.53** Air Quality
-0.02 **-0.69** Water Quantity
-0.55 **-0.61** Water Quality
0.19 **-3.07** Biodiversity
0.15 **-0.93** Terrestrial Systems

Reducing Stresses

0.57 **0.33** Reducing Air Pollution
0.57 **0.52** Reducing Water Pollution
0.11 **-1.16** Reducing Ecosystem Stress
0.69 **0.92** Reducing Waste & Consumption Pressures
-1.33 **-0.70** Reducing Population Stress

Reducing Human Vulnerability

-1.38 **-1.93** Basic Human Sustenance
-1.61 **-1.37** Environmental Health

Social and International Capacity

-0.90 **-1.06** Science/Technology
-0.56 **-0.50** Capacity for Debate
-0.26 **-0.86** Regulation and Management
-0.45 **-0.48** Private Sector Responsiveness
-0.62 **-1.44** Environmental Information
0.42 **0.63** Eco-Efficiency
-0.37 **-0.28** Reducing Public Choice Distortions

Global Stewardship

-0.41 **-1.07** International Commitment
-0.49 **-1.17** Global-Scale Funding/Participation
0.62 **0.28** Protecting International Commons

-3.00   3.00

Honduras

46.9   ESI
65   Ranking
$2,433   GDP per Capita
45.3   Peer Group ESI
46   Variable Coverage (out of 67)
12   Missing Variables Imputed

0.00 Indicator value
0.00 Reference (average value for peer group)

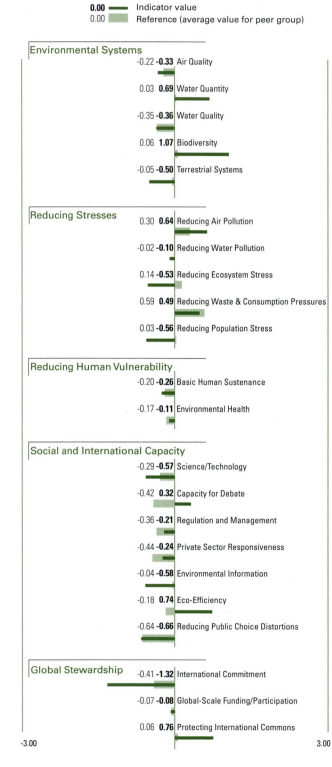

Hungary

61.0   ESI
21   Ranking
$10,232   GDP per Capita
52.2   Peer Group ESI
65   Variable Coverage (out of 67)
0   Missing Variables Imputed

0.00 Indicator value
0.00 Reference (average value for peer group)

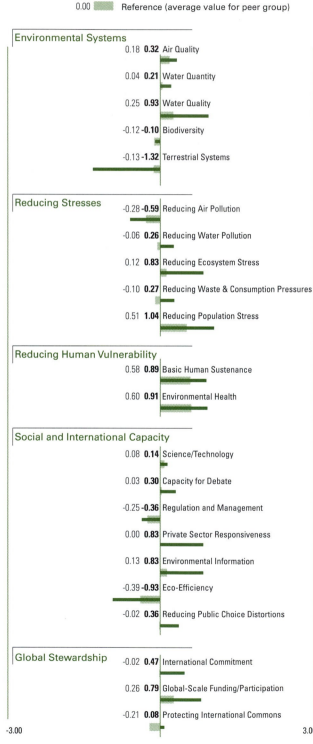

Iceland

67.3   ESI
9   Ranking
$25,110   GDP per Capita
65.2   Peer Group ESI
52   Variable Coverage (out of 67)
12   Missing Variables Imputed

0.00 ▬▬ Indicator value
0.00 ▬▬ Reference (average value for peer group)

Environmental Systems

0.81 **1.13** Air Quality
-0.08 **0.86** Water Quantity
0.77 **0.74** Water Quality
-0.04 **0.58** Biodiversity
0.07 **0.75** Terrestrial Systems

Reducing Stresses

-0.81 **-0.60** Reducing Air Pollution
-0.46 **-1.30** Reducing Water Pollution
-0.09 **0.46** Reducing Ecosystem Stress
-1.03 **-2.14** Reducing Waste & Consumption Pressures
0.77 **0.65** Reducing Population Stress

Reducing Human Vulnerability

0.88 **0.97** Basic Human Sustenance
0.96 **0.93** Environmental Health

Social and International Capacity

1.57 **1.72** Science/Technology
0.77 **2.41** Capacity for Debate
0.79 **0.19** Regulation and Management
0.54 **-0.03** Private Sector Responsiveness
0.87 **0.30** Environmental Information
0.36 **0.80** Eco-Efficiency
1.24 **1.61** Reducing Public Choice Distortions

Global Stewardship

0.89 **-0.06** International Commitment
0.36 **-0.59** Global-Scale Funding/Participation
-0.27 **0.53** Protecting International Commons

-3.00   3.00

India

40.7   ESI
93   Ranking
$2,077   GDP per Capita
45.3   Peer Group ESI
60   Variable Coverage (out of 67)
5   Missing Variables Imputed

0.00 ▬▬ Indicator value
0.00 ▬▬ Reference (average value for peer group)

Environmental Systems

-0.22 **-0.06** Air Quality
0.03 **-0.42** Water Quantity
-0.35 **-1.31** Water Quality
0.06 **-1.16** Biodiversity
-0.05 **-0.58** Terrestrial Systems

Reducing Stresses

0.30 **0.07** Reducing Air Pollution
-0.02 **-0.17** Reducing Water Pollution
0.14 **0.47** Reducing Ecosystem Stress
0.59 **0.62** Reducing Waste & Consumption Pressures
0.03 **-0.10** Reducing Population Stress

Reducing Human Vulnerability

-0.20 **-0.20** Basic Human Sustenance
-0.17 **-0.78** Environmental Health

Social and International Capacity

-0.29 **-0.42** Science/Technology
-0.42 **-0.23** Capacity for Debate
-0.36 **0.01** Regulation and Management
-0.44 **-0.44** Private Sector Responsiveness
-0.04 **0.83** Environmental Information
-0.18 **-0.03** Eco-Efficiency
-0.64 **-0.82** Reducing Public Choice Distortions

Global Stewardship

-0.41 **0.30** International Commitment
-0.07 **0.06** Global-Scale Funding/Participation
0.06 **-0.79** Protecting International Commons

-3.00   3.00

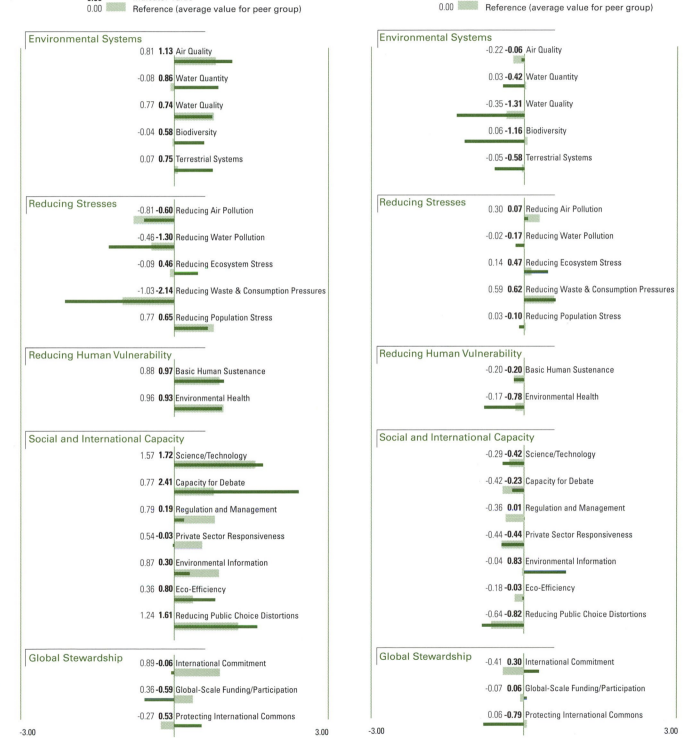
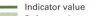

Indonesia

42.5	ESI
86	Ranking
$2,651	GDP per Capita
45.3	Peer Group ESI
60	Variable Coverage (out of 67)
6	Missing Variables Imputed

Iran

38.4	ESI
104	Ranking
$5,121	GDP per Capita
45.8	Peer Group ESI
51	Variable Coverage (out of 67)
8	Missing Variables Imputed

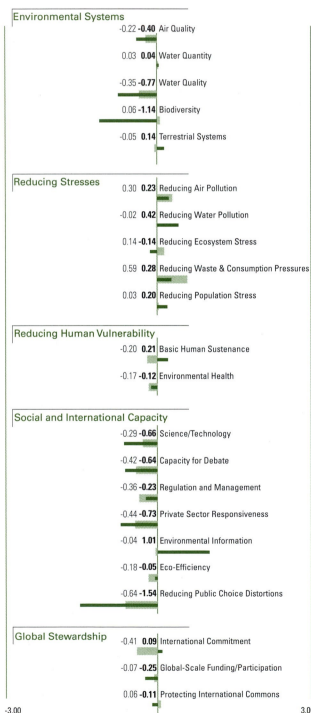

0.00 ——— Indicator value
0.00 ▬ Reference (average value for peer group)

Environmental Systems
-0.22 **-0.40** Air Quality
0.03 **0.04** Water Quantity
-0.35 **-0.77** Water Quality
0.06 **-1.14** Biodiversity
-0.05 **0.14** Terrestrial Systems

Reducing Stresses
0.30 **0.23** Reducing Air Pollution
-0.02 **0.42** Reducing Water Pollution
0.14 **-0.14** Reducing Ecosystem Stress
0.59 **0.28** Reducing Waste & Consumption Pressures
0.03 **0.20** Reducing Population Stress

Reducing Human Vulnerability
-0.20 **0.21** Basic Human Sustenance
-0.17 **-0.12** Environmental Health

Social and International Capacity
-0.29 **-0.66** Science/Technology
-0.42 **-0.64** Capacity for Debate
-0.36 **-0.23** Regulation and Management
-0.44 **-0.73** Private Sector Responsiveness
-0.04 **1.01** Environmental Information
-0.18 **-0.05** Eco-Efficiency
-0.64 **-1.54** Reducing Public Choice Distortions

Global Stewardship
-0.41 **0.09** International Commitment
-0.07 **-0.25** Global-Scale Funding/Participation
0.06 **-0.11** Protecting International Commons

-3.00   3.00

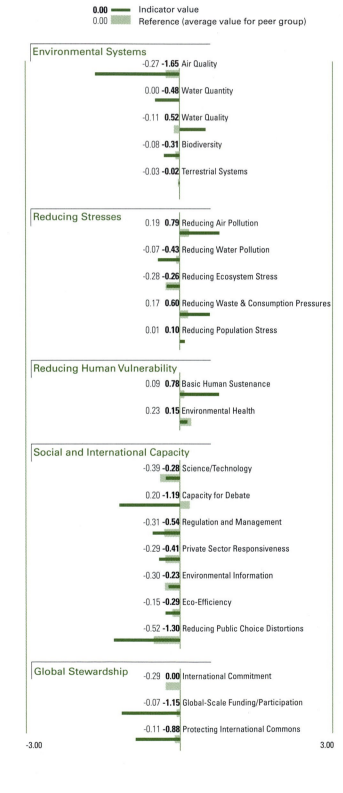

0.00 ——— Indicator value
0.00 ▬ Reference (average value for peer group)

Environmental Systems
-0.27 **-1.65** Air Quality
0.00 **-0.48** Water Quantity
-0.11 **0.52** Water Quality
-0.08 **-0.31** Biodiversity
-0.03 **-0.02** Terrestrial Systems

Reducing Stresses
0.19 **0.79** Reducing Air Pollution
-0.07 **-0.43** Reducing Water Pollution
-0.28 **-0.26** Reducing Ecosystem Stress
0.17 **0.60** Reducing Waste & Consumption Pressures
0.01 **0.10** Reducing Population Stress

Reducing Human Vulnerability
0.09 **0.78** Basic Human Sustenance
0.23 **0.15** Environmental Health

Social and International Capacity
-0.39 **-0.28** Science/Technology
0.20 **-1.19** Capacity for Debate
-0.31 **-0.54** Regulation and Management
-0.29 **-0.41** Private Sector Responsiveness
-0.30 **-0.23** Environmental Information
-0.15 **-0.29** Eco-Efficiency
-0.52 **-1.30** Reducing Public Choice Distortions

Global Stewardship
-0.29 **0.00** International Commitment
-0.07 **-1.15** Global-Scale Funding/Participation
-0.11 **-0.88** Protecting International Commons

-3.00   3.00

Ireland

64.0	ESI
17	Ranking
$21,482	GDP per Capita
65.2	Peer Group ESI
57	Variable Coverage (out of 67)
9	Missing Variables Imputed

Israel

49.6	ESI
53	Ranking
$17,301	GDP per Capita
65.2	Peer Group ESI
54	Variable Coverage (out of 67)
9	Missing Variables Imputed

0.00 —— Indicator value
0.00 ▬ Reference (average value for peer group)

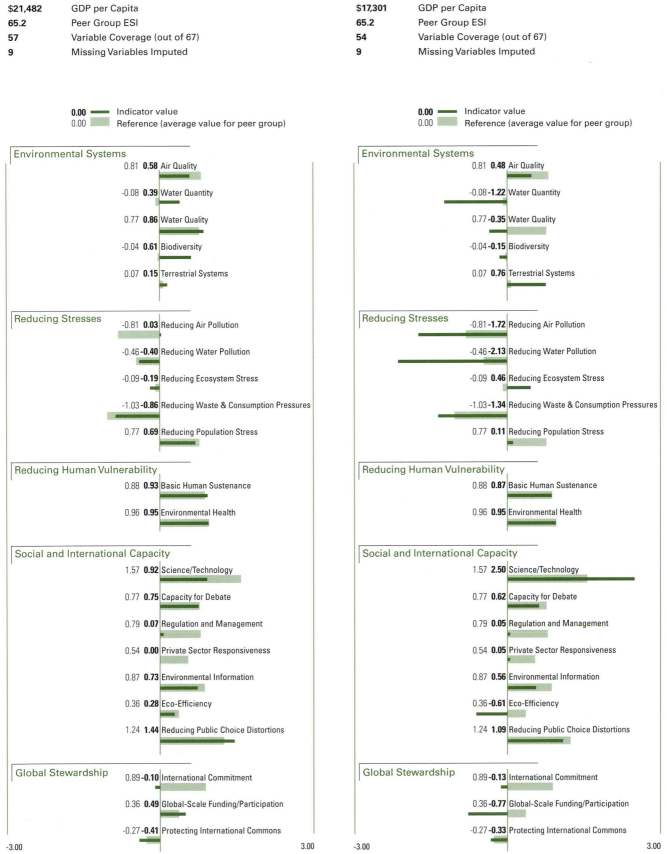

Ireland

Environmental Systems
- 0.81 **0.58** Air Quality
- -0.08 **0.39** Water Quantity
- 0.77 **0.86** Water Quality
- -0.04 **0.61** Biodiversity
- 0.07 **0.15** Terrestrial Systems

Reducing Stresses
- -0.81 **0.03** Reducing Air Pollution
- -0.46 **-0.40** Reducing Water Pollution
- -0.09 **-0.19** Reducing Ecosystem Stress
- -1.03 **-0.86** Reducing Waste & Consumption Pressures
- 0.77 **0.69** Reducing Population Stress

Reducing Human Vulnerability
- 0.88 **0.93** Basic Human Sustenance
- 0.96 **0.95** Environmental Health

Social and International Capacity
- 1.57 **0.92** Science/Technology
- 0.77 **0.75** Capacity for Debate
- 0.79 **0.07** Regulation and Management
- 0.54 **0.00** Private Sector Responsiveness
- 0.87 **0.73** Environmental Information
- 0.36 **0.28** Eco-Efficiency
- 1.24 **1.44** Reducing Public Choice Distortions

Global Stewardship
- 0.89 **-0.10** International Commitment
- 0.36 **0.49** Global-Scale Funding/Participation
- -0.27 **-0.41** Protecting International Commons

-3.00 ... 3.00

Israel

Environmental Systems
- 0.81 **0.48** Air Quality
- -0.08 **-1.22** Water Quantity
- 0.77 **-0.35** Water Quality
- -0.04 **-0.15** Biodiversity
- 0.07 **0.76** Terrestrial Systems

Reducing Stresses
- -0.81 **-1.72** Reducing Air Pollution
- -0.46 **-2.13** Reducing Water Pollution
- -0.09 **0.46** Reducing Ecosystem Stress
- -1.03 **-1.34** Reducing Waste & Consumption Pressures
- 0.77 **0.11** Reducing Population Stress

Reducing Human Vulnerability
- 0.88 **0.87** Basic Human Sustenance
- 0.96 **0.95** Environmental Health

Social and International Capacity
- 1.57 **2.50** Science/Technology
- 0.77 **0.62** Capacity for Debate
- 0.79 **0.05** Regulation and Management
- 0.54 **0.05** Private Sector Responsiveness
- 0.87 **0.56** Environmental Information
- 0.36 **-0.61** Eco-Efficiency
- 1.24 **1.09** Reducing Public Choice Distortions

Global Stewardship
- 0.89 **-0.13** International Commitment
- 0.36 **-0.77** Global-Scale Funding/Participation
- -0.27 **-0.33** Protecting International Commons

-3.00 ... 3.00

Italy

54.3   ESI
37   Ranking
$20,585 GDP per Capita
65.2   Peer Group ESI
59   Variable Coverage (out of 67)
8   Missing Variables Imputed

0.00 ━━ Indicator value
0.00 ▬ Reference (average value for peer group)

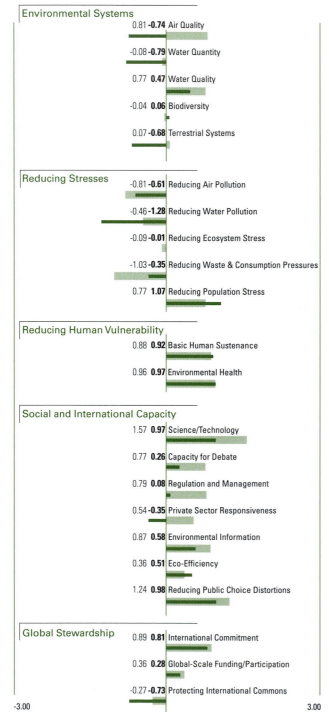

Environmental Systems

0.81 **-0.74** Air Quality

-0.08 **-0.79** Water Quantity

0.77 **0.47** Water Quality

-0.04 **0.06** Biodiversity

0.07 **-0.68** Terrestrial Systems

Reducing Stresses

-0.81 **-0.61** Reducing Air Pollution

-0.46 **-1.28** Reducing Water Pollution

-0.09 **-0.01** Reducing Ecosystem Stress

-1.03 **-0.35** Reducing Waste & Consumption Pressures

0.77 **1.07** Reducing Population Stress

Reducing Human Vulnerability

0.88 **0.92** Basic Human Sustenance

0.96 **0.97** Environmental Health

Social and International Capacity

1.57 **0.97** Science/Technology

0.77 **0.26** Capacity for Debate

0.79 **0.08** Regulation and Management

0.54 **-0.35** Private Sector Responsiveness

0.87 **0.58** Environmental Information

0.36 **0.51** Eco-Efficiency

1.24 **0.98** Reducing Public Choice Distortions

Global Stewardship

0.89 **0.81** International Commitment

0.36 **0.28** Global-Scale Funding/Participation

-0.27 **-0.73** Protecting International Commons

-3.00   3.00

Jamaica

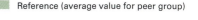

42.3   ESI
87   Ranking
$3,389   GDP per Capita
45.8   Peer Group ESI
46   Variable Coverage (out of 67)
12   Missing Variables Imputed

0.00 ━━ Indicator value
0.00 ▬ Reference (average value for peer group)

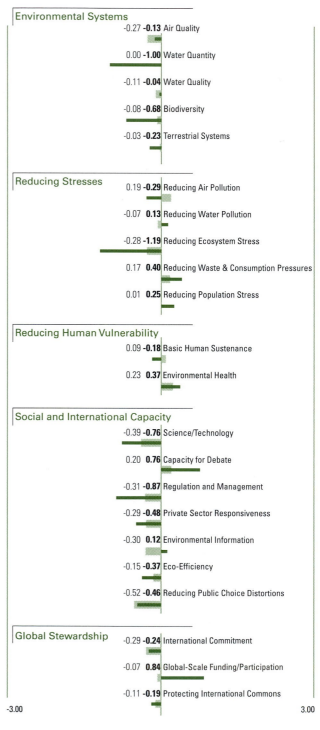

Environmental Systems

-0.27 **-0.13** Air Quality

0.00 **-1.00** Water Quantity

-0.11 **-0.04** Water Quality

-0.08 **-0.68** Biodiversity

-0.03 **-0.23** Terrestrial Systems

Reducing Stresses

0.19 **-0.29** Reducing Air Pollution

-0.07 **0.13** Reducing Water Pollution

-0.28 **-1.19** Reducing Ecosystem Stress

0.17 **0.40** Reducing Waste & Consumption Pressures

0.01 **0.25** Reducing Population Stress

Reducing Human Vulnerability

0.09 **-0.18** Basic Human Sustenance

0.23 **0.37** Environmental Health

Social and International Capacity

-0.39 **-0.76** Science/Technology

0.20 **0.76** Capacity for Debate

-0.31 **-0.87** Regulation and Management

-0.29 **-0.48** Private Sector Responsiveness

-0.30 **0.12** Environmental Information

-0.15 **-0.37** Eco-Efficiency

-0.52 **-0.46** Reducing Public Choice Distortions

Global Stewardship

-0.29 **-0.24** International Commitment

-0.07 **0.84** Global-Scale Funding/Participation

-0.11 **-0.19** Protecting International Commons

-3.00   3.00

Japan

60.6	ESI
22	Ranking
$23,257	GDP per Capita
65.2	Peer Group ESI
64	Variable Coverage (out of 67)
2	Missing Variables Imputed

0.00 ━━ Indicator value
0.00 ▬ Reference (average value for peer group)

Jordan

40.1	ESI
97	Ranking
$3,347	GDP per Capita
45.8	Peer Group ESI
54	Variable Coverage (out of 67)
9	Missing Variables Imputed

0.00 ━━ Indicator value
0.00 ▬ Reference (average value for peer group)

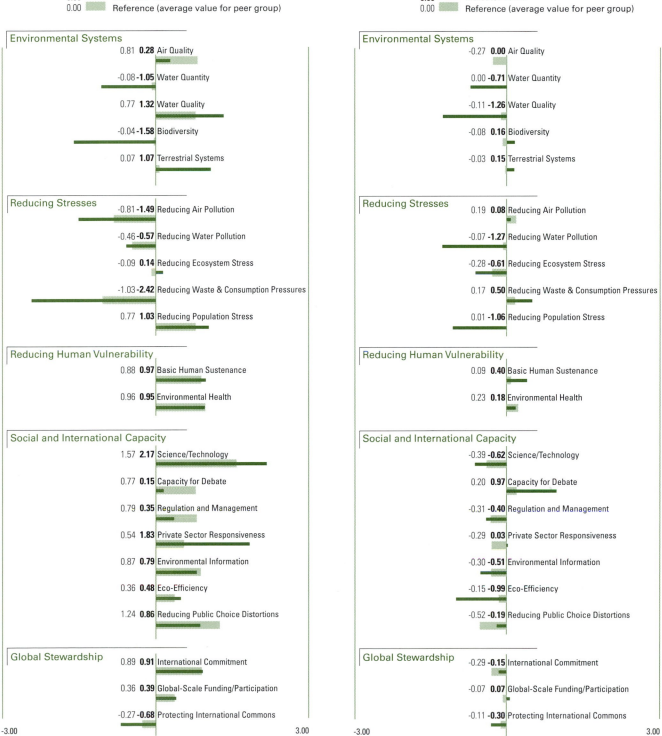

Japan

Environmental Systems
0.81 **0.28** Air Quality
-0.08 **-1.05** Water Quantity
0.77 **1.32** Water Quality
-0.04 **-1.58** Biodiversity
0.07 **1.07** Terrestrial Systems

Reducing Stresses
-0.81 **-1.49** Reducing Air Pollution
-0.46 **-0.57** Reducing Water Pollution
-0.09 **0.14** Reducing Ecosystem Stress
-1.03 **-2.42** Reducing Waste & Consumption Pressures
0.77 **1.03** Reducing Population Stress

Reducing Human Vulnerability
0.88 **0.97** Basic Human Sustenance
0.96 **0.95** Environmental Health

Social and International Capacity
1.57 **2.17** Science/Technology
0.77 **0.15** Capacity for Debate
0.79 **0.35** Regulation and Management
0.54 **1.83** Private Sector Responsiveness
0.87 **0.79** Environmental Information
0.36 **0.48** Eco-Efficiency
1.24 **0.86** Reducing Public Choice Distortions

Global Stewardship
0.89 **0.91** International Commitment
0.36 **0.39** Global-Scale Funding/Participation
-0.27 **-0.68** Protecting International Commons

-3.00   3.00

Jordan

Environmental Systems
-0.27 **0.00** Air Quality
0.00 **-0.71** Water Quantity
-0.11 **-1.26** Water Quality
-0.08 **0.16** Biodiversity
-0.03 **0.15** Terrestrial Systems

Reducing Stresses
0.19 **0.08** Reducing Air Pollution
-0.07 **-1.27** Reducing Water Pollution
-0.28 **-0.61** Reducing Ecosystem Stress
0.17 **0.50** Reducing Waste & Consumption Pressures
0.01 **-1.06** Reducing Population Stress

Reducing Human Vulnerability
0.09 **0.40** Basic Human Sustenance
0.23 **0.18** Environmental Health

Social and International Capacity
-0.39 **-0.62** Science/Technology
0.20 **0.97** Capacity for Debate
-0.31 **-0.40** Regulation and Management
-0.29 **0.03** Private Sector Responsiveness
-0.30 **-0.51** Environmental Information
-0.15 **-0.99** Eco-Efficiency
-0.52 **-0.19** Reducing Public Choice Distortions

Global Stewardship
-0.29 **-0.15** International Commitment
-0.07 **0.07** Global-Scale Funding/Participation
-0.11 **-0.30** Protecting International Commons

-3.00   3.00

Kazakhstan

41.5 ESI
91 Ranking
$4,378 GDP per Capita
45.8 Peer Group ESI
43 Variable Coverage (out of 67)
15 Missing Variables Imputed

0.00 ━━ Indicator value
0.00 ▬ Reference (average value for peer group)

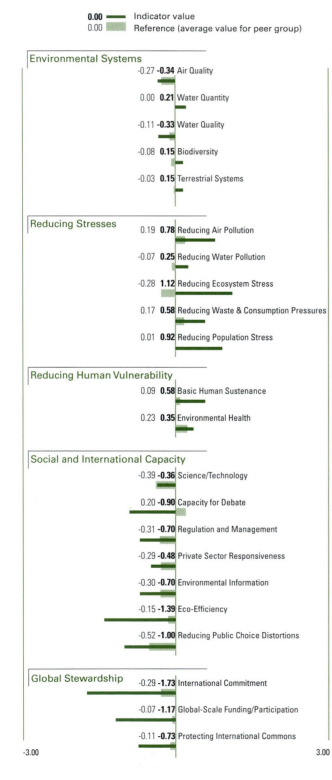

Environmental Systems
-0.27 **-0.34** Air Quality
0.00 **0.21** Water Quantity
-0.11 **-0.33** Water Quality
-0.08 **0.15** Biodiversity
-0.03 **0.15** Terrestrial Systems

Reducing Stresses
0.19 **0.78** Reducing Air Pollution
-0.07 **0.25** Reducing Water Pollution
-0.28 **1.12** Reducing Ecosystem Stress
0.17 **0.58** Reducing Waste & Consumption Pressures
0.01 **0.92** Reducing Population Stress

Reducing Human Vulnerability
0.09 **0.58** Basic Human Sustenance
0.23 **0.35** Environmental Health

Social and International Capacity
-0.39 **-0.36** Science/Technology
0.20 **-0.90** Capacity for Debate
-0.31 **-0.70** Regulation and Management
-0.29 **-0.48** Private Sector Responsiveness
-0.30 **-0.70** Environmental Information
-0.15 **-1.39** Eco-Efficiency
-0.52 **-1.00** Reducing Public Choice Distortions

Global Stewardship
-0.29 **-1.73** International Commitment
-0.07 **-1.17** Global-Scale Funding/Participation
-0.11 **-0.73** Protecting International Commons

-3.00   3.00

Kenya

43.7 ESI
84 Ranking
$980 GDP per Capita
39.2 Peer Group ESI
46 Variable Coverage (out of 67)
12 Missing Variables Imputed

0.00 ━━ Indicator value
0.00 ▬ Reference (average value for peer group)

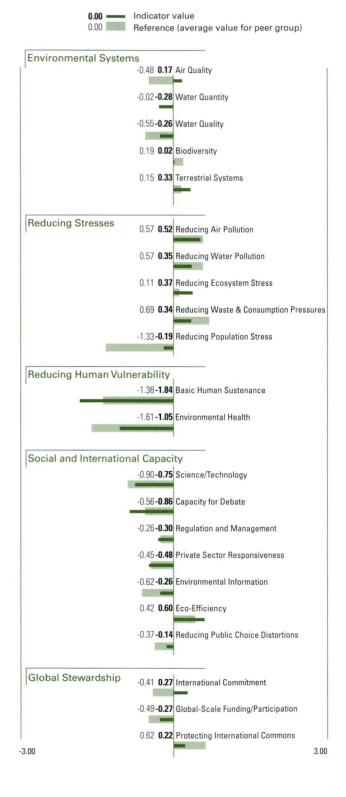

Environmental Systems
-0.48 **0.17** Air Quality
-0.02 **-0.28** Water Quantity
-0.55 **-0.26** Water Quality
0.19 **0.02** Biodiversity
0.15 **0.33** Terrestrial Systems

Reducing Stresses
0.57 **0.52** Reducing Air Pollution
0.57 **0.35** Reducing Water Pollution
0.11 **0.37** Reducing Ecosystem Stress
0.69 **0.34** Reducing Waste & Consumption Pressures
-1.33 **-0.19** Reducing Population Stress

Reducing Human Vulnerability
-1.38 **-1.84** Basic Human Sustenance
-1.61 **-1.05** Environmental Health

Social and International Capacity
-0.90 **-0.75** Science/Technology
-0.56 **-0.86** Capacity for Debate
-0.26 **-0.30** Regulation and Management
-0.45 **-0.48** Private Sector Responsiveness
-0.62 **-0.26** Environmental Information
0.42 **0.60** Eco-Efficiency
-0.37 **-0.14** Reducing Public Choice Distortions

Global Stewardship
-0.41 **0.27** International Commitment
-0.49 **-0.27** Global-Scale Funding/Participation
0.62 **0.22** Protecting International Commons

-3.00   3.00

Kuwait

31.9 ESI
116 Ranking
N/A* GDP per Capita
52.2 Peer Group ESI
43 Variable Coverage (out of 67)
14 Missing Variables Imputed

0.00 Indicator value
0.00 Reference (average value for peer group)

Environmental Systems
0.18 **0.36** Air Quality
0.04 **-1.27** Water Quantity
0.25 **0.39** Water Quality
-0.12 **-0.97** Biodiversity
-0.13 **0.20** Terrestrial Systems

Reducing Stresses
-0.28 **-1.06** Reducing Air Pollution
-0.06 **-0.82** Reducing Water Pollution
0.12 **0.46** Reducing Ecosystem Stress
-0.10 **-2.39** Reducing Waste & Consumption Pressures
0.51 **-0.40** Reducing Population Stress

Reducing Human Vulnerability
0.58 **0.75** Basic Human Sustenance
0.60 **0.91** Environmental Health

Social and International Capacity
0.08 **-0.59** Science/Technology
0.03 **0.16** Capacity for Debate
-0.25 **-0.45** Regulation and Management
0.00 **-0.43** Private Sector Responsiveness
0.13 **-0.68** Environmental Information
-0.39 **-1.38** Eco-Efficiency
-0.02 **-0.43** Reducing Public Choice Distortions

Global Stewardship
-0.02 **-0.71** International Commitment
0.26 **-1.07** Global-Scale Funding/Participation
-0.21 **-0.93** Protecting International Commons

-3.00   3.00

Kyrgyz Republic

39.6 ESI
99 Ranking
$2,317 GDP per Capita
45.3 Peer Group ESI
43 Variable Coverage (out of 67)
16 Missing Variables Imputed

0.00 Indicator value
0.00 Reference (average value for peer group)

Environmental Systems
-0.22 **-0.29** Air Quality
0.03 **-0.87** Water Quantity
-0.35 **-0.28** Water Quality
0.06 **0.35** Biodiversity
-0.05 **0.19** Terrestrial Systems

Reducing Stresses
0.30 **0.55** Reducing Air Pollution
-0.02 **-0.31** Reducing Water Pollution
0.14 **0.46** Reducing Ecosystem Stress
0.59 **1.23** Reducing Waste & Consumption Pressures
0.03 **0.38** Reducing Population Stress

Reducing Human Vulnerability
-0.20 **0.02** Basic Human Sustenance
-0.17 **0.08** Environmental Health

Social and International Capacity
-0.29 **-0.39** Science/Technology
-0.42 **-0.77** Capacity for Debate
-0.36 **-0.64** Regulation and Management
-0.44 **-0.48** Private Sector Responsiveness
-0.04 **-0.89** Environmental Information
-0.18 **-0.54** Eco-Efficiency
-0.64 **-0.62** Reducing Public Choice Distortions

Global Stewardship
-0.41 **-1.78** International Commitment
-0.07 **-1.17** Global-Scale Funding/Participation
0.06 **-0.07** Protecting International Commons

-3.00   3.00

*Not available. Peer group assigned based on Human Development Index.

Latvia

56.3	ESI	
32	Ranking	
$5,728	GDP per Capita	
45.8	Peer Group ESI	
49	Variable Coverage (out of 67)	
10	Missing Variables Imputed	

0.00 ━━ Indicator value
0.00 ▬ Reference (average value for peer group)

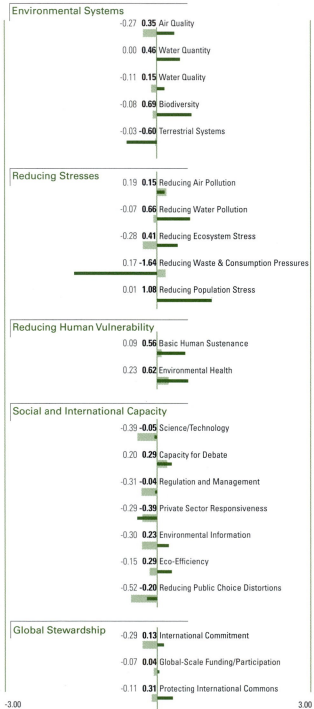

Lebanon

37.5	ESI	
108	Ranking	
$4,326	GDP per Capita	
45.8	Peer Group ESI	
43	Variable Coverage (out of 67)	
15	Missing Variables Imputed	

0.00 ━━ Indicator value
0.00 ▬ Reference (average value for peer group)

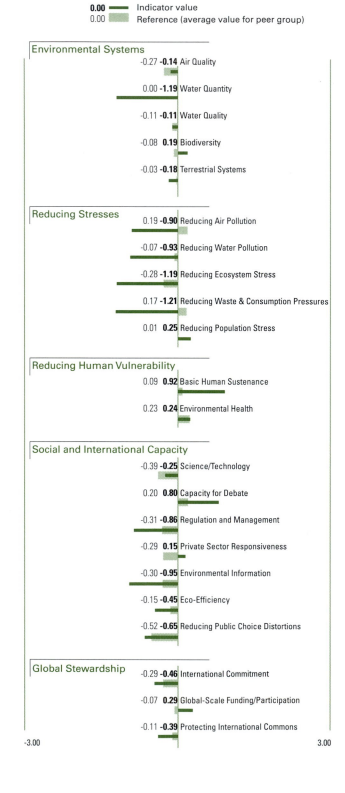

Libya

31.3   ESI
118   Ranking
N/A*   GDP per Capita
45.8   Peer Group ESI
44   Variable Coverage (out of 67)
13   Missing Variables Imputed

Lithuania

60.3   ESI
23   Ranking
$6,436   GDP per Capita
52.2   Peer Group ESI
54   Variable Coverage (out of 67)
6   Missing Variables Imputed

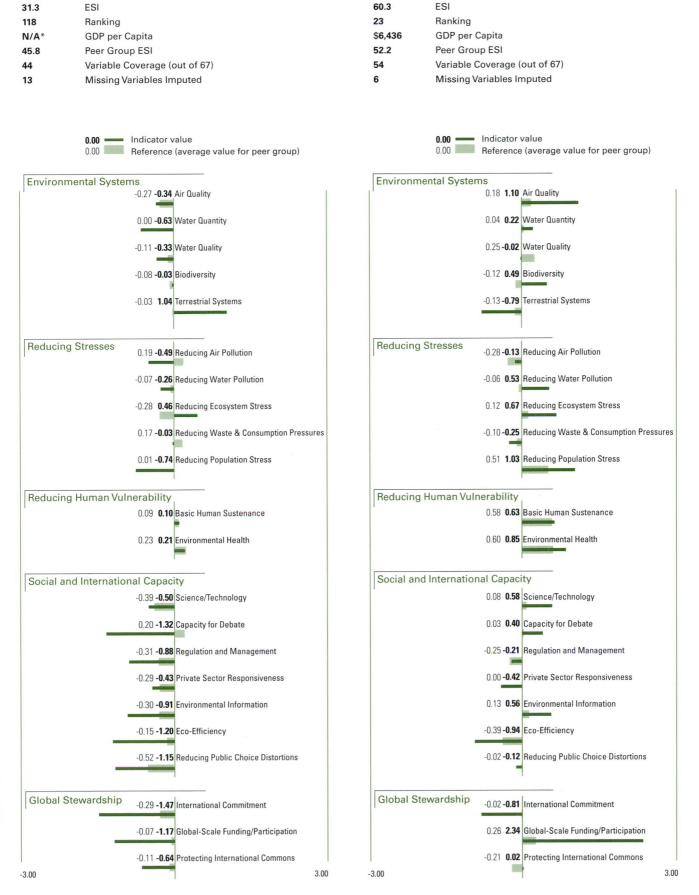

0.00 Indicator value
0.00 Reference (average value for peer group)

Libya

Environmental Systems
- -0.27 **-0.34** Air Quality
- 0.00 **-0.63** Water Quantity
- -0.11 **-0.33** Water Quality
- -0.08 **-0.03** Biodiversity
- -0.03 **1.04** Terrestrial Systems

Reducing Stresses
- 0.19 **-0.49** Reducing Air Pollution
- -0.07 **-0.26** Reducing Water Pollution
- -0.28 **0.46** Reducing Ecosystem Stress
- 0.17 **-0.03** Reducing Waste & Consumption Pressures
- 0.01 **-0.74** Reducing Population Stress

Reducing Human Vulnerability
- 0.09 **0.10** Basic Human Sustenance
- 0.23 **0.21** Environmental Health

Social and International Capacity
- -0.39 **-0.50** Science/Technology
- 0.20 **-1.32** Capacity for Debate
- -0.31 **-0.88** Regulation and Management
- -0.29 **-0.43** Private Sector Responsiveness
- -0.30 **-0.91** Environmental Information
- -0.15 **-1.20** Eco-Efficiency
- -0.52 **-1.15** Reducing Public Choice Distortions

Global Stewardship
- -0.29 **-1.47** International Commitment
- -0.07 **-1.17** Global-Scale Funding/Participation
- -0.11 **-0.64** Protecting International Commons

-3.00   3.00

Lithuania

Environmental Systems
- 0.18 **1.10** Air Quality
- 0.04 **0.22** Water Quantity
- 0.25 **-0.02** Water Quality
- -0.12 **0.49** Biodiversity
- -0.13 **-0.79** Terrestrial Systems

Reducing Stresses
- -0.28 **-0.13** Reducing Air Pollution
- -0.06 **0.53** Reducing Water Pollution
- 0.12 **0.67** Reducing Ecosystem Stress
- -0.10 **-0.25** Reducing Waste & Consumption Pressures
- 0.51 **1.03** Reducing Population Stress

Reducing Human Vulnerability
- 0.58 **0.63** Basic Human Sustenance
- 0.60 **0.85** Environmental Health

Social and International Capacity
- 0.08 **0.58** Science/Technology
- 0.03 **0.40** Capacity for Debate
- -0.25 **-0.21** Regulation and Management
- 0.00 **-0.42** Private Sector Responsiveness
- 0.13 **0.56** Environmental Information
- -0.39 **-0.94** Eco-Efficiency
- -0.02 **-0.12** Reducing Public Choice Distortions

Global Stewardship
- -0.02 **-0.81** International Commitment
- 0.26 **2.34** Global-Scale Funding/Participation
- -0.21 **0.02** Protecting International Commons

-3.00   3.00

* Not available. Peer group assigned based on Human Development Index.

Macedonia

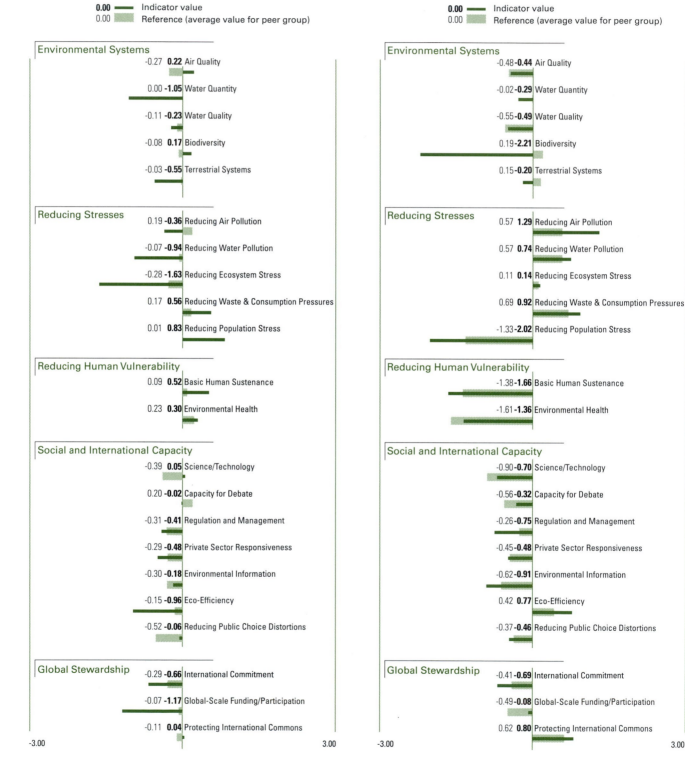

39.2	ESI
101	Ranking
$4,254	GDP per Capita
45.8	Peer Group ESI
46	Variable Coverage (out of 67)
13	Missing Variables Imputed

0.00 ——— Indicator value
0.00 ▬▬ Reference (average value for peer group)

Environmental Systems
-0.27 **0.22** Air Quality
0.00 **-1.05** Water Quantity
-0.11 **-0.23** Water Quality
-0.08 **0.17** Biodiversity
-0.03 **-0.55** Terrestrial Systems

Reducing Stresses
0.19 **-0.36** Reducing Air Pollution
-0.07 **-0.94** Reducing Water Pollution
-0.28 **-1.63** Reducing Ecosystem Stress
0.17 **0.56** Reducing Waste & Consumption Pressures
0.01 **0.83** Reducing Population Stress

Reducing Human Vulnerability
0.09 **0.52** Basic Human Sustenance
0.23 **0.30** Environmental Health

Social and International Capacity
-0.39 **0.05** Science/Technology
0.20 **-0.02** Capacity for Debate
-0.31 **-0.41** Regulation and Management
-0.29 **-0.48** Private Sector Responsiveness
-0.30 **-0.18** Environmental Information
-0.15 **-0.96** Eco-Efficiency
-0.52 **-0.06** Reducing Public Choice Distortions

Global Stewardship
-0.29 **-0.66** International Commitment
-0.07 **-1.17** Global-Scale Funding/Participation
-0.11 **0.04** Protecting International Commons

-3.00   3.00

Madagascar

35.1	ESI
113	Ranking
$756	GDP per Capita
39.2	Peer Group ESI
46	Variable Coverage (out of 67)
12	Missing Variables Imputed

0.00 ——— Indicator value
0.00 ▬▬ Reference (average value for peer group)

Environmental Systems
-0.48 **0.44** Air Quality
-0.02 **-0.29** Water Quantity
-0.55 **-0.49** Water Quality
0.19 **-2.21** Biodiversity
0.15 **-0.20** Terrestrial Systems

Reducing Stresses
0.57 **1.29** Reducing Air Pollution
0.57 **0.74** Reducing Water Pollution
0.11 **0.14** Reducing Ecosystem Stress
0.69 **0.92** Reducing Waste & Consumption Pressures
-1.33 **-2.02** Reducing Population Stress

Reducing Human Vulnerability
-1.38 **-1.66** Basic Human Sustenance
-1.61 **-1.36** Environmental Health

Social and International Capacity
-0.90 **-0.70** Science/Technology
-0.56 **-0.32** Capacity for Debate
-0.26 **-0.75** Regulation and Management
-0.45 **-0.48** Private Sector Responsiveness
-0.62 **-0.91** Environmental Information
0.42 **0.77** Eco-Efficiency
-0.37 **-0.46** Reducing Public Choice Distortions

Global Stewardship
-0.41 **-0.69** International Commitment
-0.49 **-0.08** Global-Scale Funding/Participation
0.62 **0.80** Protecting International Commons

-3.00   3.00

Malawi

41.0   ESI
92   Ranking
$523   GDP per Capita
39.2   Peer Group ESI
44   Variable Coverage (out of 67)
14   Missing Variables Imputed

Malaysia

49.8   ESI
51   Ranking
$8,137   GDP per Capita
52.2   Peer Group ESI
63   Variable Coverage (out of 67)
4   Missing Variables Imputed

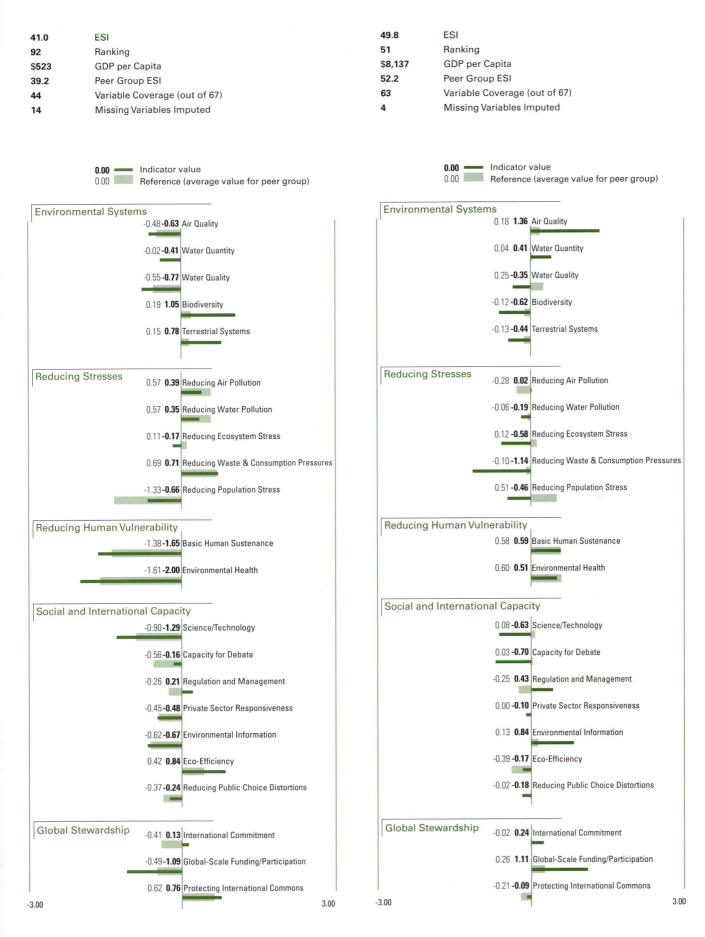

0.00 Indicator value
0.00 Reference (average value for peer group)

Malawi

Environmental Systems
- -0.48 **-0.63** Air Quality
- -0.02 **-0.41** Water Quantity
- -0.55 **-0.77** Water Quality
- 0.19 **1.05** Biodiversity
- 0.15 **0.78** Terrestrial Systems

Reducing Stresses
- 0.57 **0.39** Reducing Air Pollution
- 0.57 **0.35** Reducing Water Pollution
- 0.11 **-0.17** Reducing Ecosystem Stress
- 0.69 **0.71** Reducing Waste & Consumption Pressures
- -1.33 **-0.66** Reducing Population Stress

Reducing Human Vulnerability
- -1.38 **-1.65** Basic Human Sustenance
- -1.61 **-2.00** Environmental Health

Social and International Capacity
- -0.90 **-1.29** Science/Technology
- -0.56 **-0.16** Capacity for Debate
- -0.26 **0.21** Regulation and Management
- -0.45 **-0.48** Private Sector Responsiveness
- -0.62 **-0.67** Environmental Information
- 0.42 **0.84** Eco-Efficiency
- -0.37 **-0.24** Reducing Public Choice Distortions

Global Stewardship
- -0.41 **0.13** International Commitment
- -0.49 **-1.09** Global-Scale Funding/Participation
- 0.62 **0.76** Protecting International Commons

-3.00   3.00

Malaysia

Environmental Systems
- 0.18 **1.36** Air Quality
- 0.04 **0.41** Water Quantity
- 0.25 **-0.35** Water Quality
- -0.12 **-0.62** Biodiversity
- -0.13 **-0.44** Terrestrial Systems

Reducing Stresses
- -0.28 **0.02** Reducing Air Pollution
- -0.06 **-0.19** Reducing Water Pollution
- 0.12 **-0.58** Reducing Ecosystem Stress
- -0.10 **-1.14** Reducing Waste & Consumption Pressures
- 0.51 **-0.46** Reducing Population Stress

Reducing Human Vulnerability
- 0.58 **0.59** Basic Human Sustenance
- 0.60 **0.51** Environmental Health

Social and International Capacity
- 0.08 **-0.63** Science/Technology
- 0.03 **-0.70** Capacity for Debate
- -0.25 **0.43** Regulation and Management
- 0.00 **-0.10** Private Sector Responsiveness
- 0.13 **0.84** Environmental Information
- -0.39 **-0.17** Eco-Efficiency
- -0.02 **-0.18** Reducing Public Choice Distortions

Global Stewardship
- -0.02 **0.24** International Commitment
- 0.26 **1.11** Global-Scale Funding/Participation
- -0.21 **-0.09** Protecting International Commons

-3.00   3.00

Mali

46.1   ESI
72   Ranking
$681   GDP per Capita
39.2   Peer Group ESI
47   Variable Coverage (out of 67)
11   Missing Variables Imputed

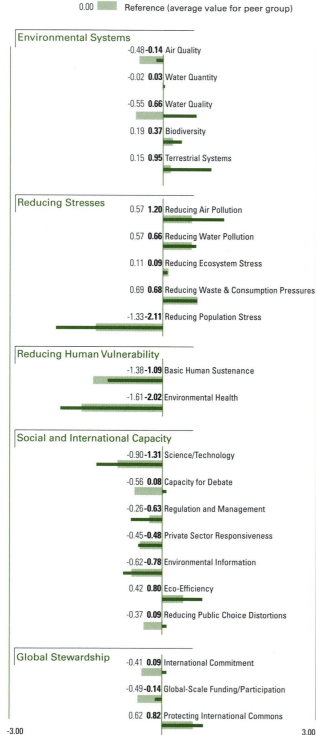

Mauritius

51.2   ESI
45   Ranking
$8,312   GDP per Capita
52.2   Peer Group ESI
50   Variable Coverage (out of 67)
13   Missing Variables Imputed

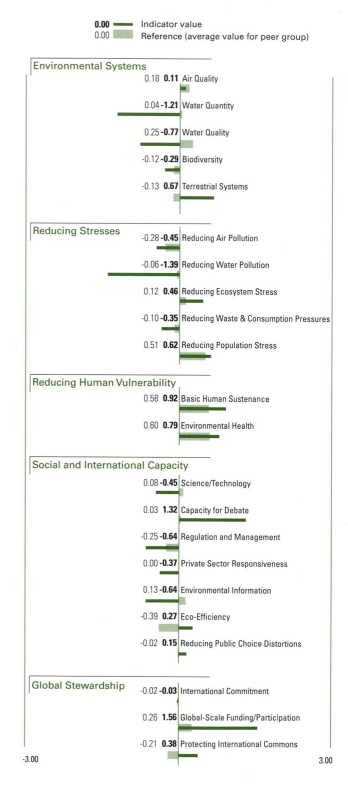

Mexico

45.3	ESI
73	Ranking
$7,704	GDP per Capita
52.2	Peer Group ESI
63	Variable Coverage (out of 67)
3	Missing Variables Imputed

0.00 ——— Indicator value
0.00 ▬▬ Reference (average value for peer group)

Moldova

47.4	ESI
60	Ranking
$1,947	GDP per Capita
45.3	Peer Group ESI
47	Variable Coverage (out of 67)
12	Missing Variables Imputed

0.00 ——— Indicator value
0.00 ▬▬ Reference (average value for peer group)

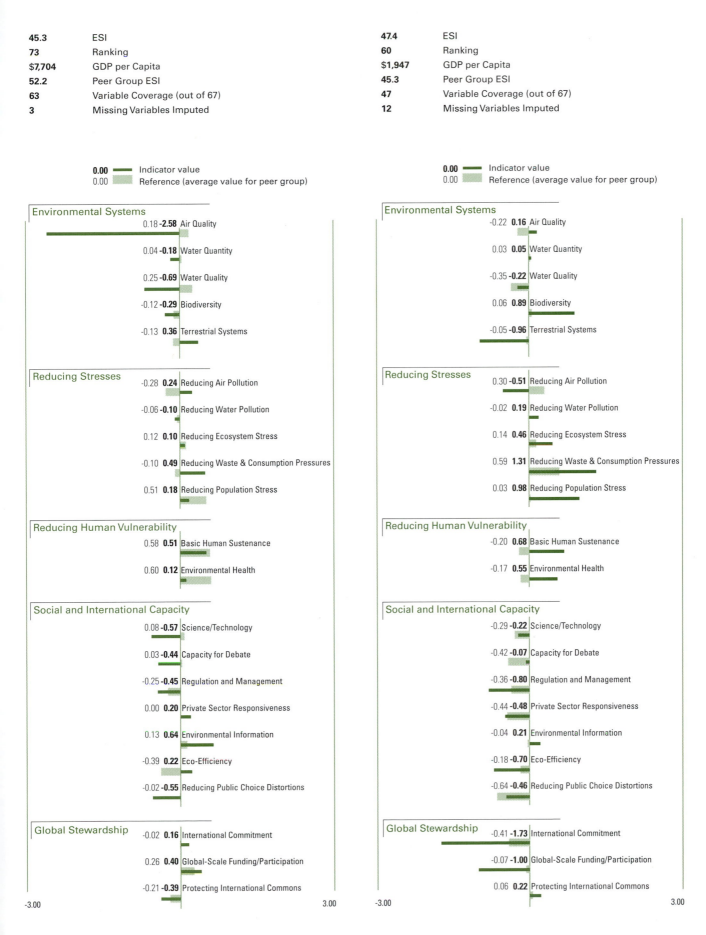

Mexico

Environmental Systems
0.18 **-2.58** Air Quality
0.04 **-0.18** Water Quantity
0.25 **-0.69** Water Quality
-0.12 **-0.29** Biodiversity
-0.13 **0.36** Terrestrial Systems

Reducing Stresses
-0.28 **0.24** Reducing Air Pollution
-0.06 **-0.10** Reducing Water Pollution
0.12 **0.10** Reducing Ecosystem Stress
-0.10 **0.49** Reducing Waste & Consumption Pressures
0.51 **0.18** Reducing Population Stress

Reducing Human Vulnerability
0.58 **0.51** Basic Human Sustenance
0.60 **0.12** Environmental Health

Social and International Capacity
0.08 **-0.57** Science/Technology
0.03 **-0.44** Capacity for Debate
-0.25 **-0.45** Regulation and Management
0.00 **0.20** Private Sector Responsiveness
0.13 **0.64** Environmental Information
-0.39 **0.22** Eco-Efficiency
-0.02 **-0.55** Reducing Public Choice Distortions

Global Stewardship
-0.02 **0.16** International Commitment
0.26 **0.40** Global-Scale Funding/Participation
-0.21 **-0.39** Protecting International Commons

-3.00   3.00

Moldova

Environmental Systems
-0.22 **0.16** Air Quality
0.03 **0.05** Water Quantity
-0.35 **-0.22** Water Quality
0.06 **0.89** Biodiversity
-0.05 **-0.96** Terrestrial Systems

Reducing Stresses
0.30 **-0.51** Reducing Air Pollution
-0.02 **0.19** Reducing Water Pollution
0.14 **0.46** Reducing Ecosystem Stress
0.59 **1.31** Reducing Waste & Consumption Pressures
0.03 **0.98** Reducing Population Stress

Reducing Human Vulnerability
-0.20 **0.68** Basic Human Sustenance
-0.17 **0.55** Environmental Health

Social and International Capacity
-0.29 **-0.22** Science/Technology
-0.42 **-0.07** Capacity for Debate
-0.36 **-0.80** Regulation and Management
-0.44 **-0.48** Private Sector Responsiveness
-0.04 **0.21** Environmental Information
-0.18 **-0.70** Eco-Efficiency
-0.64 **-0.46** Reducing Public Choice Distortions

Global Stewardship
-0.41 **-1.73** International Commitment
-0.07 **-1.00** Global-Scale Funding/Participation
0.06 **0.22** Protecting International Commons

-3.00   3.00

Mongolia

50.1   ESI
50   Ranking
$1,541   GDP per Capita
45.3   Peer Group ESI
48   Variable Coverage (out of 67)
11   Missing Variables Imputed

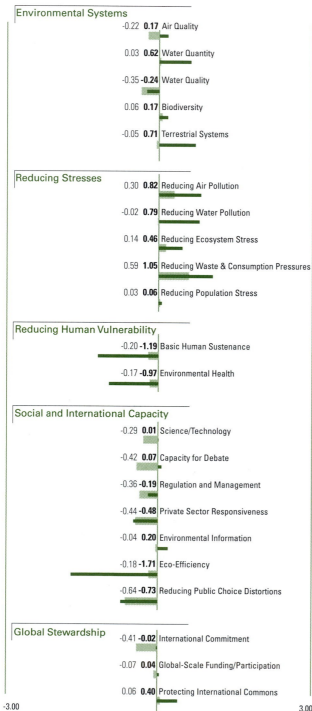

0.00 ━━ Indicator value
0.00 ▬ Reference (average value for peer group)

Environmental Systems
-0.22	**0.17**	Air Quality
0.03	**0.62**	Water Quantity
-0.35	**-0.24**	Water Quality
0.06	**0.17**	Biodiversity
-0.05	**0.71**	Terrestrial Systems

Reducing Stresses
0.30	**0.82**	Reducing Air Pollution
-0.02	**0.79**	Reducing Water Pollution
0.14	**0.46**	Reducing Ecosystem Stress
0.59	**1.05**	Reducing Waste & Consumption Pressures
0.03	**0.06**	Reducing Population Stress

Reducing Human Vulnerability
-0.20	**-1.19**	Basic Human Sustenance
-0.17	**-0.97**	Environmental Health

Social and International Capacity
-0.29	**0.01**	Science/Technology
-0.42	**0.07**	Capacity for Debate
-0.36	**-0.19**	Regulation and Management
-0.44	**-0.48**	Private Sector Responsiveness
-0.04	**0.20**	Environmental Information
-0.18	**-1.71**	Eco-Efficiency
-0.64	**-0.73**	Reducing Public Choice Distortions

Global Stewardship
-0.41	**-0.02**	International Commitment
-0.07	**0.04**	Global-Scale Funding/Participation
0.06	**0.40**	Protecting International Commons

-3.00   3.00

Morocco

41.8   ESI
89   Ranking
$3,305   GDP per Capita
45.3   Peer Group ESI
48   Variable Coverage (out of 67)
10   Missing Variables Imputed

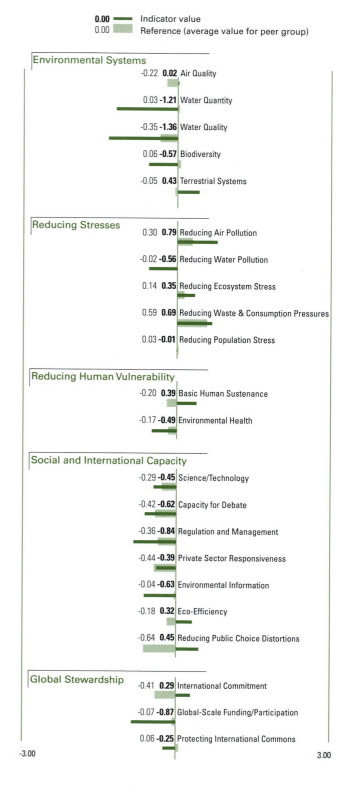

0.00 ━━ Indicator value
0.00 ▬ Reference (average value for peer group)

Environmental Systems
-0.22	**0.02**	Air Quality
0.03	**-1.21**	Water Quantity
-0.35	**-1.36**	Water Quality
0.06	**-0.57**	Biodiversity
-0.05	**0.43**	Terrestrial Systems

Reducing Stresses
0.30	**0.79**	Reducing Air Pollution
-0.02	**-0.56**	Reducing Water Pollution
0.14	**0.35**	Reducing Ecosystem Stress
0.59	**0.69**	Reducing Waste & Consumption Pressures
0.03	**-0.01**	Reducing Population Stress

Reducing Human Vulnerability
-0.20	**0.39**	Basic Human Sustenance
-0.17	**-0.49**	Environmental Health

Social and International Capacity
-0.29	**-0.45**	Science/Technology
-0.42	**-0.62**	Capacity for Debate
-0.36	**-0.84**	Regulation and Management
-0.44	**-0.39**	Private Sector Responsiveness
-0.04	**-0.63**	Environmental Information
-0.18	**0.32**	Eco-Efficiency
-0.64	**0.45**	Reducing Public Choice Distortions

Global Stewardship
-0.41	**0.29**	International Commitment
-0.07	**-0.87**	Global-Scale Funding/Participation
0.06	**-0.25**	Protecting International Commons

-3.00   3.00

Mozambique

43.8	ESI
80	Ranking
$782	GDP per Capita
39.2	Peer Group ESI
44	Variable Coverage (out of 67)
14	Missing Variables Imputed

0.00 ▬▬ Indicator value
0.00 ▬▬ Reference (average value for peer group)

Environmental Systems
-0.48 **-0.74** Air Quality
-0.02 **0.48** Water Quantity
-0.55 **-0.64** Water Quality
0.19 **0.39** Biodiversity
0.15 **0.56** Terrestrial Systems

Reducing Stresses
0.57 **1.11** Reducing Air Pollution
0.57 **0.93** Reducing Water Pollution
0.11 **0.20** Reducing Ecosystem Stress
0.69 **0.94** Reducing Waste & Consumption Pressures
-1.33 **-0.39** Reducing Population Stress

Reducing Human Vulnerability
-1.38 **-1.80** Basic Human Sustenance
-1.61 **-2.15** Environmental Health

Social and International Capacity
-0.90 **-1.46** Science/Technology
-0.56 **-0.37** Capacity for Debate
-0.26 **-0.31** Regulation and Management
-0.45 **-0.48** Private Sector Responsiveness
-0.62 **-0.62** Environmental Information
0.42 **0.71** Eco-Efficiency
-0.37 **-0.35** Reducing Public Choice Distortions

Global Stewardship
-0.41 **-0.18** International Commitment
-0.49 **-0.10** Global-Scale Funding/Participation
0.62 **0.87** Protecting International Commons

-3.00   3.00

Nepal

46.5	ESI
67	Ranking
$1,157	GDP per Capita
39.2	Peer Group ESI
45	Variable Coverage (out of 67)
14	Missing Variables Imputed

0.00 ▬▬ Indicator value
0.00 ▬▬ Reference (average value for peer group)

Environmental Systems
-0.48 **-0.03** Air Quality
-0.02 **0.08** Water Quantity
-0.55 **-0.28** Water Quality
0.19 **-0.38** Biodiversity
0.15 **0.11** Terrestrial Systems

Reducing Stresses
0.57 **0.17** Reducing Air Pollution
0.57 **-0.08** Reducing Water Pollution
0.11 **0.03** Reducing Ecosystem Stress
0.69 **0.78** Reducing Waste & Consumption Pressures
-1.33 **-0.86** Reducing Population Stress

Reducing Human Vulnerability
-1.38 **-0.45** Basic Human Sustenance
-1.61 **-1.08** Environmental Health

Social and International Capacity
-0.90 **-0.92** Science/Technology
-0.56 **-0.19** Capacity for Debate
-0.26 **0.65** Regulation and Management
-0.45 **-0.48** Private Sector Responsiveness
-0.62 **0.46** Environmental Information
0.42 **0.82** Eco-Efficiency
-0.37 **-0.39** Reducing Public Choice Distortions

Global Stewardship
-0.41 **-0.46** International Commitment
-0.49 **-0.14** Global-Scale Funding/Participation
0.62 **0.72** Protecting International Commons

-3.00   3.00

Netherlands

66.0 ESI
12 Ranking
$22,176 GDP per Capita
65.2 Peer Group ESI
66 Variable Coverage (out of 67)
1 Missing Variables Imputed

New Zealand

71.3 ESI
6 Ranking
$17,288 GDP per Capita
65.2 Peer Group ESI
62 Variable Coverage (out of 67)
2 Missing Variables Imputed

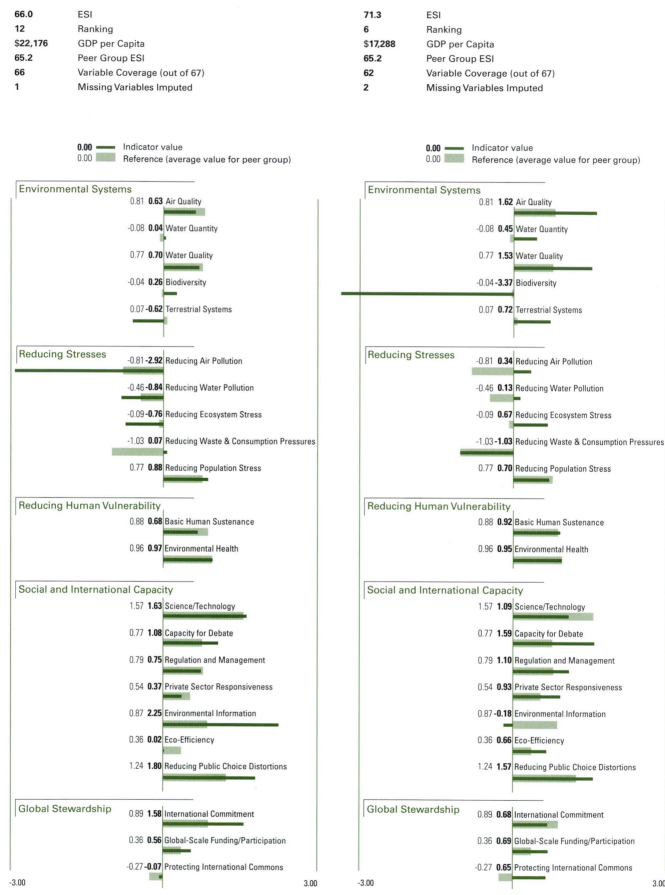

0.00 — Indicator value
0.00 ▬ Reference (average value for peer group)

Environmental Systems

0.81	**0.63**	Air Quality
-0.08	**0.04**	Water Quantity
0.77	**0.70**	Water Quality
-0.04	**0.26**	Biodiversity
0.07	**-0.62**	Terrestrial Systems

Reducing Stresses

-0.81	**-2.92**	Reducing Air Pollution
-0.46	**-0.84**	Reducing Water Pollution
-0.09	**-0.76**	Reducing Ecosystem Stress
-1.03	**0.07**	Reducing Waste & Consumption Pressures
0.77	**0.88**	Reducing Population Stress

Reducing Human Vulnerability

| 0.88 | **0.68** | Basic Human Sustenance |
| 0.96 | **0.97** | Environmental Health |

Social and International Capacity

1.57	**1.63**	Science/Technology
0.77	**1.08**	Capacity for Debate
0.79	**0.75**	Regulation and Management
0.54	**0.37**	Private Sector Responsiveness
0.87	**2.25**	Environmental Information
0.36	**0.02**	Eco-Efficiency
1.24	**1.80**	Reducing Public Choice Distortions

Global Stewardship

0.89	**1.58**	International Commitment
0.36	**0.56**	Global-Scale Funding/Participation
-0.27	**-0.07**	Protecting International Commons

-3.00   3.00

Environmental Systems

0.81	**1.62**	Air Quality
-0.08	**0.45**	Water Quantity
0.77	**1.53**	Water Quality
-0.04	**-3.37**	Biodiversity
0.07	**0.72**	Terrestrial Systems

Reducing Stresses

-0.81	**0.34**	Reducing Air Pollution
-0.46	**0.13**	Reducing Water Pollution
-0.09	**0.67**	Reducing Ecosystem Stress
-1.03	**-1.03**	Reducing Waste & Consumption Pressures
0.77	**0.70**	Reducing Population Stress

Reducing Human Vulnerability

| 0.88 | **0.92** | Basic Human Sustenance |
| 0.96 | **0.95** | Environmental Health |

Social and International Capacity

1.57	**1.09**	Science/Technology
0.77	**1.59**	Capacity for Debate
0.79	**1.10**	Regulation and Management
0.54	**0.93**	Private Sector Responsiveness
0.87	**-0.18**	Environmental Information
0.36	**0.66**	Eco-Efficiency
1.24	**1.57**	Reducing Public Choice Distortions

Global Stewardship

0.89	**0.68**	International Commitment
0.36	**0.69**	Global-Scale Funding/Participation
-0.27	**0.65**	Protecting International Commons

-3.00   3.00

Nicaragua

Niger

51.9 ESI
42 Ranking
$2,142 GDP per Capita
45.3 Peer Group ESI
48 Variable Coverage (out of 67)
10 Missing Variables Imputed

36.7 ESI
111 Ranking
$739 GDP per Capita
39.2 Peer Group ESI
43 Variable Coverage (out of 67)
15 Missing Variables Imputed

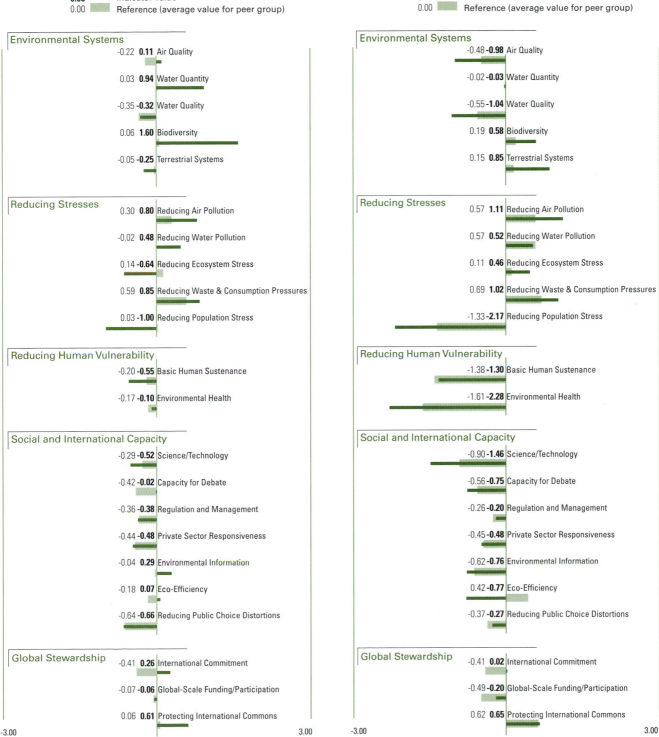

0.00 ━━ Indicator value
0.00 ▬ Reference (average value for peer group)

0.00 ━━ Indicator value
0.00 ▬ Reference (average value for peer group)

Environmental Systems
-0.22 **0.11** Air Quality
0.03 **0.94** Water Quantity
-0.35 **-0.32** Water Quality
0.06 **1.60** Biodiversity
-0.05 **-0.25** Terrestrial Systems

Reducing Stresses
0.30 **0.80** Reducing Air Pollution
-0.02 **0.48** Reducing Water Pollution
0.14 **-0.64** Reducing Ecosystem Stress
0.59 **0.85** Reducing Waste & Consumption Pressures
0.03 **-1.00** Reducing Population Stress

Reducing Human Vulnerability
-0.20 **-0.55** Basic Human Sustenance
-0.17 **-0.10** Environmental Health

Social and International Capacity
-0.29 **-0.52** Science/Technology
-0.42 **-0.02** Capacity for Debate
-0.36 **-0.38** Regulation and Management
-0.44 **-0.48** Private Sector Responsiveness
-0.04 **0.29** Environmental Information
-0.18 **0.07** Eco-Efficiency
-0.64 **-0.66** Reducing Public Choice Distortions

Global Stewardship
-0.41 **0.26** International Commitment
-0.07 **-0.06** Global-Scale Funding/Participation
0.06 **0.61** Protecting International Commons

-3.00   3.00

Environmental Systems
-0.48 **-0.98** Air Quality
-0.02 **-0.03** Water Quantity
-0.55 **-1.04** Water Quality
0.19 **0.58** Biodiversity
0.15 **0.85** Terrestrial Systems

Reducing Stresses
0.57 **1.11** Reducing Air Pollution
0.57 **0.52** Reducing Water Pollution
0.11 **0.46** Reducing Ecosystem Stress
0.69 **1.02** Reducing Waste & Consumption Pressures
-1.33 **-2.17** Reducing Population Stress

Reducing Human Vulnerability
-1.38 **-1.30** Basic Human Sustenance
-1.61 **-2.28** Environmental Health

Social and International Capacity
-0.90 **-1.46** Science/Technology
-0.56 **-0.75** Capacity for Debate
-0.26 **-0.20** Regulation and Management
-0.45 **-0.48** Private Sector Responsiveness
-0.62 **-0.76** Environmental Information
0.42 **-0.77** Eco-Efficiency
-0.37 **-0.27** Reducing Public Choice Distortions

Global Stewardship
-0.41 **0.02** International Commitment
-0.49 **-0.20** Global-Scale Funding/Participation
0.62 **0.65** Protecting International Commons

-3.00   3.00

Nigeria

31.5	ESI
117	Ranking
$795	GDP per Capita
39.2	Peer Group ESI
46	Variable Coverage (out of 67)
12	Missing Variables Imputed

0.00 — Indicator value
0.00 ▬ Reference (average value for peer group)

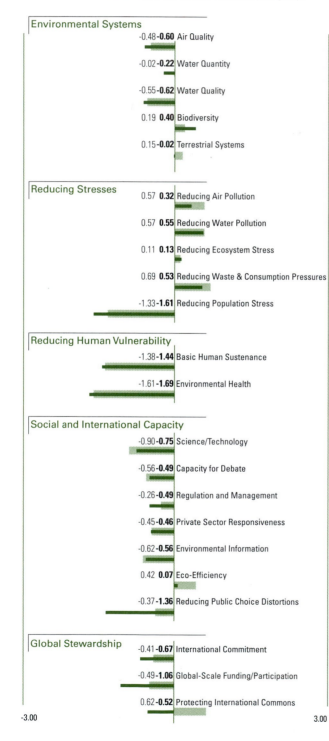

Environmental Systems

-0.48 **-0.60** Air Quality
-0.02 **-0.22** Water Quantity
-0.55 **-0.62** Water Quality
0.19 **0.40** Biodiversity
0.15 **-0.02** Terrestrial Systems

Reducing Stresses

0.57 **0.32** Reducing Air Pollution
0.57 **0.55** Reducing Water Pollution
0.11 **0.13** Reducing Ecosystem Stress
0.69 **0.53** Reducing Waste & Consumption Pressures
-1.33 **-1.61** Reducing Population Stress

Reducing Human Vulnerability

-1.38 **-1.44** Basic Human Sustenance
-1.61 **-1.69** Environmental Health

Social and International Capacity

-0.90 **-0.75** Science/Technology
-0.56 **-0.49** Capacity for Debate
-0.26 **-0.49** Regulation and Management
-0.45 **-0.46** Private Sector Responsiveness
-0.62 **-0.56** Environmental Information
0.42 **0.07** Eco-Efficiency
-0.37 **-1.36** Reducing Public Choice Distortions

Global Stewardship

-0.41 **-0.67** International Commitment
-0.49 **-1.06** Global-Scale Funding/Participation
0.62 **-0.52** Protecting International Commons

-3.00   3.00

Norway

78.2	ESI
2	Ranking
$26,342	GDP per Capita
65.2	Peer Group ESI
65	Variable Coverage (out of 67)
2	Missing Variables Imputed

0.00 — Indicator value
0.00 ▬ Reference (average value for peer group)

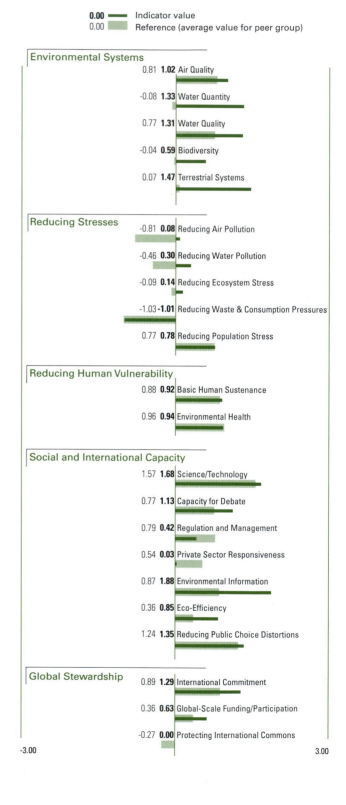

Environmental Systems

0.81 **1.02** Air Quality
-0.08 **1.33** Water Quantity
0.77 **1.31** Water Quality
-0.04 **0.59** Biodiversity
0.07 **1.47** Terrestrial Systems

Reducing Stresses

-0.81 **0.08** Reducing Air Pollution
-0.46 **0.30** Reducing Water Pollution
-0.09 **0.14** Reducing Ecosystem Stress
-1.03 **-1.01** Reducing Waste & Consumption Pressures
0.77 **0.78** Reducing Population Stress

Reducing Human Vulnerability

0.88 **0.92** Basic Human Sustenance
0.96 **0.94** Environmental Health

Social and International Capacity

1.57 **1.68** Science/Technology
0.77 **1.13** Capacity for Debate
0.79 **0.42** Regulation and Management
0.54 **0.03** Private Sector Responsiveness
0.87 **1.88** Environmental Information
0.36 **0.85** Eco-Efficiency
1.24 **1.35** Reducing Public Choice Distortions

Global Stewardship

0.89 **1.29** International Commitment
0.36 **0.63** Global-Scale Funding/Participation
-0.27 **0.00** Protecting International Commons

-3.00   3.00

Pakistan

43.4 ESI
85 Ranking
$1,715 GDP per Capita
45.3 Peer Group ESI
51 Variable Coverage (out of 67)
8 Missing Variables Imputed

0.00 ── Indicator value
0.00 ▬ Reference (average value for peer group)

Environmental Systems
-0.22 **-0.14** Air Quality
0.03 **-0.42** Water Quantity
-0.35 **-0.30** Water Quality
0.06 **-0.26** Biodiversity
-0.05 **0.29** Terrestrial Systems

Reducing Stresses
0.30 **0.57** Reducing Air Pollution
-0.02 **-0.32** Reducing Water Pollution
0.14 **-0.83** Reducing Ecosystem Stress
0.59 **1.31** Reducing Waste & Consumption Pressures
0.03 **-1.00** Reducing Population Stress

Reducing Human Vulnerability
-0.20 **-0.28** Basic Human Sustenance
-0.17 **-1.10** Environmental Health

Social and International Capacity
-0.29 **-0.34** Science/Technology
-0.42 **-1.07** Capacity for Debate
-0.36 **0.78** Regulation and Management
-0.44 **-0.46** Private Sector Responsiveness
-0.04 **0.30** Environmental Information
-0.18 **0.05** Eco-Efficiency
-0.64 **-0.65** Reducing Public Choice Distortions

Global Stewardship
-0.41 **0.24** International Commitment
-0.07 **0.11** Global-Scale Funding/Participation
0.06 **-0.17** Protecting International Commons

-3.00   3.00

Panama

55.9 ESI
34 Ranking
$5,249 GDP per Capita
45.8 Peer Group ESI
48 Variable Coverage (out of 67)
10 Missing Variables Imputed

0.00 ── Indicator value
0.00 ▬ Reference (average value for peer group)

Environmental Systems
-0.27 **0.05** Air Quality
0.00 **-0.12** Water Quantity
-0.11 **0.11** Water Quality
-0.08 **0.53** Biodiversity
-0.03 **-0.47** Terrestrial Systems

Reducing Stresses
0.19 **0.63** Reducing Air Pollution
-0.07 **0.44** Reducing Water Pollution
-0.28 **-0.44** Reducing Ecosystem Stress
0.17 **0.43** Reducing Waste & Consumption Pressures
0.01 **0.23** Reducing Population Stress

Reducing Human Vulnerability
0.09 **-0.35** Basic Human Sustenance
0.23 **0.35** Environmental Health

Social and International Capacity
-0.39 **-0.64** Science/Technology
0.20 **2.27** Capacity for Debate
-0.31 **0.40** Regulation and Management
-0.29 **-0.48** Private Sector Responsiveness
-0.30 **-0.81** Environmental Information
-0.15 **0.48** Eco-Efficiency
-0.52 **-0.57** Reducing Public Choice Distortions

Global Stewardship
-0.29 **0.52** International Commitment
-0.07 **0.67** Global-Scale Funding/Participation
-0.11 **0.05** Protecting International Commons

-3.00   3.00

Country Profiles | **Annex 3**

179

Papua New Guinea

47.1	ESI
62	Ranking
$2,359	GDP per Capita
45.3	Peer Group ESI
44	Variable Coverage (out of 67)
14	Missing Variables Imputed

Paraguay

48.8	ESI
54	Ranking
$4,288	GDP per Capita
45.8	Peer Group ESI
46	Variable Coverage (out of 67)
12	Missing Variables Imputed

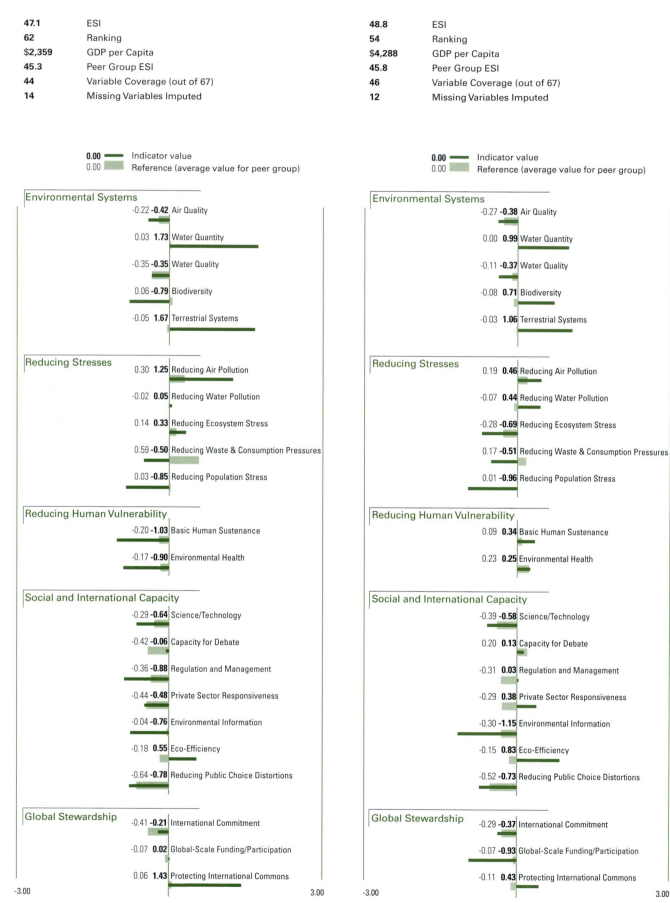

Papua New Guinea

0.00 Indicator value
0.00 Reference (average value for peer group)

Environmental Systems
-0.22 **-0.42** Air Quality
0.03 **1.73** Water Quantity
-0.35 **-0.35** Water Quality
0.06 **-0.79** Biodiversity
-0.05 **1.67** Terrestrial Systems

Reducing Stresses
0.30 **1.25** Reducing Air Pollution
-0.02 **0.05** Reducing Water Pollution
0.14 **0.33** Reducing Ecosystem Stress
0.59 **-0.50** Reducing Waste & Consumption Pressures
0.03 **-0.85** Reducing Population Stress

Reducing Human Vulnerability
-0.20 **-1.03** Basic Human Sustenance
-0.17 **-0.90** Environmental Health

Social and International Capacity
-0.29 **-0.64** Science/Technology
-0.42 **-0.06** Capacity for Debate
-0.36 **-0.88** Regulation and Management
-0.44 **-0.48** Private Sector Responsiveness
-0.04 **-0.76** Environmental Information
-0.18 **0.55** Eco-Efficiency
-0.64 **-0.78** Reducing Public Choice Distortions

Global Stewardship
-0.41 **-0.21** International Commitment
-0.07 **0.02** Global-Scale Funding/Participation
0.06 **1.43** Protecting International Commons

-3.00   3.00

Paraguay

0.00 Indicator value
0.00 Reference (average value for peer group)

Environmental Systems
-0.27 **-0.38** Air Quality
0.00 **0.99** Water Quantity
-0.11 **-0.37** Water Quality
-0.08 **0.71** Biodiversity
-0.03 **1.06** Terrestrial Systems

Reducing Stresses
0.19 **0.46** Reducing Air Pollution
-0.07 **0.44** Reducing Water Pollution
-0.28 **-0.69** Reducing Ecosystem Stress
0.17 **-0.51** Reducing Waste & Consumption Pressures
0.01 **-0.96** Reducing Population Stress

Reducing Human Vulnerability
0.09 **0.34** Basic Human Sustenance
0.23 **0.25** Environmental Health

Social and International Capacity
-0.39 **-0.58** Science/Technology
0.20 **0.13** Capacity for Debate
-0.31 **0.03** Regulation and Management
-0.29 **0.38** Private Sector Responsiveness
-0.30 **-1.15** Environmental Information
-0.15 **0.83** Eco-Efficiency
-0.52 **-0.73** Reducing Public Choice Distortions

Global Stewardship
-0.29 **-0.37** International Commitment
-0.07 **-0.93** Global-Scale Funding/Participation
-0.11 **0.43** Protecting International Commons

-3.00   3.00

Peru

54.3	ESI
38	Ranking
$4,282	GDP per Capita
45.8	Peer Group ESI
50	Variable Coverage (out of 67)
13	Missing Variables Imputed

0.00 ——— Indicator value
0.00 ▭ Reference (average value for peer group)

Philippines

35.6	ESI
112	Ranking
$3,555	GDP per Capita
45.8	Peer Group ESI
61	Variable Coverage (out of 67)
4	Missing Variables Imputed

0.00 ——— Indicator value
0.00 ▭ Reference (average value for peer group)

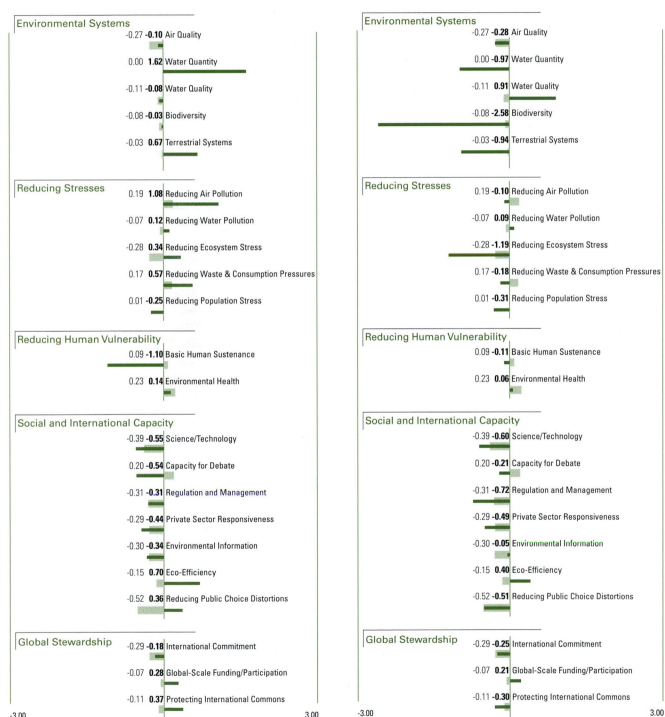

Peru

Environmental Systems
-0.27 **-0.10** Air Quality
0.00 **1.62** Water Quantity
-0.11 **-0.08** Water Quality
-0.08 **-0.03** Biodiversity
-0.03 **0.67** Terrestrial Systems

Reducing Stresses
0.19 **1.08** Reducing Air Pollution
-0.07 **0.12** Reducing Water Pollution
-0.28 **0.34** Reducing Ecosystem Stress
0.17 **0.57** Reducing Waste & Consumption Pressures
0.01 **-0.25** Reducing Population Stress

Reducing Human Vulnerability
0.09 **-1.10** Basic Human Sustenance
0.23 **0.14** Environmental Health

Social and International Capacity
-0.39 **-0.55** Science/Technology
0.20 **-0.54** Capacity for Debate
-0.31 **-0.31** Regulation and Management
-0.29 **-0.44** Private Sector Responsiveness
-0.30 **-0.34** Environmental Information
-0.15 **0.70** Eco-Efficiency
-0.52 **0.36** Reducing Public Choice Distortions

Global Stewardship
-0.29 **-0.18** International Commitment
-0.07 **0.28** Global-Scale Funding/Participation
-0.11 **0.37** Protecting International Commons

-3.00   3.00

Philippines

Environmental Systems
-0.27 **-0.28** Air Quality
0.00 **-0.97** Water Quantity
-0.11 **0.91** Water Quality
-0.08 **-2.58** Biodiversity
-0.03 **-0.94** Terrestrial Systems

Reducing Stresses
0.19 **-0.10** Reducing Air Pollution
-0.07 **0.09** Reducing Water Pollution
-0.28 **-1.19** Reducing Ecosystem Stress
0.17 **-0.18** Reducing Waste & Consumption Pressures
0.01 **-0.31** Reducing Population Stress

Reducing Human Vulnerability
0.09 **-0.11** Basic Human Sustenance
0.23 **0.06** Environmental Health

Social and International Capacity
-0.39 **-0.60** Science/Technology
0.20 **-0.21** Capacity for Debate
-0.31 **-0.72** Regulation and Management
-0.29 **-0.49** Private Sector Responsiveness
-0.30 **-0.05** Environmental Information
-0.15 **0.40** Eco-Efficiency
-0.52 **-0.51** Reducing Public Choice Distortions

Global Stewardship
-0.29 **-0.25** International Commitment
-0.07 **0.21** Global-Scale Funding/Participation
-0.11 **-0.30** Protecting International Commons

-3.00   3.00

Poland

47.6   ESI
58   Ranking
$7,619   GDP per Capita
52.2   Peer Group ESI
63   Variable Coverage (out of 67)
2   Missing Variables Imputed

Portugal

61.4   ESI
20   Ranking
$14,701   GDP per Capita
65.2   Peer Group ESI
64   Variable Coverage (out of 67)
2   Missing Variables Imputed

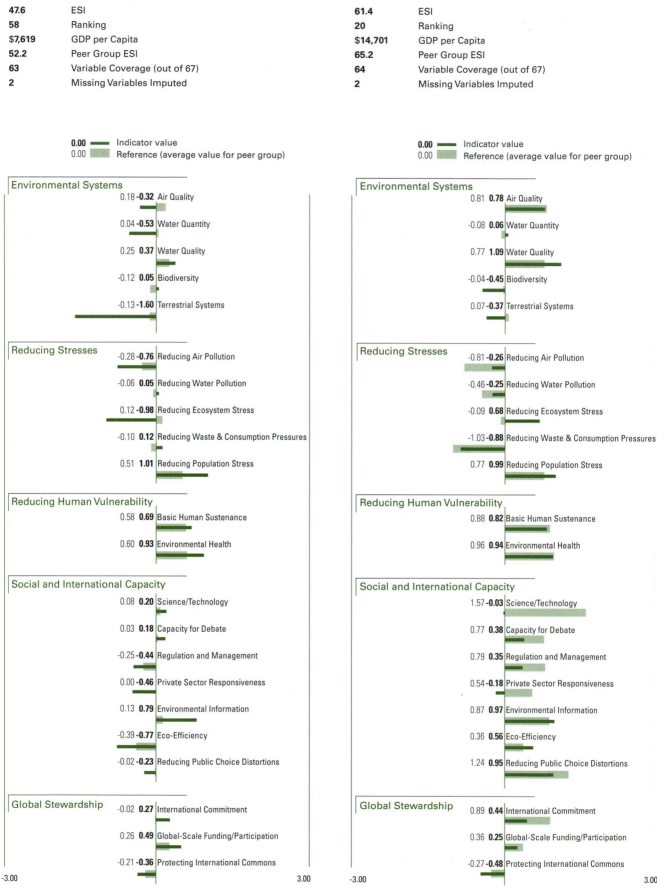

0.00 Indicator value
0.00 Reference (average value for peer group)

Environmental Systems
Poland	Portugal
0.18 **-0.32** Air Quality	0.81 **0.78** Air Quality
0.04 **-0.53** Water Quantity	-0.08 **0.06** Water Quantity
0.25 **0.37** Water Quality	0.77 **1.09** Water Quality
-0.12 **0.05** Biodiversity	-0.04 **-0.45** Biodiversity
-0.13 **-1.60** Terrestrial Systems	0.07 **-0.37** Terrestrial Systems

Reducing Stresses
Poland	Portugal
-0.28 **-0.76** Reducing Air Pollution	-0.81 **-0.26** Reducing Air Pollution
-0.06 **0.05** Reducing Water Pollution	-0.46 **-0.25** Reducing Water Pollution
0.12 **-0.98** Reducing Ecosystem Stress	-0.09 **0.68** Reducing Ecosystem Stress
-0.10 **0.12** Reducing Waste & Consumption Pressures	-1.03 **-0.88** Reducing Waste & Consumption Pressures
0.51 **1.01** Reducing Population Stress	0.77 **0.99** Reducing Population Stress

Reducing Human Vulnerability
Poland	Portugal
0.58 **0.69** Basic Human Sustenance	0.88 **0.82** Basic Human Sustenance
0.60 **0.93** Environmental Health	0.96 **0.94** Environmental Health

Social and International Capacity
Poland	Portugal
0.08 **0.20** Science/Technology	1.57 **-0.03** Science/Technology
0.03 **0.18** Capacity for Debate	0.77 **0.38** Capacity for Debate
-0.25 **-0.44** Regulation and Management	0.79 **0.35** Regulation and Management
0.00 **-0.46** Private Sector Responsiveness	0.54 **-0.18** Private Sector Responsiveness
0.13 **0.79** Environmental Information	0.87 **0.97** Environmental Information
-0.39 **-0.77** Eco-Efficiency	0.36 **0.56** Eco-Efficiency
-0.02 **-0.23** Reducing Public Choice Distortions	1.24 **0.95** Reducing Public Choice Distortions

Global Stewardship
Poland	Portugal
-0.02 **0.27** International Commitment	0.89 **0.44** International Commitment
0.26 **0.49** Global-Scale Funding/Participation	0.36 **0.25** Global-Scale Funding/Participation
-0.21 **-0.36** Protecting International Commons	-0.27 **-0.48** Protecting International Commons

-3.00   3.00

Romania

Russian Federation

	Romania			Russian Federation
ESI	44.1		ESI	56.2
Ranking	79		Ranking	33
GDP per Capita	$5,648		GDP per Capita	$6,460
Peer Group ESI	45.8		Peer Group ESI	52.2
Variable Coverage (out of 67)	53		Variable Coverage (out of 67)	60
Missing Variables Imputed	7		Missing Variables Imputed	4

0.00 ━━ Indicator value
0.00 ▬ Reference (average value for peer group)

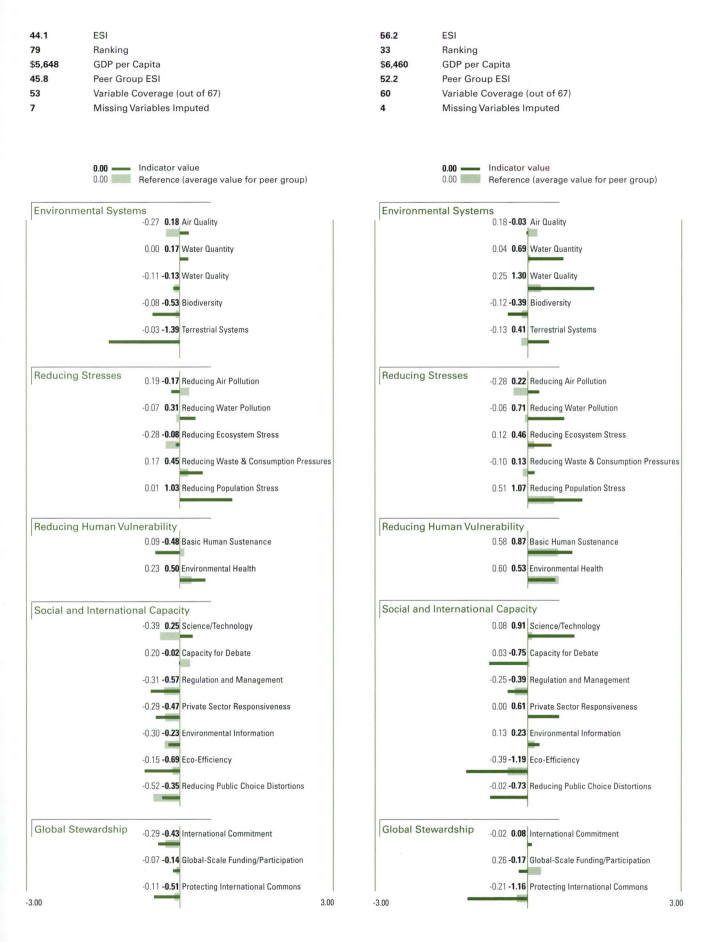

Environmental Systems

	Romania		Russian Federation
Air Quality	-0.27 **0.18**		0.18 **-0.03**
Water Quantity	0.00 **0.17**		0.04 **0.69**
Water Quality	-0.11 **-0.13**		0.25 **1.30**
Biodiversity	-0.08 **-0.53**		-0.12 **-0.39**
Terrestrial Systems	-0.03 **-1.39**		-0.13 **0.41**

Reducing Stresses

	Romania		Russian Federation
Reducing Air Pollution	0.19 **-0.17**		-0.28 **0.22**
Reducing Water Pollution	-0.07 **0.31**		-0.06 **0.71**
Reducing Ecosystem Stress	-0.28 **-0.08**		0.12 **0.46**
Reducing Waste & Consumption Pressures	0.17 **0.45**		-0.10 **0.13**
Reducing Population Stress	0.01 **1.03**		0.51 **1.07**

Reducing Human Vulnerability

	Romania		Russian Federation
Basic Human Sustenance	0.09 **-0.48**		0.58 **0.87**
Environmental Health	0.23 **0.50**		0.60 **0.53**

Social and International Capacity

	Romania		Russian Federation
Science/Technology	-0.39 **0.25**		0.08 **0.91**
Capacity for Debate	0.20 **-0.02**		0.03 **-0.75**
Regulation and Management	-0.31 **-0.57**		-0.25 **-0.39**
Private Sector Responsiveness	-0.29 **-0.47**		0.00 **0.61**
Environmental Information	-0.30 **-0.23**		0.13 **0.23**
Eco-Efficiency	-0.15 **-0.69**		-0.39 **-1.19**
Reducing Public Choice Distortions	-0.52 **-0.35**		-0.02 **-0.73**

Global Stewardship

	Romania		Russian Federation
International Commitment	-0.29 **-0.43**		-0.02 **0.08**
Global-Scale Funding/Participation	-0.07 **-0.14**		0.26 **-0.17**
Protecting International Commons	-0.11 **-0.51**		-0.21 **-1.16**

-3.00   3.00   -3.00   3.00

Rwanda

33.2	ESI
115	Ranking
N/A*	GDP per Capita
39.2	Peer Group ESI
41	Variable Coverage (out of 67)
16	Missing Variables Imputed

0.00 ━━ Indicator value
0.00 ▬ Reference (average value for peer group)

Environmental Systems

-0.48 **-0.49** Air Quality
-0.02 **-0.29** Water Quantity
-0.55 **-0.78** Water Quality
0.19 **0.76** Biodiversity
0.15 **-1.16** Terrestrial Systems

Reducing Stresses

0.57 **-0.44** Reducing Air Pollution
0.57 **1.05** Reducing Water Pollution
0.11 **0.40** Reducing Ecosystem Stress
0.69 **0.98** Reducing Waste & Consumption Pressures
-1.33 **-0.67** Reducing Population Stress

Reducing Human Vulnerability

-1.38 **-2.33** Basic Human Sustenance
-1.61 **-1.80** Environmental Health

Social and International Capacity

-0.90 **-0.74** Science/Technology
-0.56 **-1.00** Capacity for Debate
-0.26 **0.10** Regulation and Management
-0.45 **-0.43** Private Sector Responsiveness
-0.62 **-0.90** Environmental Information
0.42 **0.67** Eco-Efficiency
-0.37 **-0.32** Reducing Public Choice Distortions

Global Stewardship

-0.41 **-1.57** International Commitment
-0.49 **-1.17** Global-Scale Funding/Participation
0.62 **0.57** Protecting International Commons

-3.00   3.00

Saudi Arabia

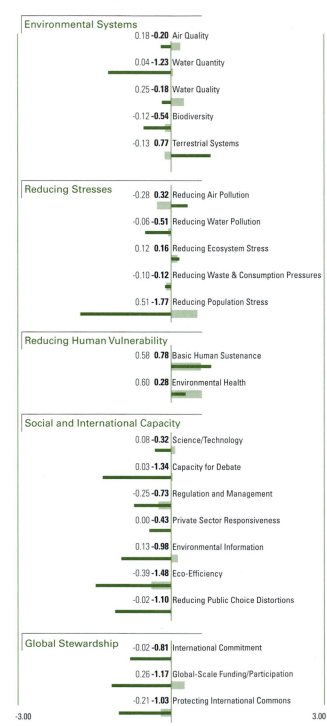

29.8	ESI
120	Ranking
$10,158	GDP per Capita
52.2	Peer Group ESI
42	Variable Coverage (out of 67)
16	Missing Variables Imputed

0.00 ━━ Indicator value
0.00 ▬ Reference (average value for peer group)

Environmental Systems

0.18 **-0.20** Air Quality
0.04 **-1.23** Water Quantity
0.25 **-0.18** Water Quality
-0.12 **-0.54** Biodiversity
-0.13 **0.77** Terrestrial Systems

Reducing Stresses

-0.28 **0.32** Reducing Air Pollution
-0.06 **-0.51** Reducing Water Pollution
0.12 **0.16** Reducing Ecosystem Stress
-0.10 **-0.12** Reducing Waste & Consumption Pressures
0.51 **-1.77** Reducing Population Stress

Reducing Human Vulnerability

0.58 **0.78** Basic Human Sustenance
0.60 **0.28** Environmental Health

Social and International Capacity

0.08 **-0.32** Science/Technology
0.03 **-1.34** Capacity for Debate
-0.25 **-0.73** Regulation and Management
0.00 **-0.43** Private Sector Responsiveness
0.13 **-0.98** Environmental Information
-0.39 **-1.48** Eco-Efficiency
-0.02 **-1.10** Reducing Public Choice Distortions

Global Stewardship

-0.02 **-0.81** International Commitment
0.26 **-1.17** Global-Scale Funding/Participation
-0.21 **-1.03** Protecting International Commons

-3.00   3.00

* Not available. Peer group assigned based on Human Development Index.

Senegal

42.3   ESI
88   Ranking
$1,307   GDP per Capita
39.2   Peer Group ESI
48   Variable Coverage (out of 67)
10   Missing Variables Imputed

Singapore

46.9   ESI
63   Ranking
$24,210   GDP per Capita
65.2   Peer Group ESI
53   Variable Coverage (out of 67)
11   Missing Variables Imputed

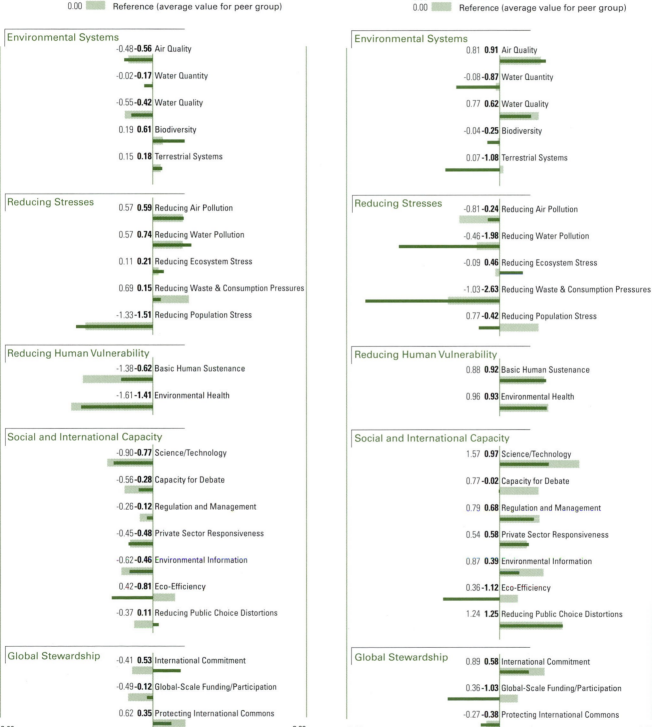

0.00 ── Indicator value
0.00 ▬ Reference (average value for peer group)

Senegal

Environmental Systems
- -0.48 **-0.56** Air Quality
- -0.02 **-0.17** Water Quantity
- -0.55 **-0.42** Water Quality
- 0.19 **0.61** Biodiversity
- 0.15 **0.18** Terrestrial Systems

Reducing Stresses
- 0.57 **0.59** Reducing Air Pollution
- 0.57 **0.74** Reducing Water Pollution
- 0.11 **0.21** Reducing Ecosystem Stress
- 0.69 **0.15** Reducing Waste & Consumption Pressures
- -1.33 **-1.51** Reducing Population Stress

Reducing Human Vulnerability
- -1.38 **-0.62** Basic Human Sustenance
- -1.61 **-1.41** Environmental Health

Social and International Capacity
- -0.90 **-0.77** Science/Technology
- -0.56 **-0.28** Capacity for Debate
- -0.26 **-0.12** Regulation and Management
- -0.45 **-0.48** Private Sector Responsiveness
- -0.62 **-0.46** Environmental Information
- 0.42 **-0.81** Eco-Efficiency
- -0.37 **0.11** Reducing Public Choice Distortions

Global Stewardship
- -0.41 **0.53** International Commitment
- -0.49 **-0.12** Global-Scale Funding/Participation
- 0.62 **0.35** Protecting International Commons

-3.00   3.00

Singapore

Environmental Systems
- 0.81 **0.91** Air Quality
- -0.08 **-0.87** Water Quantity
- 0.77 **0.62** Water Quality
- -0.04 **-0.25** Biodiversity
- 0.07 **-1.08** Terrestrial Systems

Reducing Stresses
- -0.81 **-0.24** Reducing Air Pollution
- -0.46 **-1.98** Reducing Water Pollution
- -0.09 **0.46** Reducing Ecosystem Stress
- -1.03 **-2.63** Reducing Waste & Consumption Pressures
- 0.77 **-0.42** Reducing Population Stress

Reducing Human Vulnerability
- 0.88 **0.92** Basic Human Sustenance
- 0.96 **0.93** Environmental Health

Social and International Capacity
- 1.57 **0.97** Science/Technology
- 0.77 **-0.02** Capacity for Debate
- 0.79 **0.68** Regulation and Management
- 0.54 **0.58** Private Sector Responsiveness
- 0.87 **0.39** Environmental Information
- 0.36 **-1.12** Eco-Efficiency
- 1.24 **1.25** Reducing Public Choice Distortions

Global Stewardship
- 0.89 **0.58** International Commitment
- 0.36 **-1.03** Global-Scale Funding/Participation
- -0.27 **-0.38** Protecting International Commons

-3.00   3.00

Slovak Republic

63.2   ESI
18   Ranking
$9,699   GDP per Capita
52.2   Peer Group ESI
58   Variable Coverage (out of 67)
7   Missing Variables Imputed

0.00 ▬ Indicator value
0.00 ▬ Reference (average value for peer group)

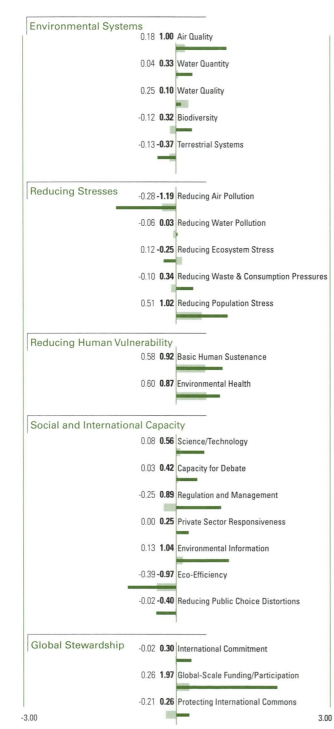

Environmental Systems
0.18 **1.00** Air Quality
0.04 **0.33** Water Quantity
0.25 **0.10** Water Quality
-0.12 **0.32** Biodiversity
-0.13 **-0.37** Terrestrial Systems

Reducing Stresses
-0.28 **-1.19** Reducing Air Pollution
-0.06 **0.03** Reducing Water Pollution
0.12 **-0.25** Reducing Ecosystem Stress
-0.10 **0.34** Reducing Waste & Consumption Pressures
0.51 **1.02** Reducing Population Stress

Reducing Human Vulnerability
0.58 **0.92** Basic Human Sustenance
0.60 **0.87** Environmental Health

Social and International Capacity
0.08 **0.56** Science/Technology
0.03 **0.42** Capacity for Debate
-0.25 **0.89** Regulation and Management
0.00 **0.25** Private Sector Responsiveness
0.13 **1.04** Environmental Information
-0.39 **-0.97** Eco-Efficiency
-0.02 **-0.40** Reducing Public Choice Distortions

Global Stewardship
-0.02 **0.30** International Commitment
0.26 **1.97** Global-Scale Funding/Participation
-0.21 **0.26** Protecting International Commons

-3.00   3.00

Slovenia

59.9   ESI
24   Ranking
$14,293   GDP per Capita
52.2   Peer Group ESI
51   Variable Coverage (out of 67)
9   Missing Variables Imputed

0.00 ▬ Indicator value
0.00 ▬ Reference (average value for peer group)

Environmental Systems
0.18 **0.52** Air Quality
0.04 **0.52** Water Quantity
0.25 **0.63** Water Quality
-0.12 **0.13** Biodiversity
-0.13 **-0.04** Terrestrial Systems

Reducing Stresses
-0.28 **-0.99** Reducing Air Pollution
-0.06 **-0.15** Reducing Water Pollution
0.12 **-0.66** Reducing Ecosystem Stress
-0.10 **-0.08** Reducing Waste & Consumption Pressures
0.51 **1.06** Reducing Population Stress

Reducing Human Vulnerability
0.58 **0.92** Basic Human Sustenance
0.60 **0.96** Environmental Health

Social and International Capacity
0.08 **0.90** Science/Technology
0.03 **0.39** Capacity for Debate
-0.25 **-0.50** Regulation and Management
0.00 **0.68** Private Sector Responsiveness
0.13 **0.26** Environmental Information
-0.39 **0.51** Eco-Efficiency
-0.02 **0.70** Reducing Public Choice Distortions

Global Stewardship
-0.02 **-0.40** International Commitment
0.26 **0.12** Global-Scale Funding/Participation
-0.21 **0.08** Protecting International Commons

-3.00   3.00

South Africa

51.2	ESI
46	Ranking
$8,488	GDP per Capita
52.2	Peer Group ESI
60	Variable Coverage (out of 67)
5	Missing Variables Imputed

0.00 ━━ Indicator value
0.00 ▬ Reference (average value for peer group)

Environmental Systems

0.18	**0.29**	Air Quality
0.04	**-0.70**	Water Quantity
0.25	**0.09**	Water Quality
-0.12	**-0.01**	Biodiversity
-0.13	**-0.50**	Terrestrial Systems

Reducing Stresses

-0.28	**-0.13**	Reducing Air Pollution
-0.06	**-0.17**	Reducing Water Pollution
0.12	**0.40**	Reducing Ecosystem Stress
-0.10	**0.30**	Reducing Waste & Consumption Pressures
0.51	**0.58**	Reducing Population Stress

Reducing Human Vulnerability

| 0.58 | **0.51** | Basic Human Sustenance |
| 0.60 | **-0.27** | Environmental Health |

Social and International Capacity

0.08	**-0.12**	Science/Technology
0.03	**0.47**	Capacity for Debate
-0.25	**0.36**	Regulation and Management
0.00	**-0.07**	Private Sector Responsiveness
0.13	**0.20**	Environmental Information
-0.39	**-0.88**	Eco-Efficiency
-0.02	**-0.02**	Reducing Public Choice Distortions

Global Stewardship

-0.02	**-0.02**	International Commitment
0.26	**0.34**	Global-Scale Funding/Participation
-0.21	**0.03**	Protecting International Commons

-3.00   3.00

South Korea

40.3	ESI
96	Ranking
$13,478	GDP per Capita
52.2	Peer Group ESI
64	Variable Coverage (out of 67)
2	Missing Variables Imputed

0.00 ━━ Indicator value
0.00 ▬ Reference (average value for peer group)

Environmental Systems

0.18	**-0.19**	Air Quality
0.04	**-0.75**	Water Quantity
0.25	**1.27**	Water Quality
-0.12	**-1.91**	Biodiversity
-0.13	**-0.33**	Terrestrial Systems

Reducing Stresses

-0.28	**-2.48**	Reducing Air Pollution
-0.06	**-1.39**	Reducing Water Pollution
0.12	**-1.25**	Reducing Ecosystem Stress
-0.10	**-1.15**	Reducing Waste & Consumption Pressures
0.51	**0.92**	Reducing Population Stress

Reducing Human Vulnerability

| 0.58 | **0.69** | Basic Human Sustenance |
| 0.60 | **0.88** | Environmental Health |

Social and International Capacity

0.08	**1.20**	Science/Technology
0.03	**-0.01**	Capacity for Debate
-0.25	**-0.28**	Regulation and Management
0.00	**0.62**	Private Sector Responsiveness
0.13	**0.23**	Environmental Information
-0.39	**-0.27**	Eco-Efficiency
-0.02	**0.31**	Reducing Public Choice Distortions

Global Stewardship

-0.02	**0.56**	International Commitment
0.26	**-1.17**	Global-Scale Funding/Participation
-0.21	**-0.90**	Protecting International Commons

-3.00   3.00

Country Profiles | **Annex 3**

187

Spain

0.00 Indicator value
0.00 Reference (average value for peer group)

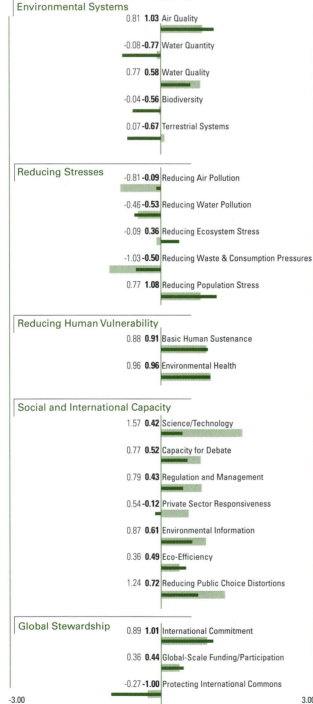

Sri Lanka

0.00 Indicator value
0.00 Reference (average value for peer group)

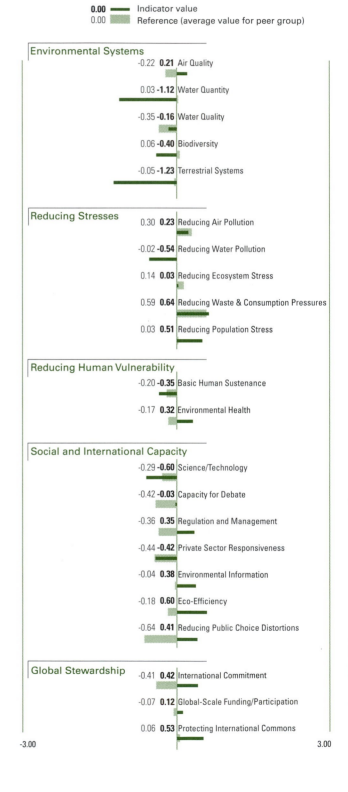

Sudan

37.6	ESI
107	Ranking
$1,394	GDP per Capita
39.2	Peer Group ESI
47	Variable Coverage (out of 67)
11	Missing Variables Imputed

0.00 ──── Indicator value
0.00 ▓▓▓ Reference (average value for peer group)

Environmental Systems

-0.48 **-0.53** Air Quality
-0.02 **-0.10** Water Quantity
-0.55 **-1.06** Water Quality
0.19 **0.53** Biodiversity
0.15 **0.90** Terrestrial Systems

Reducing Stresses

0.57 **0.95** Reducing Air Pollution
0.57 **-0.57** Reducing Water Pollution
0.11 **0.14** Reducing Ecosystem Stress
0.69 **1.08** Reducing Waste & Consumption Pressures
-1.33 **-0.80** Reducing Population Stress

Reducing Human Vulnerability

-1.38 **-1.15** Basic Human Sustenance
-1.61 **-1.15** Environmental Health

Social and International Capacity

-0.90 **-0.91** Science/Technology
-0.56 **-1.44** Capacity for Debate
-0.26 **-0.64** Regulation and Management
-0.45 **-0.48** Private Sector Responsiveness
-0.62 **-0.85** Environmental Information
0.42 **0.71** Eco-Efficiency
-0.37 **-1.02** Reducing Public Choice Distortions

Global Stewardship

-0.41 **-0.67** International Commitment
-0.49 **-0.31** Global-Scale Funding/Participation
0.62 **0.39** Protecting International Commons

-3.00   3.00

Sweden

77.1	ESI
4	Ranking
$20,659	GDP per Capita
65.2	Peer Group ESI
62	Variable Coverage (out of 67)
5	Missing Variables Imputed

0.00 ──── Indicator value
0.00 ▓▓▓ Reference (average value for peer group)

Environmental Systems

0.81 **1.45** Air Quality
-0.08 **0.41** Water Quantity
0.77 **1.19** Water Quality
-0.04 **0.45** Biodiversity
0.07 **0.59** Terrestrial Systems

Reducing Stresses

-0.81 **0.18** Reducing Air Pollution
-0.46 **0.41** Reducing Water Pollution
-0.09 **-0.50** Reducing Ecosystem Stress
-1.03 **-0.54** Reducing Waste & Consumption Pressures
0.77 **0.95** Reducing Population Stress

Reducing Human Vulnerability

0.88 **0.56** Basic Human Sustenance
0.96 **0.96** Environmental Health

Social and International Capacity

1.57 **2.61** Science/Technology
0.77 **0.73** Capacity for Debate
0.79 **0.84** Regulation and Management
0.54 **0.67** Private Sector Responsiveness
0.87 **0.51** Environmental Information
0.36 **0.80** Eco-Efficiency
1.24 **1.48** Reducing Public Choice Distortions

Global Stewardship

0.89 **1.50** International Commitment
0.36 **0.75** Global-Scale Funding/Participation
-0.27 **0.34** Protecting International Commons

-3.00   3.00

Switzerland

Country Profiles | Annex 3

190

74.6	ESI
5	Ranking
$25,512	GDP per Capita
65.2	Peer Group ESI
64	Variable Coverage (out of 67)
3	Missing Variables Imputed

0.00 ━━ Indicator value
0.00 ▬ Reference (average value for peer group)

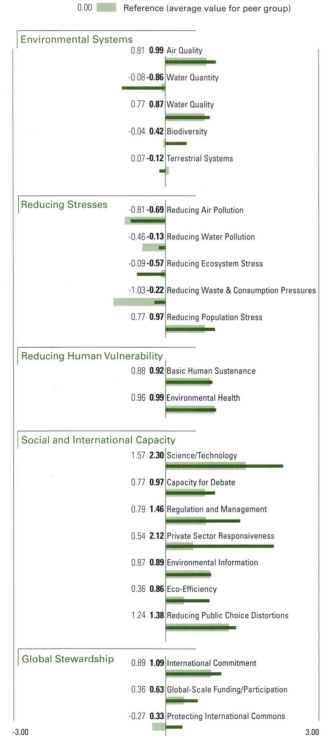

Environmental Systems
- 0.81 **0.99** Air Quality
- -0.08 **-0.86** Water Quantity
- 0.77 **0.87** Water Quality
- -0.04 **0.42** Biodiversity
- 0.07 **-0.12** Terrestrial Systems

Reducing Stresses
- -0.81 **-0.69** Reducing Air Pollution
- -0.46 **-0.13** Reducing Water Pollution
- -0.09 **-0.57** Reducing Ecosystem Stress
- -1.03 **-0.22** Reducing Waste & Consumption Pressures
- 0.77 **0.97** Reducing Population Stress

Reducing Human Vulnerability
- 0.88 **0.92** Basic Human Sustenance
- 0.96 **0.99** Environmental Health

Social and International Capacity
- 1.57 **2.30** Science/Technology
- 0.77 **0.97** Capacity for Debate
- 0.79 **1.46** Regulation and Management
- 0.54 **2.12** Private Sector Responsiveness
- 0.87 **0.89** Environmental Information
- 0.36 **0.86** Eco-Efficiency
- 1.24 **1.38** Reducing Public Choice Distortions

Global Stewardship
- 0.89 **1.09** International Commitment
- 0.36 **0.63** Global-Scale Funding/Participation
- -0.27 **0.33** Protecting International Commons

-3.00   3.00

Syria

37.9	ESI
106	Ranking
$2,892	GDP per Capita
45.3	Peer Group ESI
46	Variable Coverage (out of 67)
12	Missing Variables Imputed

0.00 ━━ Indicator value
0.00 ▬ Reference (average value for peer group)

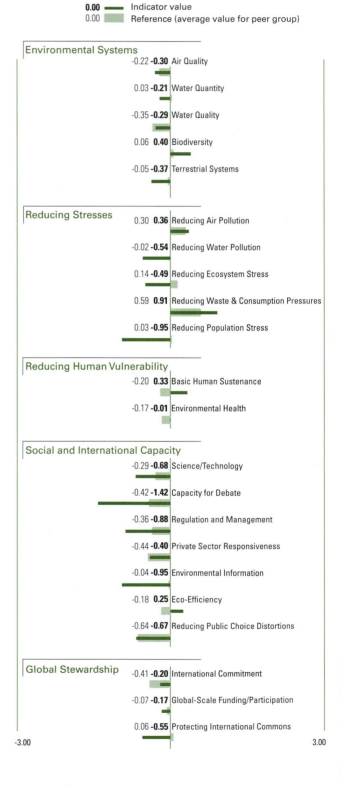

Environmental Systems
- -0.22 **-0.30** Air Quality
- 0.03 **-0.21** Water Quantity
- -0.35 **-0.29** Water Quality
- 0.06 **0.40** Biodiversity
- -0.05 **-0.37** Terrestrial Systems

Reducing Stresses
- 0.30 **0.36** Reducing Air Pollution
- -0.02 **-0.54** Reducing Water Pollution
- 0.14 **-0.49** Reducing Ecosystem Stress
- 0.59 **0.91** Reducing Waste & Consumption Pressures
- 0.03 **-0.95** Reducing Population Stress

Reducing Human Vulnerability
- -0.20 **0.33** Basic Human Sustenance
- -0.17 **-0.01** Environmental Health

Social and International Capacity
- -0.29 **-0.68** Science/Technology
- -0.42 **-1.42** Capacity for Debate
- -0.36 **-0.88** Regulation and Management
- -0.44 **-0.40** Private Sector Responsiveness
- -0.04 **-0.95** Environmental Information
- -0.18 **0.25** Eco-Efficiency
- -0.64 **-0.67** Reducing Public Choice Distortions

Global Stewardship
- -0.41 **-0.20** International Commitment
- -0.07 **-0.17** Global-Scale Funding/Participation
- 0.06 **-0.55** Protecting International Commons

-3.00   3.00

Tanzania

40.1	ESI
98	Ranking
$480	GDP per Capita
39.2	Peer Group ESI
45	Variable Coverage (out of 67)
13	Missing Variables Imputed

0.00 Indicator value
0.00 Reference (average value for peer group)

Environmental Systems
-0.48 **-0.63** Air Quality
-0.02 **-0.05** Water Quantity
-0.55 **-0.22** Water Quality
0.19 **0.01** Biodiversity
0.15 **0.16** Terrestrial Systems

Reducing Stresses
0.57 **0.86** Reducing Air Pollution
0.57 **0.47** Reducing Water Pollution
0.11 **0.09** Reducing Ecosystem Stress
0.69 **0.37** Reducing Waste & Consumption Pressures
-1.33 **-1.54** Reducing Population Stress

Reducing Human Vulnerability
-1.38 **-1.45** Basic Human Sustenance
-1.61 **-1.53** Environmental Health

Social and International Capacity
-0.90 **-1.17** Science/Technology
-0.56 **-0.57** Capacity for Debate
-0.26 **0.34** Regulation and Management
-0.45 **-0.48** Private Sector Responsiveness
-0.62 **-0.04** Environmental Information
0.42 **0.77** Eco-Efficiency
-0.37 **-0.41** Reducing Public Choice Distortions

Global Stewardship
-0.41 **0.15** International Commitment
-0.49 **-1.15** Global-Scale Funding/Participation
0.62 **0.52** Protecting International Commons

-3.00   3.00

Thailand

45.2	ESI
75	Ranking
$5,456	GDP per Capita
45.8	Peer Group ESI
61	Variable Coverage (out of 67)
4	Missing Variables Imputed

0.00 Indicator value
0.00 Reference (average value for peer group)

Environmental Systems
-0.27 **0.60** Air Quality
0.00 **0.23** Water Quantity
-0.11 **-0.59** Water Quality
-0.08 **-0.64** Biodiversity
-0.03 **-1.36** Terrestrial Systems

Reducing Stresses
0.19 **0.08** Reducing Air Pollution
-0.07 **0.11** Reducing Water Pollution
-0.28 **-0.70** Reducing Ecosystem Stress
0.17 **-0.12** Reducing Waste & Consumption Pressures
0.01 **0.73** Reducing Population Stress

Reducing Human Vulnerability
0.09 **-0.36** Basic Human Sustenance
0.23 **0.26** Environmental Health

Social and International Capacity
-0.39 **-0.67** Science/Technology
0.20 **-0.20** Capacity for Debate
-0.31 **0.12** Regulation and Management
-0.29 **0.46** Private Sector Responsiveness
-0.30 **0.56** Environmental Information
-0.15 **0.04** Eco-Efficiency
-0.52 **-0.72** Reducing Public Choice Distortions

Global Stewardship
-0.29 **-0.06** International Commitment
-0.07 **0.31** Global-Scale Funding/Participation
-0.11 **-0.76** Protecting International Commons

-3.00   3.00

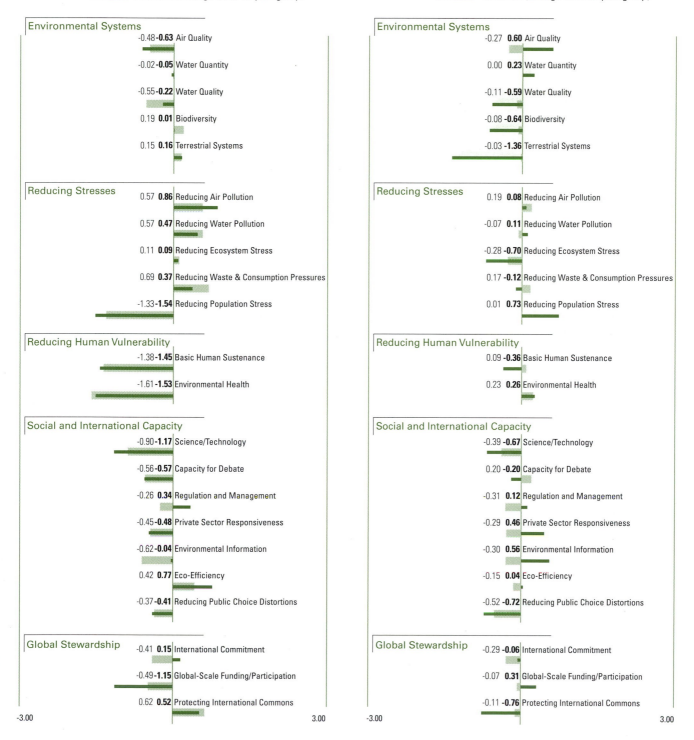

Togo

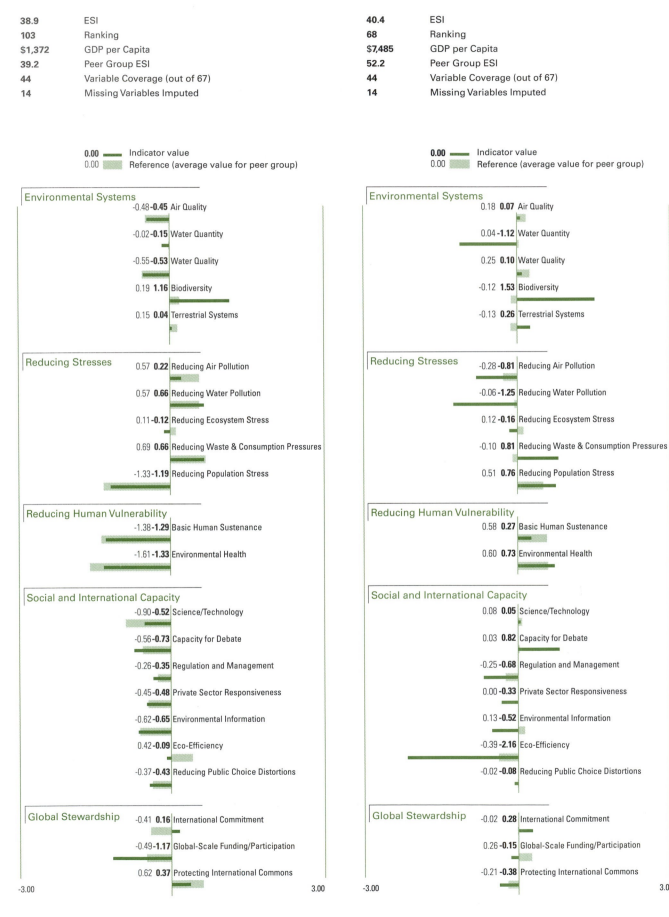

38.9	ESI	
103	Ranking	
$1,372	GDP per Capita	
39.2	Peer Group ESI	
44	Variable Coverage (out of 67)	
14	Missing Variables Imputed	

0.00 ━━ Indicator value
0.00 ▭ Reference (average value for peer group)

Environmental Systems
-0.48 **-0.45** Air Quality
-0.02 **-0.15** Water Quantity
-0.55 **-0.53** Water Quality
0.19 **1.16** Biodiversity
0.15 **0.04** Terrestrial Systems

Reducing Stresses
0.57 **0.22** Reducing Air Pollution
0.57 **0.66** Reducing Water Pollution
0.11 **-0.12** Reducing Ecosystem Stress
0.69 **0.66** Reducing Waste & Consumption Pressures
-1.33 **-1.19** Reducing Population Stress

Reducing Human Vulnerability
-1.38 **-1.29** Basic Human Sustenance
-1.61 **-1.33** Environmental Health

Social and International Capacity
-0.90 **-0.52** Science/Technology
-0.56 **-0.73** Capacity for Debate
-0.26 **-0.35** Regulation and Management
-0.45 **-0.48** Private Sector Responsiveness
-0.62 **-0.65** Environmental Information
0.42 **-0.09** Eco-Efficiency
-0.37 **-0.43** Reducing Public Choice Distortions

Global Stewardship
-0.41 **0.16** International Commitment
-0.49 **-1.17** Global-Scale Funding/Participation
0.62 **0.37** Protecting International Commons

-3.00   3.00

Trinidad and Tobago

40.4	ESI	
68	Ranking	
$7,485	GDP per Capita	
52.2	Peer Group ESI	
44	Variable Coverage (out of 67)	
14	Missing Variables Imputed	

0.00 ━━ Indicator value
0.00 ▭ Reference (average value for peer group)

Environmental Systems
0.18 **0.07** Air Quality
0.04 **-1.12** Water Quantity
0.25 **0.10** Water Quality
-0.12 **1.53** Biodiversity
-0.13 **0.26** Terrestrial Systems

Reducing Stresses
-0.28 **-0.81** Reducing Air Pollution
-0.06 **-1.25** Reducing Water Pollution
0.12 **-0.16** Reducing Ecosystem Stress
-0.10 **0.81** Reducing Waste & Consumption Pressures
0.51 **0.76** Reducing Population Stress

Reducing Human Vulnerability
0.58 **0.27** Basic Human Sustenance
0.60 **0.73** Environmental Health

Social and International Capacity
0.08 **0.05** Science/Technology
0.03 **0.82** Capacity for Debate
-0.25 **-0.68** Regulation and Management
0.00 **-0.33** Private Sector Responsiveness
0.13 **-0.52** Environmental Information
-0.39 **-2.16** Eco-Efficiency
-0.02 **-0.08** Reducing Public Choice Distortions

Global Stewardship
-0.02 **0.28** International Commitment
0.26 **-0.15** Global-Scale Funding/Participation
-0.21 **-0.38** Protecting International Commons

-3.00   3.00

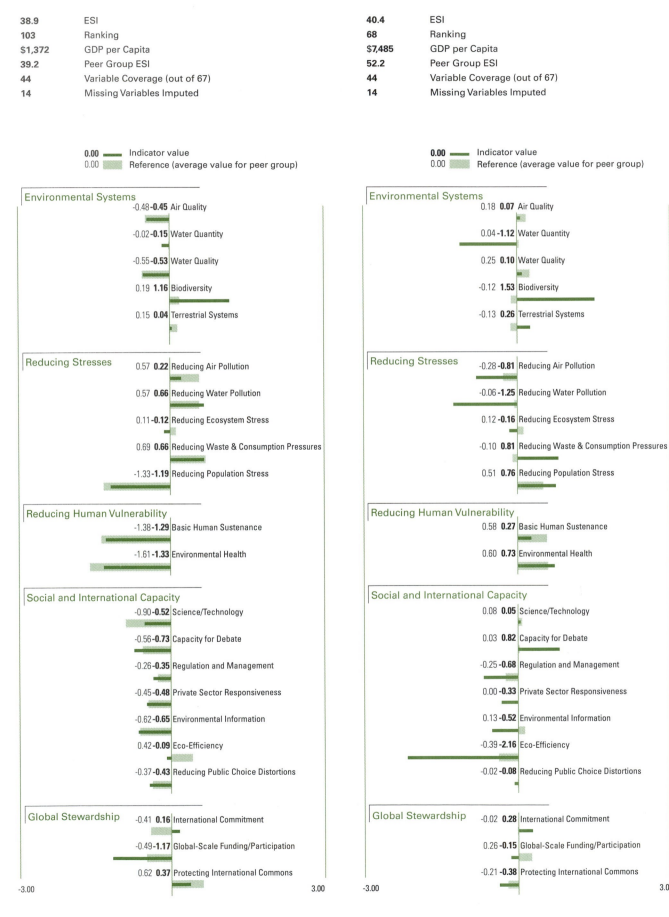

Tunisia

43.7   ESI
83   Ranking
$5,404   GDP per Capita
45.8   Peer Group ESI
48   Variable Coverage (out of 67)
11   Missing Variables Imputed

0.00 ━━ Indicator value
0.00 ▬ Reference (average value for peer group)

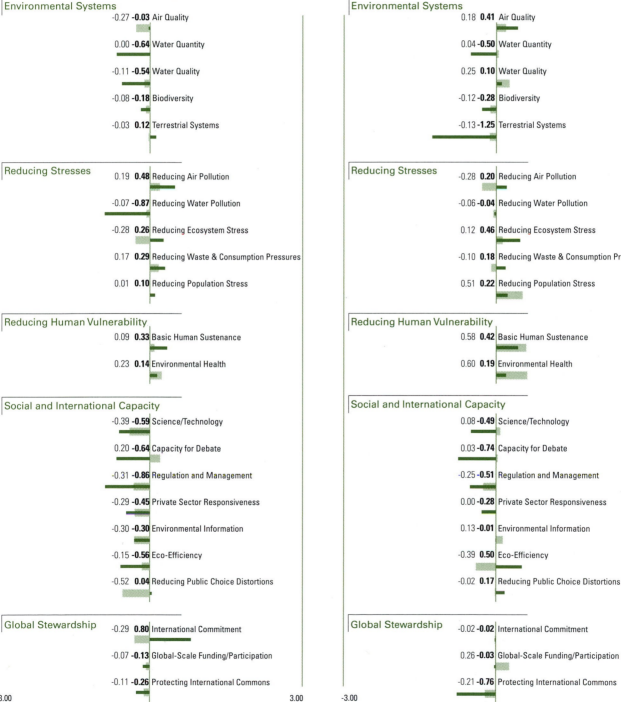

Environmental Systems

-0.27 **-0.03**	Air Quality
0.00 **-0.64**	Water Quantity
-0.11 **-0.54**	Water Quality
-0.08 **-0.18**	Biodiversity
-0.03 **0.12**	Terrestrial Systems

Reducing Stresses

0.19 **0.48**	Reducing Air Pollution
-0.07 **-0.87**	Reducing Water Pollution
-0.28 **0.26**	Reducing Ecosystem Stress
0.17 **0.29**	Reducing Waste & Consumption Pressures
0.01 **0.10**	Reducing Population Stress

Reducing Human Vulnerability

0.09 **0.33**	Basic Human Sustenance
0.23 **0.14**	Environmental Health

Social and International Capacity

-0.39 **-0.59**	Science/Technology
0.20 **-0.64**	Capacity for Debate
-0.31 **-0.86**	Regulation and Management
-0.29 **-0.45**	Private Sector Responsiveness
-0.30 **-0.30**	Environmental Information
-0.15 **-0.56**	Eco-Efficiency
-0.52 **0.04**	Reducing Public Choice Distortions

Global Stewardship

-0.29 **0.80**	International Commitment
-0.07 **-0.13**	Global-Scale Funding/Participation
-0.11 **-0.26**	Protecting International Commons

-3.00   3.00

Turkey

46.3   ESI
70   Ranking
$6,422   GDP per Capita
52.2   Peer Group ESI
59   Variable Coverage (out of 67)
6   Missing Variables Imputed

0.00 ━━ Indicator value
0.00 ▬ Reference (average value for peer group)

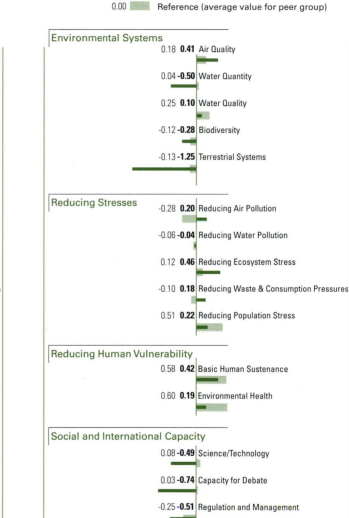

Environmental Systems

0.18 **0.41**	Air Quality
0.04 **-0.50**	Water Quantity
0.25 **0.10**	Water Quality
-0.12 **-0.28**	Biodiversity
-0.13 **-1.25**	Terrestrial Systems

Reducing Stresses

-0.28 **0.20**	Reducing Air Pollution
-0.06 **-0.04**	Reducing Water Pollution
0.12 **0.46**	Reducing Ecosystem Stress
-0.10 **0.18**	Reducing Waste & Consumption Pressures
0.51 **0.22**	Reducing Population Stress

Reducing Human Vulnerability

0.58 **0.42**	Basic Human Sustenance
0.60 **0.19**	Environmental Health

Social and International Capacity

0.08 **-0.49**	Science/Technology
0.03 **-0.74**	Capacity for Debate
-0.25 **-0.51**	Regulation and Management
0.00 **-0.28**	Private Sector Responsiveness
0.13 **-0.01**	Environmental Information
-0.39 **0.50**	Eco-Efficiency
-0.02 **0.17**	Reducing Public Choice Distortions

Global Stewardship

-0.02 **-0.02**	International Commitment
0.26 **-0.03**	Global-Scale Funding/Participation
-0.21 **-0.76**	Protecting International Commons

-3.00   3.00

Uganda

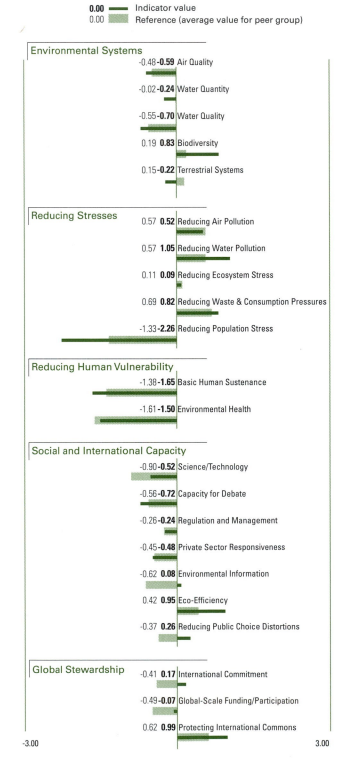

0.00 Indicator value
0.00 Reference (average value for peer group)

Environmental Systems
-0.48 **-0.59** Air Quality
-0.02 **-0.24** Water Quantity
-0.55 **-0.70** Water Quality
0.19 **0.83** Biodiversity
0.15 **-0.22** Terrestrial Systems

Reducing Stresses
0.57 **0.52** Reducing Air Pollution
0.57 **1.05** Reducing Water Pollution
0.11 **0.09** Reducing Ecosystem Stress
0.69 **0.82** Reducing Waste & Consumption Pressures
-1.33 **-2.26** Reducing Population Stress

Reducing Human Vulnerability
-1.38 **-1.65** Basic Human Sustenance
-1.61 **-1.50** Environmental Health

Social and International Capacity
-0.90 **-0.52** Science/Technology
-0.56 **-0.72** Capacity for Debate
-0.26 **-0.24** Regulation and Management
-0.45 **-0.48** Private Sector Responsiveness
-0.62 **0.08** Environmental Information
0.42 **0.95** Eco-Efficiency
-0.37 **0.26** Reducing Public Choice Distortions

Global Stewardship
-0.41 **0.17** International Commitment
-0.49 **-0.07** Global-Scale Funding/Participation
0.62 **0.99** Protecting International Commons

-3.00   3.00

Ukraine

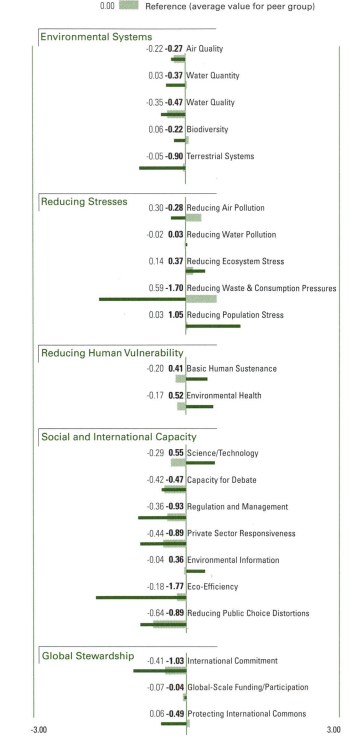

0.00 Indicator value
0.00 Reference (average value for peer group)

Environmental Systems
-0.22 **-0.27** Air Quality
0.03 **-0.37** Water Quantity
-0.35 **-0.47** Water Quality
0.06 **-0.22** Biodiversity
-0.05 **-0.90** Terrestrial Systems

Reducing Stresses
0.30 **-0.28** Reducing Air Pollution
-0.02 **0.03** Reducing Water Pollution
0.14 **0.37** Reducing Ecosystem Stress
0.59 **-1.70** Reducing Waste & Consumption Pressures
0.03 **1.05** Reducing Population Stress

Reducing Human Vulnerability
-0.20 **0.41** Basic Human Sustenance
-0.17 **0.52** Environmental Health

Social and International Capacity
-0.29 **0.55** Science/Technology
-0.42 **-0.47** Capacity for Debate
-0.36 **-0.93** Regulation and Management
-0.44 **-0.89** Private Sector Responsiveness
-0.04 **0.36** Environmental Information
-0.18 **-1.77** Eco-Efficiency
-0.64 **-0.89** Reducing Public Choice Distortions

Global Stewardship
-0.41 **-1.03** International Commitment
-0.07 **-0.04** Global-Scale Funding/Participation
0.06 **-0.49** Protecting International Commons

-3.00   3.00

United Kingdom

64.1	ESI
16	Ranking
$20,336	GDP per Capita
65.2	Peer Group ESI
65	Variable Coverage (out of 67)
2	Missing Variables Imputed

0.00 ━━ Indicator value
0.00 ▩ Reference (average value for peer group)

Environmental Systems

0.81	**0.41**	Air Quality
-0.08	**-0.80**	Water Quantity
0.77	**1.42**	Water Quality
-0.04	**0.58**	Biodiversity
0.07	**-0.60**	Terrestrial Systems

Reducing Stresses

-0.81	**-1.45**	Reducing Air Pollution
-0.46	**-0.38**	Reducing Water Pollution
-0.09	**-0.62**	Reducing Ecosystem Stress
-1.03	**-1.99**	Reducing Waste & Consumption Pressures
0.77	**0.86**	Reducing Population Stress

Reducing Human Vulnerability

0.88	**0.92**	Basic Human Sustenance
0.96	**0.94**	Environmental Health

Social and International Capacity

1.57	**1.62**	Science/Technology
0.77	**0.55**	Capacity for Debate
0.79	**1.54**	Regulation and Management
0.54	**1.02**	Private Sector Responsiveness
0.87	**1.31**	Environmental Information
0.36	**0.10**	Eco-Efficiency
1.24	**1.60**	Reducing Public Choice Distortions

Global Stewardship

0.89	**1.12**	International Commitment
0.36	**0.43**	Global-Scale Funding/Participation
-0.27	**-0.65**	Protecting International Commons

-3.00   3.00

United States

66.1	ESI
11	Ranking
$29,605	GDP per Capita
65.2	Peer Group ESI
62	Variable Coverage (out of 67)
4	Missing Variables Imputed

0.00 ━━ Indicator value
0.00 ▩ Reference (average value for peer group)

Environmental Systems

0.81	**0.61**	Air Quality
-0.08	**0.16**	Water Quantity
0.77	**1.04**	Water Quality
-0.04	**-0.36**	Biodiversity
0.07	**0.22**	Terrestrial Systems

Reducing Stresses

-0.81	**-0.64**	Reducing Air Pollution
-0.46	**0.07**	Reducing Water Pollution
-0.09	**0.18**	Reducing Ecosystem Stress
-1.03	**-1.68**	Reducing Waste & Consumption Pressures
0.77	**0.40**	Reducing Population Stress

Reducing Human Vulnerability

0.88	**0.92**	Basic Human Sustenance
0.96	**0.94**	Environmental Health

Social and International Capacity

1.57	**2.27**	Science/Technology
0.77	**0.31**	Capacity for Debate
0.79	**1.30**	Regulation and Management
0.54	**0.84**	Private Sector Responsiveness
0.87	**1.57**	Environmental Information
0.36	**0.26**	Eco-Efficiency
1.24	**0.24**	Reducing Public Choice Distortions

Global Stewardship

0.89	**0.78**	International Commitment
0.36	**0.49**	Global-Scale Funding/Participation
-0.27	**-0.79**	Protecting International Commons

-3.00   3.00

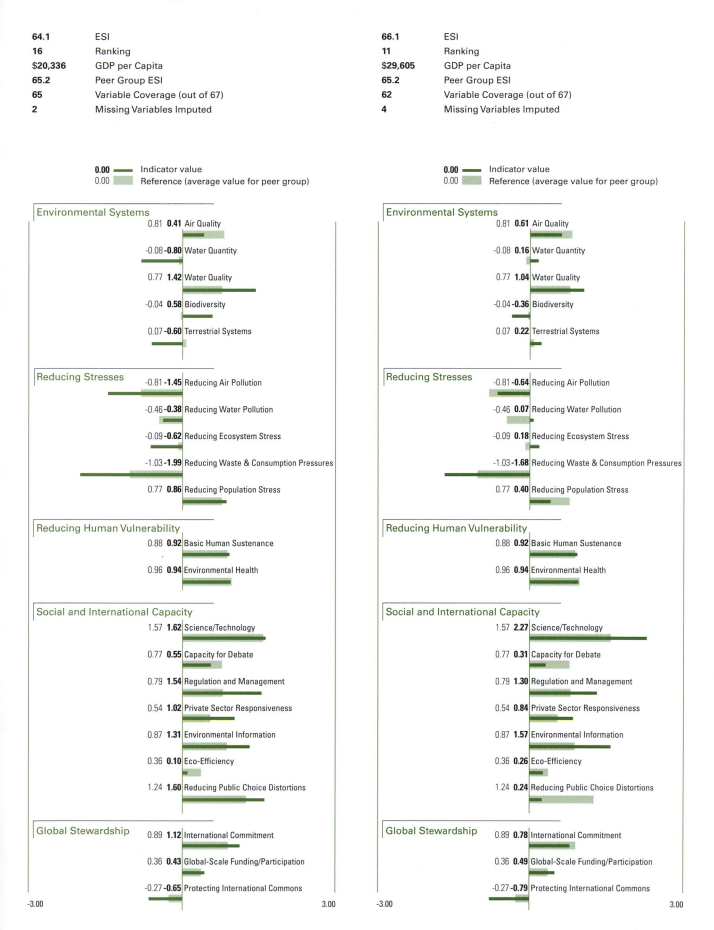

Uruguay

64.6	ESI
14	Ranking
$8,623	GDP per Capita
52.2	Peer Group ESI
48	Variable Coverage (out of 67)
10	Missing Variables Imputed

0.00 ━━ Indicator value
0.00 ▬ Reference (average value for peer group)

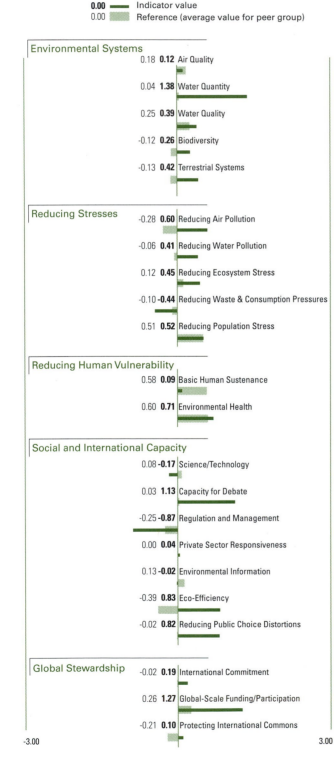

Uzbekistan

41.6	ESI
90	Ranking
$2,053	GDP per Capita
45.3	Peer Group ESI
43	Variable Coverage (out of 67)
16	Missing Variables Imputed

0.00 ━━ Indicator value
0.00 ▬ Reference (average value for peer group)

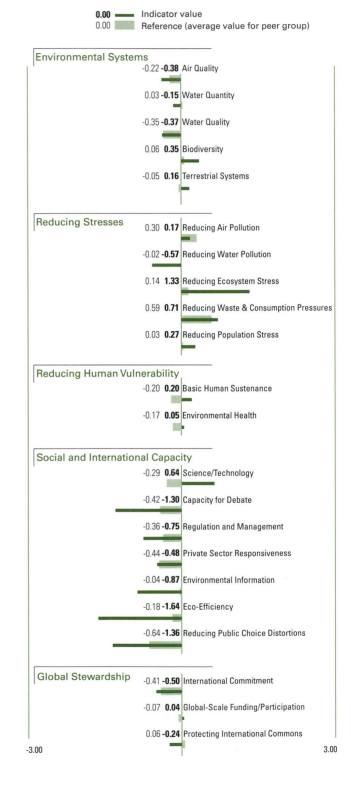

Venezuela

50.8   ESI
47   Ranking
$5,808   GDP per Capita
45.8   Peer Group ESI
58   Variable Coverage (out of 67)
6   Missing Variables Imputed

0.00 ▬▬ Indicator value
0.00 ▬▬ Reference (average value for peer group)

Environmental Systems
-0.27 **0.19** Air Quality
0.00 **1.49** Water Quantity
-0.11 **-0.01** Water Quality
-0.08 **0.53** Biodiversity
-0.03 **0.81** Terrestrial Systems

Reducing Stresses
0.19 **0.19** Reducing Air Pollution
-0.07 **0.37** Reducing Water Pollution
-0.28 **0.03** Reducing Ecosystem Stress
0.17 **0.62** Reducing Waste & Consumption Pressures
0.01 **-0.08** Reducing Population Stress

Reducing Human Vulnerability
0.09 **-0.40** Basic Human Sustenance
0.23 **0.19** Environmental Health

Social and International Capacity
-0.39 **-0.46** Science/Technology
0.20 **-0.41** Capacity for Debate
-0.31 **0.28** Regulation and Management
-0.29 **-0.79** Private Sector Responsiveness
-0.30 **-0.35** Environmental Information
-0.15 **-0.18** Eco-Efficiency
-0.52 **-1.21** Reducing Public Choice Distortions

Global Stewardship
-0.29 **-0.16** International Commitment
-0.07 **0.39** Global-Scale Funding/Participation
-0.11 **-0.59** Protecting International Commons

-3.00   3.00

Vietnam

34.2   ESI
114   Ranking
$1,689   GDP per Capita
45.3   Peer Group ESI
49   Variable Coverage (out of 67)
15   Missing Variables Imputed

0.00 ▬▬ Indicator value
0.00 ▬▬ Reference (average value for peer group)

Environmental Systems
-0.22 **0.15** Air Quality
0.03 **0.22** Water Quantity
-0.35 **-0.23** Water Quality
0.06 **-1.08** Biodiversity
-0.05 **-1.23** Terrestrial Systems

Reducing Stresses
0.30 **-0.09** Reducing Air Pollution
-0.02 **-0.26** Reducing Water Pollution
0.14 **-1.01** Reducing Ecosystem Stress
0.59 **0.64** Reducing Waste & Consumption Pressures
0.03 **0.19** Reducing Population Stress

Reducing Human Vulnerability
-0.20 **-0.70** Basic Human Sustenance
-0.17 **-0.04** Environmental Health

Social and International Capacity
-0.29 **-0.42** Science/Technology
-0.42 **-1.44** Capacity for Debate
-0.36 **-0.81** Regulation and Management
-0.44 **-0.70** Private Sector Responsiveness
-0.04 **-0.18** Environmental Information
-0.18 **-0.65** Eco-Efficiency
-0.64 **-0.77** Reducing Public Choice Distortions

Global Stewardship
-0.41 **-0.42** International Commitment
-0.07 **-0.17** Global-Scale Funding/Participation
0.06 **0.02** Protecting International Commons

-3.00   3.00

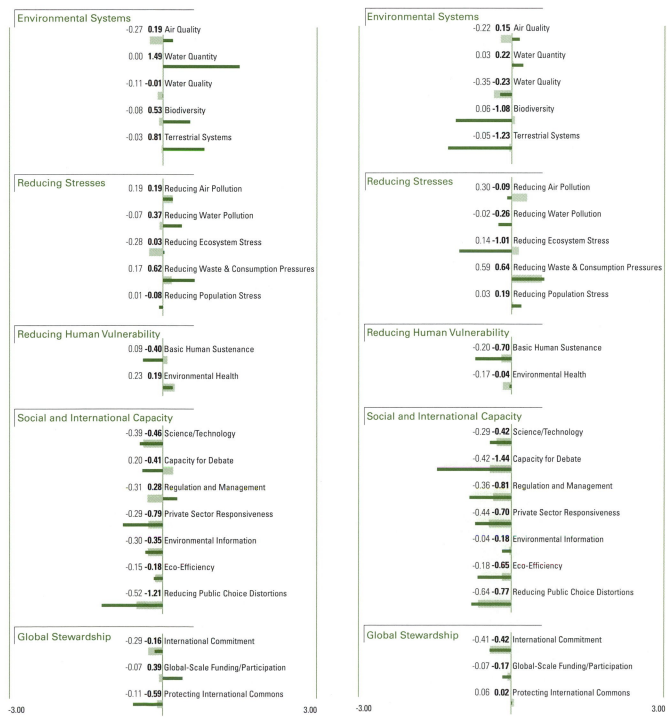

Zambia

39.5   ESI
100   Ranking
$719   GDP per Capita
39.2   Peer Group ESI
44   Variable Coverage (out of 67)
14   Missing Variables Imputed

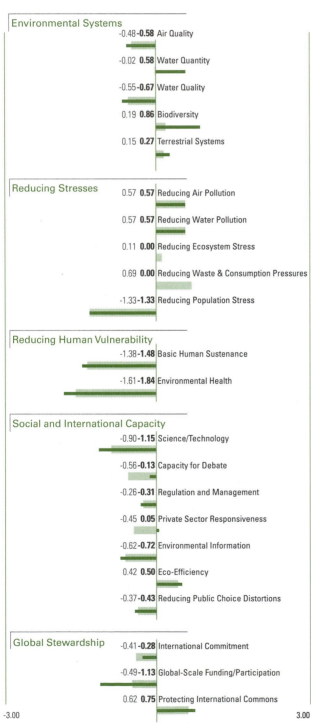

0.00 Indicator value
0.00 Reference (average value for peer group)

Environmental Systems
-0.48 **-0.58** Air Quality
-0.02 **0.58** Water Quantity
-0.55 **-0.67** Water Quality
0.19 **0.86** Biodiversity
0.15 **0.27** Terrestrial Systems

Reducing Stresses
0.57 **0.57** Reducing Air Pollution
0.57 **0.57** Reducing Water Pollution
0.11 **0.00** Reducing Ecosystem Stress
0.69 **0.00** Reducing Waste & Consumption Pressures
-1.33 **-1.33** Reducing Population Stress

Reducing Human Vulnerability
-1.38 **-1.48** Basic Human Sustenance
-1.61 **-1.84** Environmental Health

Social and International Capacity
-0.90 **-1.15** Science/Technology
-0.56 **-0.13** Capacity for Debate
-0.26 **-0.31** Regulation and Management
-0.45 **0.05** Private Sector Responsiveness
-0.62 **-0.72** Environmental Information
0.42 **0.50** Eco-Efficiency
-0.37 **-0.43** Reducing Public Choice Distortions

Global Stewardship
-0.41 **-0.28** International Commitment
-0.49 **-1.13** Global-Scale Funding/Participation
0.62 **0.75** Protecting International Commons

-3.00   3.00

Zimbabwe

51.9   ESI
43   Ranking
$2,669   GDP per Capita
45.3   Peer Group ESI
53   Variable Coverage (out of 67)
10   Missing Variables Imputed

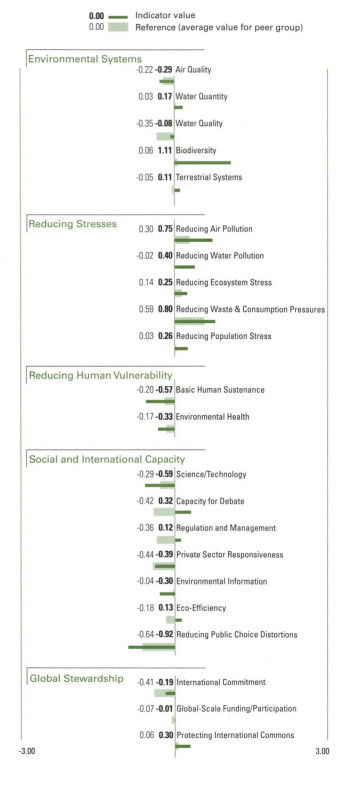

0.00 Indicator value
0.00 Reference (average value for peer group)

Environmental Systems
-0.22 **-0.29** Air Quality
0.03 **0.17** Water Quantity
-0.35 **-0.08** Water Quality
0.06 **1.11** Biodiversity
-0.05 **0.11** Terrestrial Systems

Reducing Stresses
0.30 **0.75** Reducing Air Pollution
-0.02 **0.40** Reducing Water Pollution
0.14 **0.25** Reducing Ecosystem Stress
0.59 **0.80** Reducing Waste & Consumption Pressures
0.03 **0.26** Reducing Population Stress

Reducing Human Vulnerability
-0.20 **-0.57** Basic Human Sustenance
-0.17 **-0.33** Environmental Health

Social and International Capacity
-0.29 **-0.59** Science/Technology
-0.42 **0.32** Capacity for Debate
-0.36 **0.12** Regulation and Management
-0.44 **-0.39** Private Sector Responsiveness
-0.04 **-0.30** Environmental Information
-0.18 **0.13** Eco-Efficiency
-0.64 **-0.92** Reducing Public Choice Distortions

Global Stewardship
-0.41 **-0.19** International Commitment
-0.07 **-0.01** Global-Scale Funding/Participation
0.06 **0.30** Protecting International Commons

-3.00   3.00

Annex 4
Variable Descriptions and Data

Environmental Performance Measurement:

The Global Report 2001–2002

This section contains complete variable descriptions along with the original data used to produce the 2001 Environmental Sustainability Index. Each page contains the following information:

The component and indicator in which the variable is located.

- The variable name.

- The variable code and number.

- The units for the data shown in the data table.

- The reference year (MRYA = Most Recent Year Available for the stated range).

- Data source.

- The logic for including the variable in the ESI.

- A details section summarizing the methodology used to create the variable.

- The median, minimum, and maximum data values for that variable.

- A data table containing the original data for the variable, sorted in alphabetical order by country.

Additional information on the methodology used to create several of the more innovative variables can be found in the section of the 2001 ESI Report entitled "Challenges in Measuring Environmental Sustainability".

Environmental Systems

Variable name

Urban SO2 Concentration

Variable code   **Variable number**
SO2   1

Units
Micrograms per Cubic Meter

Reference year
MRYA 1990-1996

Median	Minimum	Maximum
20.49	1	209

Source
World Bank, World Development Indicators 2000, and WHO, Air Management Information System-AMIS 2.0, 1998.

Logic
Indicator of Urban Air Quality.

Details
The values were originally collected at the city level. The number of cities with data provided by each country varied. Within each country the values have been normalized by city population for the year 1995, then summed to give the total concentration for the given country.

	Country			Country			Country
	Albania		34.00	Greece		5.47	Norway
	Algeria			Guatemala			Pakistan
1.02	Argentina			Haiti			Panama
	Armenia			Honduras			Papua New Guinea
13.17	Australia		37.33	Hungary			Paraguay
13.21	Austria		5.00	Iceland			Peru
	Azerbaijan		27.55	India		33.00	Philippines
	Bangladesh			Indonesia		54.72	Poland
21.02	Belarus		209.00	Iran		9.22	Portugal
	Belgium		18.89	Ireland		10.00	Romania
	Benin			Israel		97.55	Russian Federation
	Bhutan		15.55	Italy			Rwanda
	Bolivia			Jamaica			Saudi Arabia
	Botswana		24.33	Japan			Senegal
75.78	Brazil			Jordan		20.00	Singapore
52.45	Bulgaria			Kazakhstan		22.66	Slovak Republic
	Burkina Faso			Kenya			Slovenia
	Burundi		52.41	Korea, South		22.37	South Africa
	Cameroon			Kuwait		11.00	Spain
12.87	Canada			Kyrgyz Republic			Sri Lanka
	Central African Republic		5.36	Latvia			Sudan
29.00	Chile			Lebanon		5.23	Sweden
97.07	China			Libya		11.34	Switzerland
	Colombia		2.10	Lithuania			Syria
38.84	Costa Rica			Macedonia			Tanzania
31.00	Croatia			Madagascar		11.00	Thailand
1.00	Cuba			Malawi			Togo
27.34	Czech Republic		20.49	Malaysia			Trinidad and Tobago
7.00	Denmark			Mali			Tunisia
	Dominican Republic			Mauritius		87.02	Turkey
21.52	Ecuador		74.00	Mexico			Uganda
69.00	Egypt			Moldova			Ukraine
	El Salvador			Mongolia		21.96	United Kingdom
	Estonia			Morocco		15.43	United States
	Ethiopia			Mozambique			Uruguay
	Fiji			Nepal			Uzbekistan
4.38	Finland		10.00	Netherlands		33.00	Venezuela
13.89	France		3.49	New Zealand			Vietnam
	Gabon			Nicaragua			Zambia
12.80	Germany			Niger			Zimbabwe
	Ghana			Nigeria			

Environmental Systems

Variable name

Urban NO2 Concentration

Variable code
NO2

Variable number
2

Units
Micrograms per Cubic Meter

Reference year
MRYA 1990-1996

Median	Minimum	Maximum
45.11	0	130

Source
World Bank, World Development Indicators 2000, and WHO, Air Management Information System-AMIS 2.0,1998

Logic
Indicator of Urban Air Quality.

Details
The values were originally collected at the city level. The number of cities with data provided by each country varied. Within each country the values have been normalized by city population for the year 1995, then summed to give the total concentration for the given country.

	Albania	64.00	Greece	49.65	Norway
	Algeria	69.33	Guatemala		Pakistan
56.79	Argentina		Haiti	42.00	Panama
	Armenia	29.50	Honduras		Papua New Guinea
16.47	Australia	45.11	Hungary		Paraguay
39.75	Austria	42.00	Iceland		Peru
	Azerbaijan	29.68	India		Philippines
	Bangladesh		Indonesia	58.14	Poland
42.60	Belarus		Iran	49.57	Portugal
46.79	Belgium		Ireland	71.00	Romania
	Benin		Israel	3.44	Russian Federation
	Bhutan	124.38	Italy		Rwanda
	Bolivia		Jamaica		Saudi Arabia
	Botswana	62.01	Japan		Senegal
51.37	Brazil		Jordan	30.00	Singapore
111.14	Bulgaria		Kazakhstan	25.62	Slovak Republic
	Burkina Faso		Kenya		Slovenia
	Burundi	52.86	Korea, Rep.	44.03	South Africa
	Cameroon		Kuwait	32.36	Spain
41.24	Canada		Kyrgyz Republic		Sri Lanka
	Central African Republic	63.74	Latvia		Sudan
81.00	Chile		Lebanon	29.68	Sweden
71.72	China		Libya	42.20	Switzerland
	Colombia	28.31	Lithuania		Syria
45.75	Costa Rica		Macedonia		Tanzania
	Croatia		Madagascar	23.00	Thailand
5.00	Cuba		Malawi		Togo
28.59	Czech Republic	0.00	Malaysia		Trinidad and Tobago
54.00	Denmark		Mali		Tunisia
	Dominican Republic		Mauritius	9.45	Turkey
	Ecuador	130.00	Mexico		Uganda
	Egypt		Moldova		Ukraine
70.50	El Salvador		Mongolia	64.47	United Kingdom
	Estonia		Morocco	60.57	United States
	Ethiopia		Mozambique		Uruguay
	Fiji		Nepal		Uzbekistan
30.69	Finland	58.00	Netherlands	57.00	Venezuela
56.61	France	19.51	New Zealand		Vietnam
	Gabon	32.00	Nicaragua		Zambia
40.07	Germany		Niger		Zimbabwe
	Ghana		Nigeria		

Environmental Systems

Variable name

Urban Total Suspended Particulate Concentration

Variable code   **Variable number**
TSP   3

Units
Micrograms per Cubic Meter

Reference year
MRYA 1990-1996

Median	Minimum	Maximum
72.68	9.00	320.00

Source
World Bank, World Development Indicators 2000, and WHO, Air Management
Information System-AMIS 2.0,1998

Logic
Indicator of Urban Air Quality.

Details
The values were originally collected at the city level. The number of cities with data
provided by each country varied. Within each country the values have been
normalized by city population for the year 1995, then summed to give the total
concentration for the given country.

	Country
	Albania TSP
	Algeria
50.01	Argentina
	Armenia
43.22	Australia
45.70	Austria
	Azerbaijan
	Bangladesh
18.40	Belarus
77.91	Belgium
	Benin
	Bhutan
	Bolivia
	Botswana
106.20	Brazil
199.25	Bulgaria
	Burkina Faso
	Burundi
	Cameroon
31.26	Canada
	Central African Republic
	Chile
310.82	China
120.00	Colombia
244.48	Costa Rica
71.00	Croatia
	Cuba
58.39	Czech Republic
61.00	Denmark
	Dominican Republic
125.73	Ecuador
	Egypt
	El Salvador
	Estonia
	Ethiopia
	Fiji
49.90	Finland
14.16	France
	Gabon
43.27	Germany
137.00	Ghana

	Country
178.00	Greece
272.33	Guatemala
	Haiti
320.00	Honduras
63.74	Hungary
24.00	Iceland
277.45	India
271.00	Indonesia
248.00	Iran
	Ireland
	Israel
86.91	Italy
	Jamaica
43.63	Japan
	Jordan
	Kazakhstan
69.00	Kenya
83.79	Korea, South
	Kuwait
	Kyrgyz Republic
100.00	Latvia
	Lebanon
	Libya
114.27	Lithuania
	Macedonia
	Madagascar
	Malawi
91.58	Malaysia
	Mali
	Mauritius
279.00	Mexico
	Moldova
	Mongolia
	Morocco
	Mozambique
	Nepal
40.00	Netherlands
27.32	New Zealand
	Nicaragua
	Niger
	Nigeria

	Country
10.25	Norway
	Pakistan
	Panama
	Papua New Guinea
	Paraguay
	Peru
200.00	Philippines
	Poland
50.40	Portugal
82.00	Romania
100.00	Russian Federation
	Rwanda
	Saudi Arabia
	Senegal
	Singapore
64.49	Slovak Republic
	Slovenia
	South Africa
72.68	Spain
	Sri Lanka
	Sudan
9.00	Sweden
30.66	Switzerland
	Syria
	Tanzania
223.00	Thailand
	Togo
	Trinidad and Tobago
	Tunisia
11.35	Turkey
	Uganda
	Ukraine
	United Kingdom
	United States
	Uruguay
	Uzbekistan
53.00	Venezuela
	Vietnam
	Zambia
	Zimbabwe

Environmental Systems

Variable name

Internal Renewable Water Resources Per Capita

Variable code
WATCAP

Variable number
4

Units
Thousands Cubic meters/person

Reference year
1961-1990 (avg.)

Median	Minimum	Maximum
294.3	-7.5	2.7

Source
Center for Environmental Systems Research, University of Kassel, WaterGAP 2.1B, 2001

Logic
Logic: The per capita volume of internal renewable water resources in a country is important for a variety of environmental services and to support the needs of the population.

Details
This variable measures internal renewable water (average annual surface runoff and groundwater recharge generated from endogenous precipitation, taking into account evaporation from lakes and wetlands) per capita. These data are derived from the WaterGap 2.1 gridded hydrological model developed by the Center for Environmental Systems Research, University of Kassel, Germany. A special run of the model was performed in order to derive country-level estimates of internal renewable water resources. A logarithmic transformation of this variable was used in calculating the ESI. More details can be found in the main report.

4.09	Albania	2.96	Greece	57.71	Norway
0.39	Algeria	14.03	Guatemala	0.23	Pakistan
7.65	Argentina	0.93	Haiti	30.79	Panama
1.12	Armenia	13.09	Honduras	154.61	Papua New Guinea
27.81	Australia	1.17	Hungary	10.77	Paraguay
6.37	Austria	294.34	Iceland	47.55	Peru
0.79	Azerbaijan	1.56	India	3.79	Philippines
0.60	Bangladesh	10.96	Indonesia	1.48	Poland
2.79	Belarus	0.63	Iran	3.25	Portugal
1.19	Belgium	12.47	Ireland	1.45	Romania
2.25	Benin	0.36	Israel	22.82	Russian Federation
14.08	Bhutan	2.04	Italy	0.95	Rwanda
51.39	Bolivia	3.24	Jamaica	0.22	Saudi Arabia
-7.46	Botswana	2.60	Japan	0.96	Senegal
37.25	Brazil	0.07	Jordan		Singapore
2.00	Bulgaria	3.63	Kazakhstan	2.24	Slovak Republic
0.86	Burkina Faso	1.51	Kenya	8.04	Slovenia
0.65	Burundi	1.16	Korea, South	1.25	South Africa
17.30	Cameroon	-0.20	Kuwait	2.33	Spain
84.51	Canada	5.47	Kyrgyz Republic	1.62	Sri Lanka
37.41	Central African Republic	6.31	Latvia	-0.53	Sudan
19.56	Chile	0.66	Lebanon	15.91	Sweden
1.72	China	0.60	Libya	5.74	Switzerland
45.56	Colombia	5.10	Lithuania	0.35	Syria
23.35	Costa Rica	2.55	Macedonia	3.64	Tanzania
6.01	Croatia	22.55	Madagascar	3.50	Thailand
2.01	Cuba	1.55	Malawi	2.71	Togo
1.45	Czech Republic	20.24	Malaysia	1.58	Trinidad and Tobago
2.49	Denmark	0.40	Mali	0.22	Tunisia
1.92	Dominican Republic	0.50	Mauritius	2.59	Turkey
30.37	Ecuador	3.47	Mexico	1.00	Uganda
-0.24	Egypt	1.83	Moldova	1.26	Ukraine
1.59	El Salvador	16.32	Mongolia	3.10	United Kingdom
7.40	Estonia	0.42	Morocco	7.09	United States
2.17	Ethiopia	5.81	Mozambique	24.24	Uruguay
26.05	Fiji	5.97	Nepal	0.31	Uzbekistan
18.01	Finland	0.65	Netherlands	33.83	Venezuela
3.26	France	79.81	New Zealand	2.80	Vietnam
176.37	Gabon	29.15	Nicaragua	10.01	Zambia
1.35	Germany	-0.33	Niger	3.40	Zimbabwe
1.87	Ghana	2.26	Nigeria		

Environmental Systems

Variable name

Water Inflow from Other Countries per Capita

Variable code
WATINC

Variable number
5

Units
Thousands Cubic meters/person

Reference year
1961-1990 (avg.)

Median 1.18
Minimum 0
Maximum 235.85

Source
Center for Environmental Systems Research, University of Kassel, WaterGAP 2.1B, 2001

Logic
The sum of per capita internal water availability and the per capita volume of water flowing into a country provides a more complete assessment of a country's water resources, which are important for a variety of environmental services and to support the needs of the population.

Details
These data are derived from the WaterGap 2.1 gridded hydrological model developed by the Center for Environmental Systems Research, University of Kassel, Germany. A special run of the model was performed in order to derive country-level estimates of inflow from other countries. There are some problems, in that the size of the grid cells (0.5 x 0.5 degree) do not accurately capture small countries. A logarithmic transformation of this variable was used in calculating the ESI. More details can be found in the main report.

2.83	Albania	1.24	Greece	2.53	Norway
0.04	Algeria	1.40	Guatemala	0.68	Pakistan
18.72	Argentina	0.13	Haiti	0.00	Panama
0.56	Armenia	5.66	Honduras	0.93	Papua New Guinea
0.00	Australia	10.56	Hungary	99.41	Paraguay
4.75	Austria	0.00	Iceland	19.17	Peru
2.25	Azerbaijan	0.39	India	0.00	Philippines
9.36	Bangladesh	0.32	Indonesia	0.23	Poland
2.02	Belarus	0.42	Iran	2.33	Portugal
0.59	Belgium	1.39	Ireland	7.74	Romania
6.93	Benin	0.00	Israel	1.48	Russian Federation
5.96	Bhutan	0.05	Italy	0.95	Rwanda
29.54	Bolivia	0.00	Jamaica	0.00	Saudi Arabia
23.74	Botswana	0.00	Japan	1.68	Senegal
16.44	Brazil	0.17	Jordan		Singapore
21.88	Bulgaria	4.30	Kazakhstan	12.70	Slovak Republic
0.10	Burkina Faso	0.81	Kenya	6.53	Slovenia
0.97	Burundi	0.09	Korea, South	0.11	South Africa
2.88	Cameroon	0.00	Kuwait	0.05	Spain
4.73	Canada	0.00	Kyrgyz Republic	0.00	Sri Lanka
21.29	Central African Republic	7.10	Latvia	4.28	Sudan
1.13	Chile	0.00	Lebanon	0.91	Sweden
0.12	China	0.20	Libya	0.00	Switzerland
39.23	Colombia	2.95	Lithuania	1.83	Syria
2.25	Costa Rica	0.00	Macedonia	1.20	Tanzania
27.60	Croatia	0.00	Madagascar	5.02	Thailand
0.00	Cuba	0.41	Malawi	0.99	Togo
0.58	Czech Republic	0.50	Malaysia	0.00	Trinidad and Tobago
0.00	Denmark	5.93	Mali	0.22	Tunisia
0.13	Dominican Republic	0.00	Mauritius	0.18	Turkey
1.22	Ecuador	0.67	Mexico	1.16	Uganda
1.25	Egypt	3.66	Moldova	0.56	Ukraine
1.59	El Salvador	2.45	Mongolia	0.03	United Kingdom
5.38	Estonia	0.00	Morocco	1.36	United States
0.04	Ethiopia	8.97	Mozambique	235.85	Uruguay
0.00	Fiji	1.18	Nepal	2.54	Uzbekistan
2.35	Finland	5.50	Netherlands	27.47	Venezuela
0.79	France	0.00	New Zealand	6.07	Vietnam
22.28	Gabon	2.71	Nicaragua	5.74	Zambia
1.21	Germany	5.90	Niger	3.77	Zimbabwe
1.02	Ghana	0.83	Nigeria		

Environmental Systems

Variable name

Dissolved Oxygen Concentration

Variable code   **Variable number**
GMS_DO   6

Units
Mg/Liter

Reference year
1994-96 or MRYA

Median   **Minimum**   **Maximum**
11.2   3   8

Source
United Nations Environment Programme (UNEP), Global Environmental Monitoring System/Water Quality Monitoring System. http://www.cciw.ca/gems/

Logic
A measure of eutrophication, which has an important imapact on the health of aquatic resources and ecosystems. High levels correspond to low eutrophication.

Details
The country values represent averages of the station-level values for the three year time period 1994-96. The number of stations per country varies depending on country size, number of water bodies, and level of participation in the GEMS monitoring system.

	Albania		Greece		Norway
	Algeria		Guatemala	7.11	Pakistan
10.00	Argentina		Haiti		Panama
	Armenia		Honduras		Papua New Guinea
	Australia	10.82	Hungary		Paraguay
	Austria		Iceland		Peru
	Azerbaijan	6.38	India	8.24	Philippines
	Bangladesh	3.31	Indonesia	9.86	Poland
	Belarus	10.57	Iran	7.65	Portugal
5.62	Belgium		Ireland		Romania
	Benin		Israel	9.69	Russian Federation
	Bhutan		Italy		Rwanda
	Bolivia		Jamaica		Saudi Arabia
	Botswana	10.18	Japan	4.42	Senegal
7.27	Brazil		Jordan		Singapore
	Bulgaria		Kazakhstan		Slovak Republic
	Burkina Faso		Kenya		Slovenia
	Burundi	10.32	Korea, South		South Africa
	Cameroon		Kuwait		Spain
10.85	Canada		Kyrgyz Republic		Sri Lanka
	Central African Republic		Latvia	7.84	Sudan
	Chile		Lebanon		Sweden
7.99	China		Libya		Switzerland
5.55	Colombia	5.68	Lithuania		Syria
	Costa Rica		Macedonia	6.87	Tanzania
	Croatia		Madagascar	2.98	Thailand
8.10	Cuba		Malawi		Togo
	Czech Republic	4.54	Malaysia		Trinidad and Tobago
	Denmark	8.46	Mali		Tunisia
	Dominican Republic		Mauritius		Turkey
	Ecuador	6.10	Mexico		Uganda
	Egypt		Moldova		Ukraine
	El Salvador		Mongolia	10.40	United Kingdom
	Estonia	6.25	Morocco	9.26	United States
	Ethiopia		Mozambique		Uruguay
8.01	Fiji		Nepal		Uzbekistan
11.19	Finland	9.78	Netherlands		Venezuela
10.33	France	9.87	New Zealand		Vietnam
	Gabon		Nicaragua		Zambia
	Germany		Niger		Zimbabwe
6.80	Ghana		Nigeria		

Environmental Systems

Variable name

Phosphorus Concentration

Variable code   **Variable number**
GMS_PH   7

Units
Mg/Liter

Reference year
1994-96 or MRYA

Median	Minimum	Maximum
0.14	0.003	1.75

Source
United Nations Environment Programme (UNEP), Global Environmental Monitoring System/Water Quality Monitoring System. http://www.cciw.ca/gems/

Logic
A measure of eutrophication, which affects aquatic resources health. High levels correspond to high eutrophication.

Details
The country values represent averages of the station-level values for the three year time period 1994-96. The number of stations per country varies depending on country size, number of water bodies, and level of participation in the GEMS monitoring system.

	Country		Country		Country
	Albania		Greece	0.01	Norway
	Algeria		Guatemala	0.20	Pakistan
0.04	Argentina		Haiti		Panama
	Armenia		Honduras		Papua New Guinea
	Australia	0.21	Hungary		Paraguay
	Austria		Iceland		Peru
	Azerbaijan		India		Philippines
	Bangladesh	0.56	Indonesia	0.33	Poland
	Belarus		Iran	0.13	Portugal
1.63	Belgium		Ireland		Romania
	Benin		Israel		Russian Federation
	Bhutan		Italy		Rwanda
	Bolivia		Jamaica		Saudi Arabia
	Botswana	0.06	Japan		Senegal
0.09	Brazil	1.01	Jordan		Singapore
	Bulgaria		Kazakhstan		Slovak Republic
	Burkina Faso		Kenya		Slovenia
	Burundi		Korea, South		South Africa
	Cameroon		Kuwait		Spain
	Canada		Kyrgyz Republic		Sri Lanka
	Central African Republic		Latvia	1.75	Sudan
	Chile		Lebanon		Sweden
0.28	China		Libya	0.07	Switzerland
	Colombia	0.08	Lithuania		Syria
	Costa Rica		Macedonia		Tanzania
	Croatia		Madagascar	0.31	Thailand
0.01	Cuba		Malawi		Togo
	Czech Republic	0.04	Malaysia		Trinidad and Tobago
	Denmark	0.15	Mali		Tunisia
	Dominican Republic		Mauritius		Turkey
	Ecuador		Mexico		Uganda
	Egypt		Moldova		Ukraine
	El Salvador		Mongolia	0.09	United Kingdom
	Estonia	0.26	Morocco	0.08	United States
	Ethiopia		Mozambique		Uruguay
	Fiji		Nepal		Uzbekistan
0.01	Finland	0.27	Netherlands		Venezuela
0.17	France	0.04	New Zealand		Vietnam
	Gabon		Nicaragua		Zambia
0.32	Germany		Niger		Zimbabwe
	Ghana		Nigeria		

Environmental Systems

Variable name

Suspended Solids

Variable code   **Variable number**
GMS_SS   8

Units
Mg/Liter

Reference year
1994-96 or MRYA

Median	Minimum	Maximum
4.03	1.17	7.97

Source
Source: United Nations Environment Programme (UNEP), Global Environmental Monitoring System/Water Quality Monitoring System. http://www.cciw.ca/gems/

Logic
A measure of water quality and turbidity.

Details
The country values represent averages of the station-level values for the three year time period 1994-96. The number of stations per country varies depending on country size, number of water bodies, and level of participation in the GEMS monitoring system.

	Albania		Greece		Norway
	Algeria		Guatemala	6.76	Pakistan
4.77	Argentina		Haiti		Panama
	Armenia		Honduras		Papua New Guinea
	Australia		Iceland		Paraguay
	Austria		India		Peru
	Azerbaijan	3.42	Hungary	3.62	Philippines
4.08	Bangladesh		Iceland	3.24	Poland
	Belarus	5.37	Indonesia	1.94	Portugal
3.53	Belgium		Iran		Romania
	Benin		Ireland	3.23	Russian Federation
	Bhutan		Israel		Rwanda
	Bolivia		Italy		Saudi Arabia
	Botswana		Jamaica		Senegal
4.08	Brazil	3.27	Japan		Singapore
	Bulgaria	4.50	Jordan		Slovak Republic
	Burkina Faso		Kazakhstan		Slovenia
	Burundi		Kenya		South Africa
	Cameroon	1.69	Korea, South		Spain
2.84	Canada		Kuwait		Sri Lanka
	Central African Republic		Kyrgyz Republic	6.38	Sudan
5.10	Chile		Latvia		Sweden
7.97	China		Lebanon	3.98	Switzerland
4.77	Colombia		Libya		Syria
	Costa Rica		Lithuania		Tanzania
	Croatia		Macedonia	5.60	Thailand
	Cuba		Madagascar		Togo
	Czech Republic		Malawi		Trinidad and Tobago
	Denmark	5.70	Malaysia		Tunisia
	Dominican Republic	4.55	Mali		Turkey
	Ecuador		Mauritius		Uganda
	Egypt	5.17	Mexico		Ukraine
	El Salvador		Moldova	2.26	United Kingdom
	Estonia		Mongolia		United States
	Ethiopia	4.40	Morocco		Uruguay
	Fiji		Mozambique		Uzbekistan
1.17	Finland		Nepal		Venezuela
3.24	France	3.26	Netherlands		Vietnam
	Gabon	2.32	New Zealand		Zambia
3.06	Germany		Nicaragua		Zimbabwe
4.55	Ghana		Niger		
			Nigeria		

Environmental Systems

Variable name

Electrical Conductivity

Variable code   **Variable number**
GMS_EC   9

Units
Usie/Centimeter

Reference year
1994-96 or MRYA

Median	Minimum	Maximum
4520.2	0	355.8

Source
Source: United Nations Environment Programme (UNEP), Global Environmental Monitoring System/Water Quality Monitoring System. http://www.cciw.ca/gems/

Logic
A widely used bulk measure of metals concentration and salinity. High levels of conductivity correspond to high concentrations.

Details
The country values represent averages of the station-level values for the three year time period 1994-96. The number of stations per country varies depending on country size, number of water bodies, and level of participation in the GEMS monitoring system.

	Country			Country			Country
	Albania			Greece		0.61	Norway
	Algeria			Guatemala		410.13	Pakistan
113.68	Argentina			Haiti			Panama
	Armenia			Honduras			Papua New Guinea
	Australia		579.26	Hungary			Paraguay
	Austria			Iceland			Peru
	Azerbaijan		4,520.19	India		136.70	Philippines
231.60	Bangladesh		167.13	Indonesia		1,043.77	Poland
	Belarus		419.64	Iran		191.13	Portugal
2,626.19	Belgium			Ireland			Romania
	Benin			Israel		0.00	Russian Federation
	Bhutan			Italy			Rwanda
	Bolivia			Jamaica			Saudi Arabia
	Botswana		179.29	Japan		380.80	Senegal
145.65	Brazil		1,014.42	Jordan			Singapore
	Bulgaria			Kazakhstan			Slovak Republic
	Burkina Faso		504.00	Kenya			Slovenia
	Burundi		141.33	Korea, South			South Africa
	Cameroon			Kuwait			Spain
237.44	Canada			Kyrgyz Republic			Sri Lanka
	Central African Republic			Latvia		259.33	Sudan
667.94	Chile			Lebanon		77.56	Sweden
522.77	China			Libya		301.06	Switzerland
85.80	Colombia		598.75	Lithuania			Syria
	Costa Rica			Macedonia		363.21	Tanzania
	Croatia			Madagascar		348.33	Thailand
515.00	Cuba			Malawi			Togo
	Czech Republic		508.01	Malaysia			Trinidad and Tobago
	Denmark		120.77	Mali			Tunisia
	Dominican Republic			Mauritius			Turkey
	Ecuador		1,239.62	Mexico			Uganda
	Egypt			Moldova			Ukraine
	El Salvador			Mongolia		368.06	United Kingdom
	Estonia		3,300.63	Morocco		375.65	United States
	Ethiopia			Mozambique			Uruguay
	Fiji			Nepal			Uzbekistan
50.49	Finland		623.12	Netherlands			Venezuela
299.38	France		125.84	New Zealand			Vietnam
	Gabon			Nicaragua			Zambia
1,566.07	Germany			Niger			Zimbabwe
185.59	Ghana			Nigeria			

Environmental Systems

Variable name

Percentage of Mammals Threatened

Source

World Resources Institute, World Resources 2000-2001, Washington, DC: WRI, 2000. Original sources: World Conservation Monitoring Center, IUCN-The World Conservation Union, Food and Agriculture Organization of the United Nations and other sources.

Variable code   **Variable number**
PRTMAM   10

Logic

The percent of mammals threatened gives an estimate of a country's success at preserving its biodiversity.

Units
Percent of Mammals14

Reference year
1996

Details

Number of mammal species threatened divided by known mammal species in the country, expressed as a percentage. A logarithmic transformation of this variable was used in calculating the ESI.

Median	**Minimum**	**Maximum**
9.49	1.00	100

2.94	Albania	13.68	Greece	7.41	Norway
16.30	Algeria	3.20	Guatemala	8.61	Pakistan
8.44	Argentina	100.00	Haiti	7.80	Panama
4.76	Armenia	4.05	Honduras	25.68	Papua New Guinea
22.31	Australia	9.64	Hungary	3.28	Paraguay
8.43	Austria	9.09	Iceland	10.00	Peru
11.11	Azerbaijan	23.73	India	31.01	Philippines
16.51	Bangladesh	28.01	Indonesia	11.90	Poland
5.41	Belarus	14.29	Iran	20.63	Portugal
10.34	Belgium	8.00	Ireland	19.05	Romania
4.79	Benin	11.21	Israel	11.52	Russian Federation
20.20	Bhutan	11.11	Italy	5.96	Rwanda
7.59	Bolivia	16.67	Jamaica	11.69	Saudi Arabia
3.05	Botswana	15.43	Japan	6.77	Senegal
17.03	Brazil	9.86	Jordan	7.06	Singapore
16.05	Bulgaria	8.43	Kazakhstan	9.41	Slovak Republic
4.08	Burkina Faso	11.98	Kenya	13.33	Slovenia
4.67	Burundi	12.24	Korea, South	12.94	South Africa
7.82	Cameroon	4.76	Kuwait	23.17	Spain
3.63	Canada	7.23	Kyrgyz Republic	15.91	Sri Lanka
5.26	Central African Republic	4.82	Latvia	7.87	Sudan
17.58	Chile	8.77	Lebanon	8.33	Sweden
18.75	China	14.47	Libya	8.00	Switzerland
9.75	Colombia	7.35	Lithuania	6.35	Syria
6.83	Costa Rica	12.82	Macedonia	10.44	Tanzania
13.16	Croatia	32.62	Madagascar	12.83	Thailand
29.03	Cuba	3.59	Malawi	4.08	Togo
8.64	Czech Republic	14.00	Malaysia	1.00	Trinidad and Tobago
6.98	Denmark	9.49	Mali	14.10	Tunisia
20.00	Dominican Republic		Mauritius	12.93	Turkey
9.27	Ecuador	13.03	Mexico	5.33	Uganda
15.31	Egypt	2.94	Moldova	13.89	Ukraine
1.48	El Salvador	9.02	Mongolia	8.00	United Kingdom
6.15	Estonia	17.14	Morocco	8.10	United States
13.73	Ethiopia	7.26	Mozambique	6.17	Uruguay
100.00	Fiji	15.47	Nepal	7.22	Uzbekistan
6.67	Finland	10.91	Netherlands	7.43	Venezuela
13.98	France	100.00	New Zealand	17.84	Vietnam
6.32	Gabon	2.00	Nicaragua	4.72	Zambia
10.53	Germany	8.40	Niger	3.33	Zimbabwe
5.86	Ghana	9.49	Nigeria		

Environmental Systems

Variable name

Percentage of Breeding Birds Threatened

Variable code   **Variable number**
PRTBRD   11

Units
Percent of Breeding Birds

Reference year
1996

Median	Minimum	Maximum
3.08	0.00	43.88

Source
World Resources Institute, *World Resources 2000-2001*, Washington, DC: WRI, 2000. Original sources: World Conservation Monitoring Center, IUCN-The World Conservation Union, Food and Agriculture Organization of the United Nations and other sources.

Logic
The percent of breeding birds threatened gives an estimate of a country's success at preserving its biodiversity.

Details
Number of bird species threatened divided by known bird species in the country, expressed as a percentage.

3.04	Albania	3.98	Greece	1.23	Norway
4.17	Algeria	0.87	Guatemala	6.67	Pakistan
4.57	Argentina	14.67	Haiti	1.37	Panama
2.07	Armenia	0.95	Honduras	4.75	Papua New Guinea
6.93	Australia	4.88	Hungary	4.68	Paraguay
2.35	Austria	0.00	Iceland	4.15	Peru
3.23	Azerbaijan	7.88	India	43.88	Philippines
10.17	Bangladesh	6.80	Indonesia	2.64	Poland
1.81	Belarus	4.33	Iran	3.38	Portugal
1.67	Belgium	0.70	Ireland	4.45	Romania
0.33	Benin	4.44	Israel	6.05	Russian Federation
3.13	Bhutan	2.99	Italy	1.17	Rwanda
	Bolivia	6.19	Jamaica	7.10	Saudi Arabia
1.81	Botswana	13.20	Japan	1.56	Senegal
6.87	Brazil	2.84	Jordan	7.63	Singapore
5.00	Bulgaria	3.79	Kazakhstan	1.91	Slovak Republic
0.30	Burkina Faso	2.83	Kenya	1.45	Slovenia
1.33	Burundi	16.96	Korea, South	2.68	South Africa
2.03	Cameroon	15.00	Kuwait	3.60	Spain
1.17	Canada		Kyrgyz Republic	4.40	Sri Lanka
0.37	Central African Republic	2.76	Latvia	1.32	Sudan
6.08	Chile	3.25	Lebanon	1.61	Sweden
8.16	China	2.20	Libya	2.07	Switzerland
3.76	Colombia	1.98	Lithuania	3.43	Syria
2.17	Costa Rica	1.43	Macedonia	3.63	Tanzania
1.79	Croatia	13.86	Madagascar	7.31	Thailand
9.49	Cuba	1.73	Malawi	0.26	Togo
3.02	Czech Republic	6.69	Malaysia	1.15	Trinidad and Tobago
1.02	Denmark	1.51	Mali	3.47	Tunisia
8.09	Dominican Republic		Mauritius	4.64	Turkey
3.82	Ecuador	4.66	Mexico	1.20	Uganda
7.19	Egypt	3.95	Moldova	3.80	Ukraine
0.00	El Salvador	3.29	Mongolia	0.87	United Kingdom
0.94	Estonia	5.24	Morocco	7.69	United States
3.19	Ethiopia	2.81	Mozambique	4.64	Uruguay
12.16	Fiji	4.42	Nepal		Uzbekistan
1.61	Finland	1.57	Netherlands	1.64	Venezuela
2.60	France	29.33	New Zealand	8.79	Vietnam
0.86	Gabon	0.62	Nicaragua	1.65	Zambia
2.09	Germany	0.67	Niger	1.69	Zimbabwe
1.89	Ghana	1.32	Nigeria		

Environmental Systems

Variable name

Severity of Human Induced Soil Degradation

Variable code   **Variable number**
SOIL   12

Units
Index Ranging from 0 (Low Levels of Degradation) to 3.66 (High Levels)

Reference year
1990

Median	Minimum	Maximum
1.75	0.00	3.66

Source
UNEP, Global Assessment of Human Induced Soil Degradation (GLASOD), database, 1990.

Logic
A measure of the degree of soil degradation within a country, which affects biological productivity and sedimentation of water bodies.

Details
The original data classify countries' territories into 4 classes of degradation.
We calculated the fraction of each country's territory falling into each class, and then computed a single weighted composite, using the degradation class as the weight.

3.66	Albania	1.94	Greece	0.34	Norway
1.06	Algeria	2.26	Guatemala	1.66	Pakistan
1.65	Argentina	3.10	Haiti	2.66	Panama
	Armenia	2.62	Honduras	0.22	Papua New Guinea
1.22	Australia	2.51	Hungary	0.94	Paraguay
2.37	Austria		Iceland	1.67	Peru
	Azerbaijan	1.92	India	2.04	Philippines
2.06	Bangladesh	1.90	Indonesia	2.86	Poland
	Belarus	2.52	Iran	1.91	Portugal
2.57	Belgium	0.28	Ireland	3.07	Romania
1.74	Benin	0.44	Israel		Russian Federation
1.42	Bhutan	2.17	Italy	3.35	Rwanda
1.19	Bolivia	2.39	Jamaica	1.63	Saudi Arabia
1.13	Botswana	0.16	Japan	2.12	Senegal
1.62	Brazil	2.43	Jordan	1.99	Singapore
2.94	Bulgaria		Kazakhstan		Slovak Republic
2.77	Burkina Faso	1.84	Kenya		Slovenia
3.21	Burundi	2.23	Korea, South	2.54	South Africa
1.73	Cameroon	1.86	Kuwait	2.09	Spain
0.52	Canada		Kyrgyz Republic	2.51	Sri Lanka
0.64	Central African Republic		Latvia	1.40	Sudan
1.05	Chile	1.48	Lebanon	1.57	Sweden
1.83	China	1.28	Libya	1.73	Switzerland
1.43	Colombia		Lithuania	2.70	Syria
3.42	Costa Rica		Macedonia	1.63	Tanzania
	Croatia	2.80	Madagascar	3.22	Thailand
1.79	Cuba	0.97	Malawi	2.49	Togo
	Czech Republic	2.76	Malaysia	1.59	Trinidad and Tobago
0.47	Denmark	1.40	Mali	2.27	Tunisia
2.16	Dominican Republic	0.79	Mauritius	3.21	Turkey
1.30	Ecuador	1.76	Mexico	2.27	Uganda
0.58	Egypt		Moldova		Ukraine
2.39	El Salvador	1.75	Mongolia	1.48	United Kingdom
	Estonia	2.06	Morocco	1.72	United States
2.30	Ethiopia	1.08	Mozambique	0.75	Uruguay
0.00	Fiji	1.51	Nepal		Uzbekistan
1.26	Finland	1.40	Netherlands	1.32	Venezuela
1.47	France	1.51	New Zealand	3.20	Vietnam
0.39	Gabon	2.31	Nicaragua	1.59	Zambia
	Germany	1.55	Niger	1.29	Zimbabwe
1.67	Ghana	2.36	Nigeria		

Environmental Systems

Land Area Impacted by Human Activities as a Percentage of Total Land Area

Variable code ANTHRO
Variable number 13

Units
Percent of Land Area

Reference year
1992/93 (agriculture) and October 1994 to March 1995 (lit area)

Median	Minimum	Maximum
33.91	0.98	100

Source
NOAA/NGDC World Stable Lights Images - October 1994 to March 1995. Derived from DMSP OLS Nighttime Imagery during the dark half of each lunar cycle. 30 Arc Second Grid and USGS EDCDAAC Version 2.0 Global Land Cover Characteristics Data Base (USGS legend)

Logic
Agricultural activities and the built environment have high impacts on the natural environment. The clearing of natural vegetation for anthropogenic activity has important ecological implications.

Details
This variable measures urbanized (as indicated by lights at night) and agricultural area as a percentage of a country's total area. A complete description of the methodology is included in the main report.

81.19	Albania	65.65	Greece	14.82	Norway
2.23	Algeria	35.70	Guatemala	30.03	Pakistan
32.90	Argentina	43.29	Haiti	34.32	Panama
57.09	Armenia	37.39	Honduras	7.65	Papua New Guinea
7.07	Australia	88.47	Hungary	14.65	Paraguay
46.41	Austria	2.81	Iceland	8.12	Peru
67.35	Azerbaijan	69.56	India	85.03	Philippines
74.77	Bangladesh	29.56	Indonesia	91.20	Poland
95.81	Belarus	13.78	Iran	57.92	Portugal
97.88	Belgium	92.47	Ireland	70.61	Romania
4.02	Benin	50.77	Israel	16.02	Russian Federation
5.76	Bhutan	65.62	Italy	47.42	Rwanda
9.31	Bolivia	31.34	Jamaica	4.14	Saudi Arabia
20.46	Botswana	41.93	Japan	18.20	Senegal
26.44	Brazil	7.84	Jordan	100.00	Singapore
74.55	Bulgaria	29.26	Kazakhstan	63.93	Slovak Republic
14.52	Burkina Faso	20.49	Kenya	47.03	Slovenia
60.83	Burundi	43.24	Korea, South	40.69	South Africa
20.97	Cameroon	27.61	Kuwait	68.37	Spain
8.51	Canada	27.89	Kyrgyz Republic	83.54	Sri Lanka
12.82	Central African Republic	75.31	Latvia	5.39	Sudan
10.27	Chile	64.15	Lebanon	16.52	Sweden
30.06	China	0.98	Libya	51.01	Switzerland
18.90	Colombia	87.90	Lithuania	27.03	Syria
33.47	Costa Rica	71.22	Macedonia	38.79	Tanzania
57.28	Croatia	13.48	Madagascar	62.99	Thailand
46.56	Cuba	29.09	Malawi	12.11	Togo
79.98	Czech Republic	28.70	Malaysia	34.82	Trinidad and Tobago
91.28	Denmark	2.90	Mali	15.82	Tunisia
29.98	Dominican Republic		Mauritius	57.16	Turkey
27.35	Ecuador	21.94	Mexico	35.26	Uganda
4.72	Egypt	94.14	Moldova	89.43	Ukraine
70.09	El Salvador	2.47	Mongolia	87.75	United Kingdom
71.89	Estonia	6.63	Morocco	31.86	United States
15.75	Ethiopia	37.80	Mozambique	58.18	Uruguay
5.66	Fiji	46.17	Nepal	28.46	Uzbekistan
14.61	Finland	92.41	Netherlands	14.25	Venezuela
82.83	France	11.60	New Zealand	56.63	Vietnam
8.84	Gabon	35.54	Nicaragua	33.91	Zambia
85.01	Germany	1.46	Niger	54.87	Zimbabwe
23.37	Ghana	20.62	Nigeria		

Reducing Stresses

Variable name

NOx Emissions per Populated Land Area

Variable code
NOXKM

Variable number
14

Units
Metric Tons/Populated Land Area

Reference year
1990

Median	Minimum	Maximum
0.51	0.03	12.06

Source
RIVM, Emission Database for Global Atmospheric Research (EDGAR)-version 2.0, 1996

Logic
Indicator of air pollution: emissions contribute to declines in air quality.

Details
The gridded emissions data, originally available as 1by1 degree cells (approximately 100 km by 100 km at the equator, decreasing to approximately 100 km by 50 km at a latitude of 60 degrees), were summarized at the country level to give the total emissions for each country. Air pollution is generally greatest in densely populated areas. To take this into account, we used the Gridded Population of the World dataset available from CIESIN and calculated the total land area in each country inhabited with a population density of greater than 5 persons per sq. km. We then utilized this land area as the denominator for the emissions data. A logarithmic transformation of this variable was used in calculating the ESI.

0.76	Albania	1.39	Greece	0.78	Norway		
0.22	Algeria	0.33	Guatemala	0.25	Pakistan		
0.12	Argentina	0.26	Haiti	0.20	Panama		
3.43	Armenia	0.19	Honduras	0.03	Papua New Guinea		
2.09	Australia	1.34	Hungary	0.46	Paraguay		
2.13	Austria	2.00	Iceland	0.07	Peru		
1.38	Azerbaijan	0.51	India	0.73	Philippines		
1.68	Bangladesh	0.43	Indonesia	1.08	Poland		
0.79	Belarus	0.14	Iran	1.11	Portugal		
10.42	Belgium	0.72	Ireland	0.97	Romania		
0.42	Benin	3.13	Israel	0.44	Russian Federation		
0.06	Bhutan	1.79	Italy	2.00	Rwanda		
0.18	Bolivia	0.56	Jamaica	0.18	Saudi Arabia		
1.98	Botswana	3.57	Japan	0.35	Senegal		
0.32	Brazil	0.74	Jordan		Singapore		
1.11	Bulgaria	0.23	Kazakhstan	2.53	Slovak Republic		
0.15	Burkina Faso	0.29	Kenya	2.43	Slovenia		
1.43	Burundi	5.06	Korea, South	0.63	South Africa		
0.22	Cameroon	2.58	Kuwait	0.75	Spain		
1.39	Canada	0.47	Kyrgyz Republic	0.32	Sri Lanka		
0.49	Central African Republic	0.72	Latvia	0.15	Sudan		
0.19	Chile	2.90	Lebanon	0.63	Sweden		
0.49	China	0.98	Libya	3.44	Switzerland		
0.24	Colombia	1.20	Lithuania	0.43	Syria		
0.23	Costa Rica	1.06	Macedonia	0.23	Tanzania		
2.94	Croatia	0.08	Madagascar	0.54	Thailand		
0.32	Cuba	0.55	Malawi	0.51	Togo		
2.50	Czech Republic	0.54	Malaysia	2.24	Trinidad and Tobago		
7.19	Denmark	0.12	Mali	0.26	Tunisia		
0.26	Dominican Republic	0.86	Mauritius	0.52	Turkey		
0.35	Ecuador	0.40	Mexico	0.36	Uganda		
1.18	Egypt	2.72	Moldova	1.17	Ukraine		
0.47	El Salvador	0.23	Mongolia	4.03	United Kingdom		
0.74	Estonia	0.16	Morocco	1.79	United States		
0.15	Ethiopia	0.15	Mozambique	0.25	Uruguay		
0.09	Fiji	0.23	Nepal	0.69	Uzbekistan		
0.86	Finland	12.06	Netherlands	0.44	Venezuela		
1.77	France	0.31	New Zealand	0.46	Vietnam		
0.15	Gabon	0.16	Nicaragua	0.34	Zambia		
2.79	Germany	0.13	Niger	0.21	Zimbabwe		
0.38	Ghana	0.28	Nigeria				

Reducing Stresses

Variable name

SO2 Emissions per Populated Land Area

Variable code   **Variable number**

SO2KM   15

Units

Metric Tons/Populated Land Area

Reference year

1990

Median   **Minimum**   **Maximum**

1.17   0.02   59.12

Source

RIVM, Emission Database for Global Atmospheric Research (EDGAR)-version 2.0, 1996

Logic

Indicator of air pollution: emissions contribute to declines in air quality.

Details

The gridded emissions data, originally available as 1x1 degree cells (approximately 100 km by 100 km at the equator, decreasing to approximately 100 km by 50 km at a latitude of 60 degrees), were summarized at the country level to give the total emissions for each country. Air pollution is generally greatest in densely populated areas. To take this into account, we used the Gridded Population of the World dataset available from CIESIN and calculated the total land area in each country inhabited with a population density of greater than 5 persons per sq. km. We then utilized this land area as the denominator for the emissions data.

8.36	Albania	0.13	Ghana	0.05	Niger
0.19	Algeria	8.21	Greece	0.11	Nigeria
0.09	Argentina	0.22	Guatemala	1.19	Norway
11.17	Armenia	0.30	Haiti	0.24	Pakistan
3.85	Australia	0.14	Honduras	0.23	Panama
5.90	Austria	7.15	Hungary	0.02	Papua New Guinea
3.90	Azerbaijan	1.54	Iceland	0.20	Paraguay
1.22	Bangladesh	0.70	India	0.17	Peru
2.15	Belarus	0.68	Indonesia	1.82	Philippines
37.84	Belgium	0.35	Iran	6.59	Poland
0.15	Benin	1.92	Ireland	3.50	Portugal
0.05	Bhutan	8.24	Israel	4.08	Romania
0.06	Bolivia	4.10	Italy	1.40	Russian Federation
1.29	Botswana	14.19	Jamaica	0.69	Rwanda
0.26	Brazil	4.02	Japan	0.25	Saudi Arabia
10.83	Bulgaria	1.65	Jordan	0.15	Senegal
0.05	Burkina Faso	1.17	Kazakhstan		Singapore
0.48	Burundi	0.12	Kenya	16.39	Slovak Republic
0.08	Cameroon	25.50	Korea, South	8.12	Slovenia
3.02	Canada	2.81	Kuwait	1.26	South Africa
0.17	Central African Republic	2.16	Kyrgyz Republic	1.82	Spain
3.67	Chile	1.90	Latvia	0.14	Sri Lanka
2.15	China	5.89	Lebanon	0.06	Sudan
0.15	Colombia	1.61	Libya	1.30	Sweden
0.21	Costa Rica	3.22	Lithuania	2.59	Switzerland
9.00	Croatia	6.24	Macedonia	0.86	Syria
1.30	Cuba	0.03	Madagascar	0.08	Tanzania
35.15	Czech Republic	0.19	Malawi	0.76	Thailand
12.86	Denmark	0.87	Malaysia	0.19	Togo
0.51	Dominican Republic	0.04	Mali	0.76	Trinidad and Tobago
0.24	Ecuador	2.08	Mauritius	0.94	Tunisia
2.92	Egypt	0.68	Mexico	1.57	Turkey
0.43	El Salvador	7.70	Moldova	0.13	Uganda
2.39	Estonia	0.63	Mongolia	3.39	Ukraine
0.06	Ethiopia	0.38	Morocco	8.95	United Kingdom
0.15	Fiji	0.06	Mozambique	2.48	United States
2.62	Finland	0.19	Nepal	0.19	Uruguay
3.81	France	59.12	Netherlands	2.43	Uzbekistan
0.07	Gabon	0.41	New Zealand	0.27	Venezuela
9.55	Germany	0.11	Nicaragua	0.77	Vietnam

Reducing Stresses

Variable name

VOCs emissions per populated land area

Variable code	Variable number
VOCKM	16

Units
Metric Tons/Populated Land Area

Reference year
1990

Median	Minimum	Maximum
3.30	0.25	60.53

Source
RIVM, Emission Database for Global Atmospheric Research (EDGAR)-version 2.0, 1996

Logic
Indicator of air pollution: emissions contribute to declines in air quality.

Details
The gridded emissions data, originally available as 1x1 degree cells (approximately 100 km by 100 km at the equator, decreasing to approximately 100 km North-South side by 50 km East-West side at a latitude of 60 degrees), were summarized at the country level to give the total emissions for each country. Air pollution is generally greatest in densely populated areas. To take this into account, we used the Gridded Population of the World dataset available from CIESIN and calculated the total land area in each country inhabited with a population density of greater than 5 persons per sq. km. We then utilized this land area as the denominator for the emissions data. A logarithmic transformation of this variable was used in calculating the ESI.

3.30	Albania	2.99	Ghana	1.05	Niger
2.45	Algeria	5.59	Greece	5.15	Nigeria
0.79	Argentina	4.02	Guatemala	2.24	Norway
8.36	Armenia	5.05	Haiti	2.50	Pakistan
6.33	Australia	1.83	Honduras	1.35	Panama
6.88	Austria	6.24	Hungary	0.25	Papua New Guinea
4.46	Azerbaijan	6.39	Iceland	3.39	Paraguay
20.78	Bangladesh	4.87	India	0.77	Peru
2.25	Belarus	4.73	Indonesia	7.22	Philippines
23.06	Belgium	1.56	Iran	2.38	Poland
5.54	Benin	2.15	Ireland	2.99	Portugal
0.80	Bhutan	9.34	Israel	3.35	Romania
1.24	Bolivia	4.92	Italy	2.52	Russian Federation
10.39	Botswana	3.20	Jamaica	21.06	Rwanda
1.89	Brazil	14.23	Japan	5.79	Saudi Arabia
3.92	Bulgaria	2.48	Jordan	2.10	Senegal
1.23	Burkina Faso	0.86	Kazakhstan		Singapore
14.66	Burundi	3.01	Kenya	7.17	Slovak Republic
2.05	Cameroon	14.31	Korea, South	9.38	Slovenia
5.94	Canada	60.53	Kuwait	1.81	South Africa
3.55	Central African Republic	1.65	Kyrgyz Republic	2.40	Spain
1.10	Chile	2.31	Latvia	4.05	Sri Lanka
2.49	China	14.76	Lebanon	1.47	Sudan
2.78	Colombia	20.13	Libya	1.71	Sweden
1.49	Costa Rica	3.39	Lithuania	9.25	Switzerland
11.41	Croatia	3.82	Macedonia	3.16	Syria
1.23	Cuba	0.78	Madagascar	2.08	Tanzania
7.29	Czech Republic	4.78	Malawi	3.60	Thailand
25.10	Denmark	5.47	Malaysia	4.33	Togo
2.13	Dominican Republic	0.81	Mali	18.11	Trinidad and Tobago
4.05	Ecuador	8.27	Mauritius	1.72	Tunisia
8.90	Egypt	2.78	Mexico	2.21	Turkey
5.07	El Salvador	6.47	Moldova	3.38	Uganda
1.54	Estonia	0.88	Mongolia	3.30	Ukraine
1.27	Ethiopia	0.77	Morocco	13.50	United Kingdom
0.83	Fiji	1.24	Mozambique	3.53	United States
1.71	Finland	4.04	Nepal	1.42	Uruguay
5.27	France	31.63	Netherlands	1.84	Uzbekistan
2.47	Gabon	1.32	New Zealand	4.98	Venezuela
8.62	Germany	1.24	Nicaragua	4.73	Vietnam

Reducing Stresses

Variable name

Coal Consumption per Populated Land Area

Variable code
COALKM

Variable number
17

Units
Billion Btu/Populated Land Area

Reference year
1998

Median	Minimum	Maximum
0.28	0.01	15.43

Source
US Energy Information Agency, available at
http://www.eia.doe.gov/emeu/international/contents.html

Logic
Coal fired power plants emit higher levels of SO2 and other air pollutants than natural gas or oil fired plants, and the energy produced is more carbon-intensive.

Details
Air pollution is generally greatest in densely populated areas. To take this into account, we used the Gridded Population of the World dataset available from CIESIN and calculated the total land area in each country inhabited with a population density of greater than 5 persons per sq. km. We then utilized this land area as the denominator for the coal consumption data. A logarithmic transformation of this variable was used in calculating the ESI.

0.01	Albania	2.57	Greece	0.38	Norway
0.06	Algeria		Guatemala	0.12	Pakistan
0.02	Argentina		Haiti	0.03	Panama
	Armenia		Honduras		Papua New Guinea
7.20	Australia	1.84	Hungary	0.01	Paraguay
1.55	Austria	1.55	Iceland	0.02	Peru
	Azerbaijan	2.31	India	0.38	Philippines
	Bangladesh	0.27	Indonesia	7.51	Poland
0.07	Belarus	0.03	Iran	1.22	Portugal
10.70	Belgium	1.11	Ireland	1.22	Romania
	Benin	13.31	Israel	1.07	Russian Federation
0.01	Bhutan	1.55	Italy		Rwanda
	Bolivia	0.18	Jamaica		Saudi Arabia
0.49	Botswana	7.82	Japan		Senegal
0.14	Brazil		Jordan	0.06	Singapore
2.95	Bulgaria	0.67	Kazakhstan	3.96	Slovak Republic
	Burkina Faso	0.01	Kenya	3.57	Slovenia
	Burundi	15.43	Korea, South	5.50	South Africa
	Cameroon		Kuwait	1.33	Spain
2.32	Canada	0.11	Kyrgyz Republic		Sri Lanka
	Central African Republic	0.12	Latvia		Sudan
0.43	Chile	0.56	Lebanon	0.34	Sweden
3.86	China		Libya	0.05	Switzerland
0.22	Colombia	0.07	Lithuania		Syria
0.01	Costa Rica	2.32	Macedonia		Tanzania
0.23	Croatia		Madagascar	0.57	Thailand
0.02	Cuba	0.02	Malawi		Togo
12.04	Czech Republic	0.19	Malaysia		Trinidad and Tobago
3.16	Denmark		Mali	0.03	Tunisia
0.08	Dominican Republic	1.10	Mauritius	1.00	Turkey
	Ecuador	0.18	Mexico		Uganda
0.37	Egypt	0.25	Moldova	3.22	Ukraine
	El Salvador	0.60	Mongolia	5.81	United Kingdom
0.29	Estonia	0.28	Morocco	4.89	United States
	Ethiopia		Mozambique		Uruguay
0.03	Fiji	0.02	Nepal	0.09	Uzbekistan
0.82	Finland	10.67	Netherlands		Venezuela
1.06	France	0.56	New Zealand	0.51	Vietnam
	Gabon		Nicaragua	0.01	Zambia
8.76	Germany	0.01	Niger	0.38	Zimbabwe
	Ghana		Nigeria		

Reducing Stresses

Variable name

Vehicles Per Populated Land Area

Variable code   **Variable number**
CARSKM   18

Units
Vehicles/Populated Land Area

Reference year
MRYA 1996-1998

Median   **Minimum**   **Maximum**
5.26   0.01   1041.12

Source
World Bank, World Development Indicators 2000

Logic
Proxy for environmental impacts associated with production, use and disposal of motor vehicles and the transportation infrastructure that supports them.

Details
Air pollution is generally greatest in densely populated areas. To take this into account, we used the Gridded Population of the World dataset available from CIESIN and calculated the total land area in each country inhabited with a population density of greater than 5 persons per sq. km. We then utilized this land area as the denominator for the vehicles data. A logarithmic transformation of this variable was used in calculating the ESI.

Value	Country	Value	Country	Value	Country
4.57	Albania	26.89	Greece	19.13	Norway
4.11	Algeria	2.12	Guatemala	1.46	Pakistan
4.54	Argentina	2.07	Haiti	5.04	Panama
0.19	Armenia	2.39	Honduras	0.27	Papua New Guinea
48.23	Australia	29.48	Hungary	0.78	Paraguay
50.97	Austria	106.51	Iceland	1.39	Peru
4.17	Azerbaijan	2.21	India	7.43	Philippines
0.94	Bangladesh	2.90	Indonesia	34.69	Poland
5.60	Belarus	1.42	Iran	37.68	Portugal
149.66	Belgium	16.66	Ireland	13.17	Romania
0.42	Benin	76.64	Israel	5.71	Russian Federation
0.08	Bhutan	115.24	Italy	0.79	Rwanda
0.89	Bolivia	11.15	Jamaica	3.89	Saudi Arabia
1.54	Botswana	187.46	Japan	0.64	Senegal
3.94	Brazil	7.78	Jordan	1,041.12	Singapore
19.11	Bulgaria	1.07	Kazakhstan	28.27	Slovak Republic
0.22	Burkina Faso	1.34	Kenya	43.63	Slovenia
	Burundi	104.45	Korea, South	7.72	South Africa
0.43	Cameroon	44.91	Kuwait	37.06	Spain
33.63	Canada	0.77	Kyrgyz Republic	9.70	Sri Lanka
0.01	Central African Republic	9.40	Latvia	0.21	Sudan
5.21	Chile		Lebanon	18.44	Sweden
1.74	China	30.23	Libya	94.59	Switzerland
2.66	Colombia	16.80	Lithuania	2.23	Syria
9.53	Costa Rica	12.21	Macedonia	0.16	Tanzania
	Croatia	0.16	Madagascar	12.04	Thailand
3.26	Cuba	0.51	Malawi	2.14	Togo
53.52	Czech Republic	10.99	Malaysia	26.85	Trinidad and Tobago
51.16	Denmark	0.11	Mali	6.03	Tunisia
7.55	Dominican Republic	70.72	Mauritius	6.70	Turkey
3.60	Ecuador	10.42	Mexico	0.42	Uganda
16.98	Egypt	8.65	Moldova	8.24	Ukraine
17.49	El Salvador	0.90	Mongolia	106.38	United Kingdom
12.74	Estonia	4.11	Morocco	47.02	United States
0.10	Ethiopia	0.02	Mozambique	4.29	Uruguay
3.15	Fiji		Nepal		Uzbekistan
15.19	Finland	193.80	Netherlands	5.26	Venezuela
56.34	France	27.08	New Zealand		Vietnam
0.34	Gabon	1.33	Nicaragua	0.53	Zambia
	Germany	0.19	Niger	0.98	Zimbabwe
0.61	Ghana	2.99	Nigeria		

Reducing Stresses

Variable name

Fertilizer Consumption per Hectare of Arable Land

Variable code   **Variable number**
FERTHA   19

Units
Hundreds Grams/Hectare of Arable Land

Reference year
1997

Median   **Minimum**   **Maximum**
788.92   1.19   32133.33

Source
World Bank, World Development Indicators 2000

Logic
Logic: Excessive use of fertilizers from agricultural activities has a negative impact on soil and water, altering chemistry and levels of nutrients and leading to eutrophication problems.

Details
A logarithmic transformation of this variable was used in calculating the ESI.

88.39	Albania	1,792.42	Greece	2,308.20	Norway
128.90	Algeria	1,571.22	Guatemala	1,264.30	Pakistan
332.88	Argentina	173.21	Haiti	672.00	Panama
161.94	Armenia	781.71	Honduras	2,166.67	Papua New Guinea
427.42	Australia	891.32	Hungary	154.55	Paraguay
1,664.28	Austria	32,133.33	Iceland	470.81	Peru
135.77	Azerbaijan	999.99	India	1,582.79	Philippines
1,354.75	Bangladesh	1,372.83	Indonesia	1,140.54	Poland
1,190.28	Belarus	649.25	Iran	1,110.08	Portugal
4,166.67	Belgium	5,026.06	Ireland	338.71	Romania
256.33	Benin	3,410.26	Israel	132.51	Russian Federation
7.14	Bhutan	2,222.62	Italy	4.71	Rwanda
67.52	Bolivia	1,258.62	Jamaica	883.24	Saudi Arabia
109.33	Botswana	3,856.96	Japan	102.69	Senegal
1,030.17	Brazil	890.00	Jordan	20,630.00	Singapore
439.80	Bulgaria	42.17	Kazakhstan	796.14	Slovak Republic
125.60	Burkina Faso	333.25	Kenya	3,202.25	Slovenia
59.74	Burundi	5,257.54	Korea, South	507.75	South Africa
56.21	Cameroon	2,000.00	Kuwait	1,437.19	Spain
604.35	Canada	229.63	Kyrgyz Republic	2,428.87	Sri Lanka
1.55	Central African Republic	216.67	Latvia	46.35	Sudan
2,194.75	Chile	3,344.72	Lebanon	1,104.17	Sweden
2,898.88	China	340.50	Libya	2,509.52	Switzerland
2,885.95	Colombia	466.06	Lithuania	772.07	Syria
9,017.91	Costa Rica	766.83	Macedonia	174.10	Tanzania
1,776.23	Croatia	36.97	Madagascar	865.55	Thailand
552.81	Cuba	358.36	Malawi	67.15	Togo
1,015.19	Czech Republic	6,593.41	Malaysia	1,413.33	Trinidad and Tobago
1,882.45	Denmark	103.78	Mali	329.31	Tunisia
955.88	Dominican Republic	3,645.60	Mauritius	686.90	Turkey
1,064.80	Ecuador	636.11	Mexico	1.19	Uganda
3,565.63	Egypt	678.25	Moldova	268.32	Ukraine
1,633.63	El Salvador	15.16	Mongolia	3,299.37	United Kingdom
237.59	Estonia	347.35	Morocco	1,141.83	United States
133.79	Ethiopia	22.03	Mozambique	1,022.22	Uruguay
960.00	Fiji	375.78	Nepal	1,229.05	Uzbekistan
1,453.43	Finland	5,566.67	Netherlands	1,121.21	Venezuela
2,770.83	France	4,443.73	New Zealand	2,773.29	Vietnam
6.15	Gabon	201.25	Nicaragua	111.13	Zambia
2,414.54	Germany	19.02	Niger	587.66	Zimbabwe
74.95	Ghana	48.76	Nigeria		

Reducing Stresses

Variable name

Pesticide Use per Hectare of Crop Land

Source
World Resource Institute, *World Resources 2000-2001*, Washington, DC: WRI, 2000.

Logic
Excessive use of pesticides in agricultural activities has a negative impact on soil, water, humans and wildlife.

Variable code	**Variable number**
PESTHA	20

Units
Kg/Hectare of Cropland

Reference year
1996

Median	**Minimum**	**Maximum**
1217.5	1	19288

435.00	Albania		Greece	941.00	Norway
835.00	Algeria	574.00	Guatemala	365.00	Pakistan
1,266.00	Argentina	23.00	Haiti		Panama
	Armenia	6,521.00	Honduras	1,750.00	Papua New Guinea
2,535.00	Australia	2,863.00	Hungary	1,542.00	Paraguay
2,710.00	Austria		Iceland		Peru
	Azerbaijan	436.00	India		Philippines
176.00	Bangladesh	88.00	Indonesia	490.00	Poland
	Belarus	1,881.00	Iran	2,584.00	Portugal
	Belgium		Ireland	1,617.00	Romania
	Benin		Israel	407.00	Russian Federation
670.00	Bhutan	19,288.00	Italy	260.00	Rwanda
1,514.00	Bolivia		Jamaica		Saudi Arabia
40.00	Botswana		Japan	183.00	Senegal
836.00	Brazil	1,495.00	Jordan		Singapore
966.00	Bulgaria		Kazakhstan	4,148.00	Slovak Republic
1.00	Burkina Faso		Kenya	6,389.00	Slovenia
268.00	Burundi	13,829.00	Korea, South	57.00	South Africa
253.00	Cameroon		Kuwait		Spain
644.00	Canada	1,860.00	Kyrgyz Republic	6,271.00	Sri Lanka
12.00	Central African Republic	208.00	Latvia	106.00	Sudan
3,240.00	Chile		Lebanon	509.00	Sweden
	China		Libya	4,576.00	Switzerland
6,134.00	Colombia	312.00	Lithuania		Syria
18,726.00	Costa Rica	7,718.00	Macedonia		Tanzania
3,060.00	Croatia	28.00	Madagascar	1,116.00	Thailand
	Cuba		Malawi	95.00	Togo
1,169.00	Czech Republic	5,982.00	Malaysia	11,827.00	Trinidad and Tobago
2,200.00	Denmark	136.00	Mali		Tunisia
	Dominican Republic		Mauritius	1,145.00	Turkey
1,696.00	Ecuador		Mexico	17.00	Uganda
1,293.00	Egypt	1,434.00	Moldova	2,001.00	Ukraine
2,642.00	El Salvador		Mongolia	4,745.00	United Kingdom
105.00	Estonia		Morocco	1,599.00	United States
34.00	Ethiopia		Mozambique	1,316.00	Uruguay
2,333.00	Fiji	21.00	Nepal		Uzbekistan
410.00	Finland	11,842.00	Netherlands	1,403.00	Venezuela
	France	2,215.00	New Zealand		Vietnam
	Gabon	357.00	Nicaragua	317.00	Zambia
2,085.00	Germany		Niger	531.00	Zimbabwe
2,333.00	Ghana		Nigeria		

Reducing Stresses

Variable name

Industrial Organic Pollutants per Available Freshwater

Variable code   **Variable number**
BODWAT   21

Units

Kg of Biochemical Oxygen Demand (BOD) Emissions/Cubic Km of Water

Reference year
1996

Median	**Minimum**	**Maximum**
892.38	30.72	18083.68

Source

World Bank, World Development Indicators 2000 and Center for Environmental Systems Research, University of Kassel, WaterGap 2.1, 2000

Logic

Emissions of organic pollutants from industrial activities cause water quality degradation. Given these considerations, BOD emissions have been normalized per amount of freshwater availability.

Details

These are modeled emissions data from the World Bank, measuring organic pollutants in terms of Biochemical Oxygen Demand (BOD).

233.77	Albania	1,273.13	Greece	188.19	Norway
8,580.76	Algeria		Guatemala		Pakistan
	Argentina		Haiti	138.97	Panama
2,143.00	Armenia		Honduras		Papua New Guinea
	Australia	1,118.41	Hungary		Paraguay
876.85	Austria		Iceland		Peru
	Azerbaijan	892.38	India	691.06	Philippines
	Bangladesh	324.63	Indonesia	5,669.83	Poland
	Belarus		Iran	2,462.13	Portugal
	Belgium	666.54	Ireland		Romania
	Benin	18,083.68	Israel	475.66	Russian Federation
	Bhutan		Italy		Rwanda
	Bolivia	2,188.43	Jamaica		Saudi Arabia
243.67	Botswana	4,504.89	Japan	601.34	Senegal
	Brazil	15,225.22	Jordan		Singapore
422.52	Bulgaria		Kazakhstan		Slovak Republic
	Burkina Faso		Kenya	1,216.60	Slovenia
	Burundi	5,965.52	Korea, South	3,719.32	South Africa
48.84	Cameroon		Kuwait	3,566.38	Spain
116.28	Canada		Kyrgyz Republic		Sri Lanka
	Central African Republic	759.91	Latvia		Sudan
252.08	Chile		Lebanon	604.04	Sweden
3,248.13	China		Libya	3,018.35	Switzerland
36.68	Colombia		Lithuania		Syria
336.47	Costa Rica	4,697.95	Macedonia		Tanzania
322.67	Croatia		Madagascar		Thailand
	Cuba		Malawi		Togo
	Czech Republic	391.01	Malaysia		Trinidad and Tobago
5,674.21	Denmark		Mali	11,451.51	Tunisia
	Dominican Republic	17,424.19	Mauritius	1,011.98	Turkey
91.96	Ecuador		Mexico		Uganda
	Egypt		Moldova	5,800.97	Ukraine
910.26	El Salvador		Mongolia	3,571.12	United Kingdom
	Estonia	7,691.03	Morocco	1,122.61	United States
160.24	Ethiopia	30.72	Mozambique	32.54	Uruguay
	Fiji	175.83	Nepal		Uzbekistan
608.04	Finland	1,349.91	Netherlands	68.63	Venezuela
2,469.97	France		New Zealand		Vietnam
	Gabon		Nicaragua		Zambia
	Germany		Niger	415.29	Zimbabwe
	Ghana		Nigeria		

Reducing Stresses

Variable name

Percentage of Country's Territory Under Severe Water Availability Stress

Variable code **Variable number**
WATSTR   22

Units
Percent of Land Area

Reference year
1961-1990 (avg.)

Median	Minimum	Maximum
4.50	0	100

Source
Center for Environmental Systems Research, University of Kassel, WaterGAP 2.1B, 2001

Logic
The regional distribution of water availability relative to population and consumption needs is more important than its overall water availability. This variable captures the percent of the territory that is under water stress, which will affect the availability of water for environmental services and human well-being.

Details
These data are derived from the WaterGap 2.1 gridded hydrological model developed by the Center for Environmental Systems Research, University of Kassel, Germany. The modelers identified grid cells in which water consumption exceeds 40 percent of the water available in that particular grid cell. These were then converted to land area equivalents, and the percentage of the territory under severe water stress was calculated.

19.50	Albania	58.00	Greece	0.40	Norway
71.00	Algeria	0.00	Guatemala	76.30	Pakistan
23.30	Argentina	0.00	Haiti	0.00	Panama
84.60	Armenia	0.00	Honduras	0.00	Papua New Guinea
8.00	Australia	0.00	Hungary	0.00	Paraguay
0.00	Austria	0.00	Iceland	23.60	Peru
95.40	Azerbaijan	80.20	India	10.40	Philippines
22.10	Bangladesh	1.40	Indonesia	0.00	Poland
0.00	Belarus	87.50	Iran	54.70	Portugal
93.90	Belgium	0.00	Ireland	1.70	Romania
0.00	Benin	100.00	Israel	3.80	Russian Federation
0.00	Bhutan	26.30	Italy	0.00	Rwanda
14.00	Bolivia	0.00	Jamaica	88.30	Saudi Arabia
14.20	Botswana	9.50	Japan	5.00	Senegal
0.30	Brazil	82.60	Jordan		Singapore
45.90	Bulgaria	60.40	Kazakhstan	0.00	Slovak Republic
0.00	Burkina Faso	1.10	Kenya	0.00	Slovenia
0.00	Burundi	49.80	Korea, South	68.50	South Africa
0.00	Cameroon	97.70	Kuwait	72.30	Spain
0.90	Canada	93.00	Kyrgyz Republic	39.50	Sri Lanka
0.00	Central African Republic	0.00	Latvia	31.10	Sudan
41.10	Chile	82.10	Lebanon	0.60	Sweden
44.70	China	83.70	Libya	0.00	Switzerland
1.00	Colombia	0.40	Lithuania	99.60	Syria
0.00	Costa Rica	91.60	Macedonia	0.00	Tanzania
0.00	Croatia	1.70	Madagascar	0.60	Thailand
24.60	Cuba	0.00	Malawi	0.00	Togo
0.00	Czech Republic	1.60	Malaysia	100.00	Trinidad and Tobago
7.70	Denmark	2.70	Mali	89.00	Tunisia
4.50	Dominican Republic	0.00	Mauritius	61.70	Turkey
1.20	Ecuador	43.80	Mexico	0.00	Uganda
88.10	Egypt	6.30	Moldova	17.00	Ukraine
0.00	El Salvador	8.10	Mongolia	21.00	United Kingdom
0.30	Estonia	81.50	Morocco	31.30	United States
24.70	Ethiopia	13.60	Mozambique	0.00	Uruguay
0.00	Fiji	98.10	Nepal	87.10	Uzbekistan
2.10	Finland	36.00	Netherlands	2.40	Venezuela
19.40	France	0.00	New Zealand	2.80	Vietnam
0.00	Gabon	0.30	Nicaragua	0.00	Zambia
1.10	Germany	40.50	Niger	16.20	Zimbabwe
0.00	Ghana	17.80	Nigeria		

Reducing Stresses

Variable name

Percentage Change in Forest Cover 1990-1995

Variable code   **Variable number**
FOREST   23

Units
Percent Change

Reference year
1995

Median   **Minimum**   **Maximum**
-0.02   -0.33   0.14

Source
Source: Forest Resources Assessment Programme 2000, Working Paper 1, Rome 1998, Forest Department FAO. Originally published in the *State of the World's Forests 1997* (FAO, 1997).

Logic
When forests are lost or severely degraded, their capacity to function as regulators for the enviornment is also lost, increasing flood and erosion hazards, reducing soil fertility, and contributing to the loss of plant and animal life. As a result, the sustainable provision of goods and services from forests is jeopardized (Forest Resources Assessme

Details
Values for Croatia, Czech Republic, Estonia, Hungary and Latvia disagreed with national reports so values were replaced with those from World Resources 2000-2001, FG1. The 1995 figures are the most current global data set on forest cover and forest cover change available according to the FAO. A logarithmic transformation of this variable was used in calculating the ESI.

	Albania	0.12	Greece	0.02	Norway
-0.06	Algeria	-0.10	Guatemala	-0.14	Pakistan
-0.01	Argentina	-0.16	Haiti	-0.10	Panama
0.14	Armenia	-0.11	Honduras	-0.02	Papua New Guinea
	Australia	0.07	Hungary	-0.12	Paraguay
	Austria		Iceland	-0.02	Peru
	Azerbaijan		India	-0.16	Philippines
-0.04	Bangladesh	-0.05	Indonesia	0.01	Poland
0.05	Belarus	-0.08	Iran	0.04	Portugal
	Belgium	0.14	Ireland		Romania
-0.06	Benin		Israel		Russian Federation
-0.02	Bhutan		Italy	-0.01	Rwanda
-0.06	Bolivia	-0.31	Jamaica	-0.04	Saudi Arabia
-0.02	Botswana		Japan	-0.03	Senegal
-0.02	Brazil	-0.12	Jordan		Singapore
	Bulgaria	0.10	Kazakhstan	0.01	Slovak Republic
-0.04	Burkina Faso	-0.01	Kenya		Slovenia
-0.02	Burundi	-0.01	Korea, South	-0.01	South Africa
-0.03	Cameroon		Kuwait		Spain
	Canada		Kyrgyz Republic	-0.05	Sri Lanka
-0.02	Central African Republic		Latvia	-0.04	Sudan
-0.02	Chile	-0.33	Lebanon		Sweden
	China		Libya		Switzerland
-0.02	Colombia	0.03	Lithuania	-0.11	Syria
-0.14	Costa Rica		Macedonia	-0.05	Tanzania
	Croatia	-0.04	Madagascar	-0.12	Thailand
-0.06	Cuba	-0.08	Malawi	-0.07	Togo
0.04	Czech Republic	-0.11	Malaysia	-0.07	Trinidad and Tobago
	Denmark	-0.05	Mali	-0.03	Tunisia
-0.08	Dominican Republic		Mauritius		Turkey
-0.08	Ecuador	-0.04	Mexico	-0.05	Uganda
	Egypt		Moldova		Ukraine
-0.15	El Salvador		Mongolia	0.03	United Kingdom
0.02	Estonia	-0.02	Morocco	0.01	United States
-0.02	Ethiopia	-0.03	Mozambique		Uruguay
-0.02	Fiji	-0.05	Nepal	0.14	Uzbekistan
	Finland		Netherlands	-0.05	Venezuela
0.06	France	0.03	New Zealand	-0.07	Vietnam
-0.02	Gabon	-0.12	Nicaragua	-0.04	Zambia
	Germany		Niger	-0.03	Zimbabwe
-0.06	Ghana	-0.04	Nigeria		

Reducing Stresses

Variable name

Percentage of Country's Territory with Acidification Exceedance

Variable code
AC_EXC

Variable number
24

Units
Percent Land Area

Reference year
1990

Median	Minimum	Maximum
0	0	97.48

Source
Stockholm Environment Institute at York, Acidification in developing countries: ecosystem sensitivity and the critical loads approach at the global scale, 2000

Logic
Exceedance of critical SO2 loading represents an indicator for ecosystems under stress due to acidification from anthropogenic sulphur deposition. Since it takes into account both the deposition and the ability of the ecosystem to respond to stress, it is a good indicator of the ecosystems' "sustainability".

Details
From a map of acidification exceedance, the areas at risk were summed within each country and then the percentage of a country at risk of exceedance was calculated. See the main report for more details on how the acidification exceedance map was produced.

2.54	Albania	2.77	Greece	15.96	Norway
0.00	Algeria	0.00	Guatemala	0.00	Pakistan
0.00	Argentina	0.00	Haiti	0.00	Panama
0.00	Armenia	0.00	Honduras	0.00	Papua New Guinea
0.00	Australia	4.93	Hungary	0.00	Paraguay
50.81	Austria	0.00	Iceland	0.00	Peru
0.00	Azerbaijan	0.00	India	0.00	Philippines
0.00	Bangladesh	8.15	Indonesia	53.45	Poland
4.91	Belarus	0.00	Iran	3.24	Portugal
75.83	Belgium	54.16	Ireland	19.27	Romania
0.00	Benin	0.00	Israel	0.33	Russian Federation
0.00	Bhutan	17.94	Italy	0.00	Rwanda
0.00	Bolivia	0.00	Jamaica	0.00	Saudi Arabia
0.00	Botswana	10.99	Japan	0.00	Senegal
0.00	Brazil	0.00	Jordan	0.00	Singapore
14.10	Bulgaria	0.00	Kazakhstan	27.23	Slovak Republic
0.00	Burkina Faso	0.00	Kenya	40.11	Slovenia
0.00	Burundi	58.90	Korea, South	0.00	South Africa
0.00	Cameroon	0.00	Kuwait	3.65	Spain
5.39	Canada	0.00	Kyrgyz Republic	0.00	Sri Lanka
0.00	Central African Republic	1.95	Latvia	0.00	Sudan
0.00	Chile	0.00	Lebanon	34.37	Sweden
15.66	China	0.00	Libya	36.90	Switzerland
0.00	Colombia	0.00	Lithuania	0.00	Syria
0.00	Costa Rica	97.48	Macedonia	0.00	Tanzania
4.69	Croatia	0.00	Madagascar	0.27	Thailand
0.00	Cuba	0.00	Malawi	0.00	Togo
89.22	Czech Republic	0.00	Malaysia	0.00	Trinidad and Tobago
54.88	Denmark	0.00	Mali	0.00	Tunisia
0.00	Dominican Republic	0.00	Mauritius	0.02	Turkey
0.00	Ecuador	0.68	Mexico	0.00	Uganda
0.00	Egypt	0.00	Moldova	4.27	Ukraine
0.00	El Salvador	0.00	Mongolia	45.75	United Kingdom
0.00	Estonia	0.00	Morocco	13.74	United States
0.00	Ethiopia	0.00	Mozambique	0.00	Uruguay
0.00	Fiji	0.00	Nepal	0.00	Uzbekistan
1.19	Finland	43.81	Netherlands	0.00	Venezuela
18.84	France	0.00	New Zealand	32.17	Vietnam
0.00	Gabon	0.00	Nicaragua	5.13	Zambia
51.88	Germany	0.00	Niger	0.00	Zimbabwe
0.00	Ghana	0.00	Nigeria		

Reducing Stresses

Variable name

Consumption Pressure per Capita

Source

World Wide Fund for Nature, *Living Planet Report*. Gland, Switzerland: WWF, 1999.

Variable code	**Variable number**
PRESS	25

Logic

Higher level of consumption pressure produce higher levels of environmental stress, both in terms of resource depletion and waste disposal.

Units

Consumption as a Proportion of Global Average

Reference year

1996

Details

The Consumption Pressure Index was calculated using the same methodology used by WWF for the 1998 Living Planet report, but using only grain-equivalent, fish, wood-equivalent and cement consumption per person. For each commodity, a country's per capita average was divided by the the global per person average, giving a relative score. The relative scores for all the 4 components were then averaged to give the consumption pressure per person for that country.

Median	**Minimum**	**Maximum**
0.89	0.30	5.70

224

0.56	Albania	1.41	Greece	2.47	Norway
0.65	Algeria	0.87	Guatemala	0.35	Pakistan
0.86	Argentina	0.58	Haiti	0.86	Panama
0.39	Armenia	0.82	Honduras	1.39	Papua New Guinea
1.53	Australia	1.00	Hungary	1.39	Paraguay
1.80	Austria		Iceland	0.78	Peru
0.30	Azerbaijan	0.44	India	1.21	Philippines
0.46	Bangladesh	0.99	Indonesia	1.17	Poland
1.19	Belarus	0.62	Iran	2.33	Portugal
	Belgium	1.59	Ireland	0.78	Romania
0.92	Benin	1.87	Israel	1.03	Russian Federation
0.90	Bhutan	1.62	Italy	0.54	Rwanda
0.49	Bolivia	0.88	Jamaica	1.17	Saudi Arabia
1.02	Botswana	2.49	Japan	1.02	Senegal
1.16	Brazil	0.82	Jordan	5.70	Singapore
0.75	Bulgaria	0.77	Kazakhstan	0.86	Slovak Republic
0.65	Burkina Faso	0.91	Kenya	1.40	Slovenia
0.59	Burundi	2.62	Korea, South	0.91	South Africa
0.85	Cameroon	2.47	Kuwait	1.83	Spain
1.80	Canada	0.40	Kyrgyz Republic	0.74	Sri Lanka
0.64	Central African Republic	2.04	Latvia	0.48	Sudan
1.26	Chile	1.79	Lebanon	1.87	Sweden
1.11	China	1.12	Libya	1.55	Switzerland
0.65	Colombia	1.46	Lithuania	0.58	Syria
1.24	Costa Rica	0.78	Macedonia	0.89	Tanzania
0.89	Croatia	0.57	Madagascar	1.45	Thailand
0.57	Cuba	0.69	Malawi	0.72	Togo
1.22	Czech Republic	2.69	Malaysia	0.64	Trinidad and Tobago
1.90	Denmark	0.72	Mali	0.97	Tunisia
0.64	Dominican Republic	1.30	Mauritius	1.12	Turkey
0.97	Ecuador	0.74	Mexico	0.63	Uganda
0.73	Egypt	0.32	Moldova	0.66	Ukraine
0.74	El Salvador	0.50	Mongolia	1.17	United Kingdom
1.55	Estonia	0.71	Morocco	2.10	United States
0.55	Ethiopia	0.57	Mozambique	1.36	Uruguay
	Fiji	0.65	Nepal	0.49	Uzbekistan
2.14	Finland	1.22	Netherlands	0.75	Venezuela
1.57	France	1.69	New Zealand	0.74	Vietnam
1.22	Gabon	0.62	Nicaragua	1.10	Zambia
1.45	Germany	0.52	Niger	0.65	Zimbabwe
1.18	Ghana	0.80	Nigeria		

Reducing Stresses

Variable name

Radioactive Waste

Variable code
NUKE

Variable number
26

Units
Standardized Scale (Z score)

Reference year
1996

Median | **Minimum** | **Maximum**
-0.33 | -0.36 | 4.36

Source
International Atomic Energy Agency, Waste Management Database, 1997

Logic
Radioactive waste, as a source of ionizing radiation, has long been recognized as a potential hazard to human health. Many practices in the fields of research, medicine, industry and generation of electricity generate waste that requires management to ensure the protection of human health and the environment now and in the future, without imposing undue burdens on future generations (The Principle of Radioactive Waste Management, IAEA, 1997).

Details
Two variables were initially available for Radioactive Waste: Accumulated Quantity (cubic meters) as generated and Accumulated Quantity (cubic meters) after treatment. We calculated the Z scores for the two variables, in order to make them comparable, and took the one available for each country. For the three countries (Australia, Canada and Czech Republic) which had both variables, we took the higher.

-0.33	Albania		Ghana		Niger	
	Algeria		Greece		Nigeria	
-0.35	Argentina	-0.33	Guatemala	-0.35	Norway	
	Armenia		Haiti		Pakistan	
-0.34	Australia		Honduras		Panama	
	Austria	-0.34	Hungary		Papua New Guinea	
	Azerbaijan		Iceland		Paraguay	
	Bangladesh	-0.06	India		Peru	
-0.32	Belarus	-0.36	Indonesia		Philippines	
-0.31	Belgium	-0.33	Iran	-0.35	Poland	
	Benin		Ireland	-0.36	Portugal	
	Bhutan		Israel	-0.31	Romania	
	Bolivia	-0.19	Italy		Russian Federation	
	Botswana		Jamaica		Rwanda	
-0.34	Brazil		Japan		Saudi Arabia	
-0.20	Bulgaria		Jordan		Senegal	
	Burkina Faso		Kazakhstan		Singapore	
	Burundi		Kenya	-0.24	Slovak Republic	
	Cameroon	-0.30	Korea, South	-0.35	Slovenia	
0.66	Canada		Kuwait	-0.23	South Africa	
	Central African Republic		Kyrgyz Republic	-0.26	Spain	
-0.36	Chile		Latvia		Sri Lanka	
	China		Lebanon		Sudan	
	Colombia		Libya	-0.23	Sweden	
	Costa Rica	-0.10	Lithuania	-0.32	Switzerland	
	Croatia		Macedonia		Syria	
-0.33	Cuba		Madagascar		Tanzania	
-0.28	Czech Republic		Malawi	-0.36	Thailand	
-0.35	Denmark	-0.33	Malaysia		Togo	
	Dominican Republic		Mali		Trinidad and Tobago	
	Ecuador		Mauritius	-0.33	Tunisia	
-0.33	Egypt	-0.33	Mexico	-0.36	Turkey	
	El Salvador		Moldova		Uganda	
-0.36	Estonia		Mongolia	4.36	Ukraine	
	Ethiopia		Morocco	3.98	United Kingdom	
	Fiji		Mozambique	1.67	United States	
-0.34	Finland		Nepal		Uruguay	
2.18	France	-0.32	Netherlands	-0.33	Uzbekistan	
	Gabon		New Zealand		Venezuela	
0.19	Germany		Nicaragua		Vietnam	

Reducing Stresses

Variable name

Total Fertility Rate

Source

Population Reference Bureau, *2000 World Population Data Sheet*,
Washington, DC: PRB, 2000.

Variable code	**Variable number**
TFR	27

Units

Average Number of Births Per Woman

Logic

Fertility affects population growth and thus resource and consumption pressure.
High levels of fertility are environmentally unsustainable.

Reference year

2000

Median	**Minimum**	**Maximum**
2.74	1.11	7.5

2.16	Albania	1.29	Greece	1.81	Norway
3.81	Algeria	5.00	Guatemala	5.60	Pakistan
2.62	Argentina	4.72	Haiti	2.60	Panama
1.30	Armenia	4.41	Honduras	4.84	Papua New Guinea
1.70	Australia	1.30	Hungary	4.30	Paraguay
1.32	Austria	2.05	Iceland	3.40	Peru
1.87	Azerbaijan	3.34	India	3.69	Philippines
3.27	Bangladesh	2.78	Indonesia	1.39	Poland
1.28	Belarus	2.92	Iran	1.46	Portugal
1.56	Belgium	1.93	Ireland	1.32	Romania
6.32	Benin	2.92	Israel	1.18	Russian Federation
5.60	Bhutan	1.19	Italy	6.50	Rwanda
4.20	Bolivia	2.58	Jamaica	6.37	Saudi Arabia
4.10	Botswana	1.35	Japan	5.70	Senegal
2.44	Brazil	4.40	Jordan	1.48	Singapore
1.11	Bulgaria	1.70	Kazakhstan	1.38	Slovak Republic
6.80	Burkina Faso	4.70	Kenya	1.23	Slovenia
6.48	Burundi	1.48	Korea, South	2.90	South Africa
5.20	Cameroon	3.19	Kuwait	1.15	Spain
1.48	Canada	2.83	Kyrgyz Republic	2.06	Sri Lanka
5.07	Central African Republic	1.16	Latvia	4.57	Sudan
2.44	Chile	2.35	Lebanon	1.49	Sweden
1.80	China	4.10	Libya	1.46	Switzerland
3.00	Colombia	1.35	Lithuania	4.70	Syria
3.20	Costa Rica	1.93	Macedonia	5.63	Tanzania
1.45	Croatia	5.97	Madagascar	1.90	Thailand
1.55	Cuba	5.90	Malawi	6.12	Togo
1.14	Czech Republic	3.20	Malaysia	1.68	Trinidad and Tobago
1.72	Denmark	6.70	Mali	2.84	Tunisia
3.10	Dominican Republic	2.03	Mauritius	2.46	Turkey
3.30	Ecuador	2.65	Mexico	6.86	Uganda
3.30	Egypt	1.51	Moldova	1.28	Ukraine
3.58	El Salvador	2.69	Mongolia	1.70	United Kingdom
1.21	Estonia	3.10	Morocco	2.07	United States
6.70	Ethiopia	5.62	Mozambique	2.26	Uruguay
3.30	Fiji	4.60	Nepal	2.85	Uzbekistan
1.72	Finland	1.59	Netherlands	2.90	Venezuela
1.77	France	1.97	New Zealand	2.50	Vietnam
5.40	Gabon	4.40	Nicaragua	6.08	Zambia
1.32	Germany	7.50	Niger	4.00	Zimbabwe
4.48	Ghana	5.99	Nigeria		

Reducing Stresses

Variable name

Percentage Change in Projected Population Between 2000 and 2050

Variable code   **Variable number**
GR2050   28

Units
Percent Change in Population

Reference year
2000

Median   **Minimum**   **Maximum**
47.17   -34.96   260.7

Source
Population Reference Bureau, *2000 World Population Data Sheet*, Washington, DC: PRB, 2000.

Logic
The projected change in population between 2000 and 2050 provides an indication of the trajectory of population change, which has an impact on a country's per capita natural resource availability and environmental conditions.

Details
A lower threshold of 0 was applied in calculating the ESI. All countries with growth rates of 0 or below received the same score.

51.71	Albania	-8.93	Greece	13.02	Norway
83.44	Algeria	154.03	Guatemala	89.20	Pakistan
47.17	Argentina	84.52	Haiti	49.21	Panama
-0.50	Armenia	79.46	Honduras	97.83	Papua New Guinea
29.72	Australia	-19.80	Hungary	128.25	Paraguay
-5.10	Austria	19.22	Iceland	76.51	Peru
48.69	Azerbaijan	62.45	India	73.80	Philippines
64.53	Bangladesh	46.96	Indonesia	-12.26	Poland
-14.73	Belarus	52.70	Iran	-18.23	Portugal
-2.40	Belgium	19.34	Ireland	-20.56	Romania
182.32	Benin	51.60	Israel	-12.07	Russian Federation
132.04	Bhutan	-27.46	Italy	23.20	Rwanda
87.24	Bolivia	47.18	Jamaica	152.05	Saudi Arabia
-25.95	Botswana	-20.79	Japan	144.01	Senegal
43.57	Brazil	135.49	Jordan	159.21	Singapore
-34.96	Bulgaria	-12.60	Kazakhstan	-12.89	Slovak Republic
187.15	Burkina Faso	27.42	Kenya	-16.57	Slovenia
165.73	Burundi	8.19	Korea, South	-25.06	South Africa
124.76	Cameroon	101.10	Kuwait	-22.04	Spain
30.81	Canada	24.53	Kyrgyz Republic	35.09	Sri Lanka
81.30	Central African Republic	-27.69	Latvia	100.66	Sudan
46.05	Chile	54.62	Lebanon	3.86	Sweden
8.26	China	109.31	Libya	3.00	Switzerland
83.20	Colombia	-15.77	Lithuania	113.89	Syria
95.63	Costa Rica	3.69	Macedonia	150.04	Tanzania
-14.89	Croatia	215.97	Madagascar	15.83	Thailand
-4.89	Cuba	41.82	Malawi	93.68	Togo
-9.43	Czech Republic	107.27	Malaysia	19.15	Trinidad and Tobago
15.05	Denmark	179.09	Mali	56.29	Tunisia
76.75	Dominican Republic	23.30	Mauritius	54.13	Turkey
67.56	Ecuador	52.63	Mexico	260.70	Uganda
71.37	Egypt	-0.75	Moldova	-22.54	Ukraine
117.01	El Salvador	65.21	Mongolia	7.38	United Kingdom
-27.36	Estonia	60.25	Morocco	46.48	United States
193.05	Ethiopia	20.07	Mozambique	27.71	Uruguay
61.53	Fiji	106.10	Nepal	36.64	Uzbekistan
-7.67	Finland	8.22	Netherlands	74.40	Venezuela
9.68	France	17.05	New Zealand	57.13	Vietnam
118.76	Gabon	128.62	Nicaragua	111.74	Zambia
-10.76	Germany	182.42	Niger	-18.21	Zimbabwe
63.58	Ghana	146.14	Nigeria		

227

Reducing Human Vulnerability

Variable name

Daily Per Capita Calorie Supply as a Percentage of Total Requirements

Variable code
CALOR

Variable number
29

Units
Percent of Total Calorie Requirements

Reference year
MRYA 1988-90

Median	Minimum	Maximum
112.5	73	157

Source
World Resource Institute, World Development Indicators 1998-1999

Logic
This indicator represents a measure of the population vulnerability to malnutrition, famine or diseases, in addition to showing the incapacity of an economy to supply an adequate amount of food and to manage food resources.

Details
An upper treshold of 120 was applied in calculating the ESI. Countries with values higher than 120 were assigned a value of 120, based on considerations in the background paper: Bender, W. H., "An end use analysis of global food requirements", Food Policy, vol.19, n.4, August 1994.

107.00	Albania	151.00	Greece	120.00	Norway
123.00	Algeria	103.00	Guatemala	99.00	Pakistan
131.00	Argentina	89.00	Haiti	98.00	Panama
	Armenia	98.00	Honduras	114.00	Papua New Guinea
124.00	Australia	137.00	Hungary	116.00	Paraguay
133.00	Austria		Iceland	87.00	Peru
	Azerbaijan	101.00	India	104.00	Philippines
88.00	Bangladesh	121.00	Indonesia	131.00	Poland
	Belarus	125.00	Iran	136.00	Portugal
149.00	Belgium	157.00	Ireland	116.00	Romania
104.00	Benin	125.00	Israel		Russian Federation
128.00	Bhutan	139.00	Italy	82.00	Rwanda
84.00	Bolivia	114.00	Jamaica	121.00	Saudi Arabia
97.00	Botswana	125.00	Japan	98.00	Senegal
114.00	Brazil	110.00	Jordan	136.00	Singapore
148.00	Bulgaria		Kazakhstan		Slovak Republic
94.00	Burkina Faso	89.00	Kenya		Slovenia
84.00	Burundi	120.00	Korea, South	128.00	South Africa
95.00	Cameroon		Kuwait	141.00	Spain
122.00	Canada		Kyrgyz Republic	101.00	Sri Lanka
82.00	Central African Republic		Latvia	87.00	Sudan
102.00	Chile	127.00	Lebanon	111.00	Sweden
112.00	China	140.00	Libya	130.00	Switzerland
106.00	Colombia		Lithuania	126.00	Syria
121.00	Costa Rica		Macedonia	95.00	Tanzania
	Croatia	95.00	Madagascar	103.00	Thailand
135.00	Cuba	88.00	Malawi	99.00	Togo
	Czech Republic	120.00	Malaysia	114.00	Trinidad and Tobago
135.00	Denmark	96.00	Mali	131.00	Tunisia
102.00	Dominican Republic	128.00	Mauritius	127.00	Turkey
105.00	Ecuador	131.00	Mexico	93.00	Uganda
132.00	Egypt		Moldova		Ukraine
102.00	El Salvador	97.00	Mongolia	130.00	United Kingdom
	Estonia	125.00	Morocco	138.00	United States
73.00	Ethiopia	77.00	Mozambique	101.00	Uruguay
	Fiji	100.00	Nepal		Uzbekistan
113.00	Finland	114.00	Netherlands	99.00	Venezuela
143.00	France	131.00	New Zealand		Vietnam
104.00	Gabon	99.00	Nicaragua	87.00	Zambia
	Germany	95.00	Niger	94.00	Zimbabwe
93.00	Ghana	93.00	Nigeria		

Reducing Human Vulnerability

Variable name

Percentage of Population with Access to Improved Drinking Water Supply

Variable code	**Variable number**
WATSUP	30

Units

Percent of Population

Reference year

2000

Median	**Minimum**	**Maximum**
83.5	24	100

Source

World Health Organization and the United Nations Children's Fund, *Global Water Supply and Sanitation Assessment 2000*, New York: WHO and UNICEF, 2000.

Logic

The percentage of population with access to improved sources of drinking water supply is directly related to the capacity of a country to provide a healthy environment, reducing the risks associated with water-related diseases and exposure to pollutants.

	Albania		Greece	100.00	Norway
94.00	Algeria	92.00	Guatemala	88.00	Pakistan
79.00	Argentina	46.00	Haiti	87.00	Panama
	Armenia	90.00	Honduras	42.00	Papua New Guinea
100.00	Australia	99.00	Hungary		Paraguay
100.00	Austria		Iceland	77.00	Peru
	Azerbaijan	88.00	India	87.00	Philippines
97.00	Bangladesh	76.00	Indonesia		Poland
100.00	Belarus	95.00	Iran		Portugal
	Belgium		Ireland	58.00	Romania
63.00	Benin		Israel	99.00	Russian Federation
62.00	Bhutan		Italy	41.00	Rwanda
79.00	Bolivia	71.00	Jamaica	95.00	Saudi Arabia
95.00	Botswana		Japan	78.00	Senegal
87.00	Brazil	96.00	Jordan	100.00	Singapore
100.00	Bulgaria	91.00	Kazakhstan	100.00	Slovak Republic
53.00	Burkina Faso	49.00	Kenya	100.00	Slovenia
65.00	Burundi	92.00	Korea, South	86.00	South Africa
62.00	Cameroon		Kuwait		Spain
100.00	Canada	77.00	Kyrgyz Republic	83.00	Sri Lanka
60.00	Central African Republic		Latvia	75.00	Sudan
94.00	Chile	100.00	Lebanon	100.00	Sweden
75.00	China	72.00	Libya	100.00	Switzerland
91.00	Colombia		Lithuania	80.00	Syria
98.00	Costa Rica		Macedonia	54.00	Tanzania
	Croatia	47.00	Madagascar	80.00	Thailand
95.00	Cuba	57.00	Malawi	54.00	Togo
	Czech Republic		Malaysia	86.00	Trinidad and Tobago
100.00	Denmark	65.00	Mali	80.00	Tunisia
79.00	Dominican Republic	100.00	Mauritius	83.00	Turkey
71.00	Ecuador	86.00	Mexico	50.00	Uganda
95.00	Egypt	100.00	Moldova		Ukraine
74.00	El Salvador	60.00	Mongolia	100.00	United Kingdom
	Estonia	82.00	Morocco	100.00	United States
24.00	Ethiopia	60.00	Mozambique	98.00	Uruguay
47.00	Fiji	81.00	Nepal	85.00	Uzbekistan
100.00	Finland	100.00	Netherlands	84.00	Venezuela
	France		New Zealand	56.00	Vietnam
70.00	Gabon	79.00	Nicaragua	64.00	Zambia
	Germany	59.00	Niger	85.00	Zimbabwe
64.00	Ghana	57.00	Nigeria		

Reducing Human Vulnerability

Variable name

Child Death Rate from Respiratory Diseases

Variable code	Variable number
DISRES	31

Units
Deaths/100,000 population

Reference year
MRYA 1990-1998

Median	Minimum	Maximum
3.53	0.24	179.57

Source
World Health Organisation. *1997-1999 World Health Statistics Annual*, Geneva: WHO, 2000, available at http://www.who.int/whosis/mort/download.htm

Logic
Indicator of the degree to which children are impacted by poor air quality.

Details
The final number is based on an aggregation of deaths recorded for WHO codes B31, B320, and B321, by sex and by age. These were then combined with UN Population Division population data broken down by age group to produce rates. See the main report for more details on the methodology.

40.92	Albania		1.63	Greece		0.24	Norway
	Algeria			Guatemala			Pakistan
10.34	Argentina			Haiti			Panama
	Armenia			Honduras			Papua New Guinea
1.37	Australia		4.04	Hungary		20.03	Paraguay
0.28	Austria		3.07	Iceland			Peru
	Azerbaijan			India		46.49	Philippines
	Bangladesh			Indonesia		2.67	Poland
	Belarus			Iran		1.87	Portugal
0.94	Belgium		1.43	Ireland		48.44	Romania
	Benin		1.45	Israel			Russian Federation
	Bhutan		0.70	Italy			Rwanda
	Bolivia			Jamaica			Saudi Arabia
	Botswana		1.52	Japan			Senegal
	Brazil			Jordan		3.14	Singapore
19.52	Bulgaria		46.00	Kazakhstan		10.63	Slovak Republic
	Burkina Faso			Kenya		1.39	Slovenia
	Burundi		2.55	Korea, South		19.57	South Africa
	Cameroon		3.53	Kuwait		0.64	Spain
0.62	Canada			Kyrgyz Republic			Sri Lanka
	Central African Republic			Latvia			Sudan
11.86	Chile			Lebanon		1.03	Sweden
	China			Libya			Switzerland
12.73	Colombia		3.11	Lithuania			Syria
6.35	Costa Rica			Macedonia			Tanzania
2.77	Croatia			Madagascar			Thailand
5.11	Cuba			Malawi			Togo
2.35	Czech Republic			Malaysia		6.38	Trinidad and Tobago
	Denmark			Mali			Tunisia
	Dominican Republic		4.70	Mauritius			Turkey
32.80	Ecuador		27.97	Mexico			Uganda
120.86	Egypt		33.59	Moldova			Ukraine
17.69	El Salvador		179.57	Mongolia		1.78	United Kingdom
5.12	Estonia			Morocco			United States
	Ethiopia			Mozambique		11.00	Uruguay
	Fiji			Nepal			Uzbekistan
0.41	Finland		0.88	Netherlands		19.07	Venezuela
0.78	France		1.75	New Zealand			Vietnam
	Gabon		26.20	Nicaragua			Zambia
0.51	Germany			Niger		44.52	Zimbabwe
	Ghana			Nigeria			

Reducing Human Vulnerability

Variable name

Death Rate from Intestinal Infectious Diseases

Variable code DISINT
Variable number 32

Units
Deaths/100,000 population

Reference year
MRYA 1990-1999

Median	Minimum	Maximum
0.97	0.00	36.17

Source
World Health Organisation. *1997-1999 World Health Statistics Annual*, Geneva: WHO, 2000, available at http://www.who.int/whosis/mort/download.htm

Logic
Indicator of the degree to which the population is affected by poor sanitation and water quality, which are related to environmental conditions.

Details
The final number is based on an aggregation of deaths recorded for WHO code B01 for all age groups by sex. These were then combined with UN Population Division population data for the country in that particular year. The death rates were standardized to a common age structure. See the main report for more details on the methodology.

Value	Country	Value	Country	Value	Country
0.33	Albania	0.00	Greece	1.33	Norway
	Algeria		Guatemala		Pakistan
1.95	Argentina		Haiti		Panama
3.15	Armenia		Honduras		Papua New Guinea
0.62	Australia	0.25	Hungary	16.00	Paraguay
0.13	Austria	1.11	Iceland		Peru
5.05	Azerbaijan		India	13.78	Philippines
	Bangladesh		Indonesia	0.11	Poland
0.43	Belarus		Iran	0.17	Portugal
0.84	Belgium	0.57	Ireland	1.08	Romania
	Benin	0.45	Israel	0.90	Russian Federation
	Bhutan	0.12	Italy		Rwanda
	Bolivia		Jamaica		Saudi Arabia
	Botswana	0.88	Japan		Senegal
	Brazil		Jordan	1.24	Singapore
0.56	Bulgaria	3.24	Kazakhstan	0.24	Slovak Republic
	Burkina Faso		Kenya	0.29	Slovenia
	Burundi	2.62	Korea, South	24.99	South Africa
	Cameroon	0.26	Kuwait	0.56	Spain
0.30	Canada	8.28	Kyrgyz Republic		Sri Lanka
	Central African Republic	0.23	Latvia		Sudan
3.21	Chile		Lebanon	0.39	Sweden
	China		Libya		Switzerland
6.42	Colombia	0.34	Lithuania		Syria
9.28	Costa Rica		Macedonia		Tanzania
0.38	Croatia		Madagascar		Thailand
9.51	Cuba		Malawi		Togo
0.43	Czech Republic		Malaysia	4.97	Trinidad and Tobago
	Denmark		Mali		Tunisia
	Dominican Republic	2.15	Mauritius		Turkey
14.28	Ecuador	18.48	Mexico		Uganda
19.65	Egypt	1.04	Moldova	0.54	Ukraine
36.17	El Salvador	2.06	Mongolia	0.75	United Kingdom
0.31	Estonia		Morocco		United States
	Ethiopia		Mozambique	4.30	Uruguay
	Fiji		Nepal	9.58	Uzbekistan
0.97	Finland	0.28	Netherlands	20.16	Venezuela
0.97	France	0.51	New Zealand		Vietnam
	Gabon	24.07	Nicaragua		Zambia
0.34	Germany		Niger	19.43	Zimbabwe
	Ghana		Nigeria		

Reducing Human Vulnerability

Variable name

Under-5 Mortality Rate

Variable code   **Variable number**
U5MORT   33

Units
Deaths Per 1,000 Live Births

Reference year
1998

Median	Minimum	Maximum
35	4	280

Source
The United Nations Chidren's Fund (UNICEF), *The State of the World's Children 2000*, New York: UNICEF, 2000.

Logic
Under-5 mortality rate is a measure of the vulnerability of the most vulnerable population group.

Details
Deaths between birth and age five, divided by 1,000 live births.

Value	Country		Value	Country		Value	Country
37.00	Albania		7.00	Greece		4.00	Norway
40.00	Algeria		52.00	Guatemala		136.00	Pakistan
22.00	Argentina		130.00	Haiti		20.00	Panama
30.00	Armenia		44.00	Honduras		112.00	Papua New Guinea
5.00	Australia		11.00	Hungary		33.00	Paraguay
5.00	Austria		5.00	Iceland		54.00	Peru
46.00	Azerbaijan		105.00	India		44.00	Philippines
106.00	Bangladesh		56.00	Indonesia		11.00	Poland
27.00	Belarus		33.00	Iran		9.00	Portugal
6.00	Belgium		7.00	Ireland		24.00	Romania
165.00	Benin		6.00	Israel		25.00	Russian Federation
116.00	Bhutan		6.00	Italy		170.00	Rwanda
85.00	Bolivia		11.00	Jamaica		26.00	Saudi Arabia
48.00	Botswana		4.00	Japan		121.00	Senegal
42.00	Brazil		36.00	Jordan		5.00	Singapore
17.00	Bulgaria		43.00	Kazakhstan		10.00	Slovak Republic
165.00	Burkina Faso		117.00	Kenya		5.00	Slovenia
176.00	Burundi		5.00	Korea, South		83.00	South Africa
153.00	Cameroon		13.00	Kuwait		6.00	Spain
6.00	Canada		66.00	Kyrgyz Republic		19.00	Sri Lanka
173.00	Central African Republic		22.00	Latvia		115.00	Sudan
12.00	Chile		35.00	Lebanon		4.00	Sweden
47.00	China		24.00	Libya		5.00	Switzerland
30.00	Colombia		23.00	Lithuania		32.00	Syria
16.00	Costa Rica		27.00	Macedonia		142.00	Tanzania
9.00	Croatia		157.00	Madagascar		37.00	Thailand
8.00	Cuba		213.00	Malawi		144.00	Togo
6.00	Czech Republic		10.00	Malaysia		18.00	Trinidad and Tobago
5.00	Denmark		237.00	Mali		32.00	Tunisia
51.00	Dominican Republic		23.00	Mauritius		42.00	Turkey
39.00	Ecuador		34.00	Mexico		134.00	Uganda
69.00	Egypt		35.00	Moldova		22.00	Ukraine
34.00	El Salvador		150.00	Mongolia		6.00	United Kingdom
22.00	Estonia		70.00	Morocco		8.00	United States
173.00	Ethiopia		206.00	Mozambique		19.00	Uruguay
23.00	Fiji		100.00	Nepal		58.00	Uzbekistan
5.00	Finland		5.00	Netherlands		25.00	Venezuela
5.00	France		6.00	New Zealand		42.00	Vietnam
144.00	Gabon		48.00	Nicaragua		202.00	Zambia
5.00	Germany		280.00	Niger		89.00	Zimbabwe
105.00	Ghana		187.00	Nigeria			

Social and Institutional Capacity

Variable name

Research & Development Scientists and Engineers per Million Population

Variable code   **Variable number**
RDPERS   34

Units
Scientists & Engineers/Million Population

Reference year
MRYA 1980-1997

Median	Minimum	Maximum
663.5	3	4909

Source
United Nations Educational, Scientific and Cultural Organization (UNESCO), *Statistical Yearbook 1999*, Paris: UNESCO, 1999.

Logic
The greater the proportion of a country's population that is dedicated to research and development in a variety of scientific fields, the more capacity it has to respond effectively to environmental threats.

	Albania	773.00	Greece	3,664.00	Norway
	Algeria	104.00	Guatemala	72.00	Pakistan
660.00	Argentina		Haiti	252.00	Panama
1,485.00	Armenia		Honduras		Papua New Guinea
3,357.00	Australia	1,099.00	Hungary		Paraguay
1,627.00	Austria	4,131.00	Iceland	233.00	Peru
2,791.00	Azerbaijan	149.00	India	157.00	Philippines
52.00	Bangladesh	182.00	Indonesia	1,358.00	Poland
2,248.00	Belarus	560.00	Iran	1,182.00	Portugal
2,272.00	Belgium	2,319.00	Ireland	1,387.00	Romania
176.00	Benin	4,828.00	Israel	3,587.00	Russian Federation
	Bhutan	1,318.00	Italy	35.00	Rwanda
172.00	Bolivia	8.00	Jamaica		Saudi Arabia
	Botswana	4,909.00	Japan	3.00	Senegal
168.00	Brazil	94.00	Jordan	2,318.00	Singapore
1,747.00	Bulgaria		Kazakhstan	1,866.00	Slovak Republic
17.00	Burkina Faso		Kenya	2,251.00	Slovenia
33.00	Burundi	2,193.00	Korea, South	1,031.00	South Africa
	Cameroon	230.00	Kuwait	1,305.00	Spain
2,719.00	Canada	584.00	Kyrgyz Republic	191.00	Sri Lanka
56.00	Central African Republic	1,049.00	Latvia		Sudan
455.00	Chile		Lebanon	3,826.00	Sweden
454.00	China	362.00	Libya	3,006.00	Switzerland
37.00	Colombia	2,028.00	Lithuania	30.00	Syria
532.00	Costa Rica	1,335.00	Macedonia		Tanzania
1,916.00	Croatia	12.00	Madagascar	103.00	Thailand
1,612.00	Cuba		Malawi	98.00	Togo
1,222.00	Czech Republic	93.00	Malaysia		Trinidad and Tobago
3,259.00	Denmark		Mali	125.00	Tunisia
	Dominican Republic	361.00	Mauritius	291.00	Turkey
146.00	Ecuador	214.00	Mexico	21.00	Uganda
459.00	Egypt	330.00	Moldova	2,171.00	Ukraine
20.00	El Salvador	910.00	Mongolia	2,448.00	United Kingdom
2,017.00	Estonia		Morocco	3,676.00	United States
	Ethiopia		Mozambique	667.00	Uruguay
	Fiji		Nepal	1,763.00	Uzbekistan
2,799.00	Finland	2,219.00	Netherlands	209.00	Venezuela
2,659.00	France	1,663.00	New Zealand	334.00	Vietnam
234.00	Gabon	204.00	Nicaragua		Zambia
2,831.00	Germany		Niger		Zimbabwe
	Ghana	15.00	Nigeria		

Social and Institutional Capacity

Variable name

Expenditure for Research & Development as a Percentage of GNP

Variable code	Variable number
RDEXP	35

Units
Percent of Gross National Product

Reference year
MRYA 1980-1997

Median	Minimum	Maximum
0.64	0.01	3.76

Source

Source: United Nations Educational, Scientific and Cultural Organization (UNESCO), *Statistical Yearbook 1999*, Paris: UNESCO, 1999.

Logic

The greater the proportion of a country's annual GNP that is dedicated to research and development in a variety of scientific fields, the more capacity it has to respond effectively to environmental threats.

	Albania	0.47	Greece	1.58	Norway
	Algeria	0.16	Guatemala	0.92	Pakistan
0.38	Argentina		Haiti	0.01	Panama
	Armenia		Honduras		Papua New Guinea
1.80	Australia	0.68	Hungary		Paraguay
1.53	Austria	1.55	Iceland	0.25	Peru
0.21	Azerbaijan	0.73	India	0.22	Philippines
0.03	Bangladesh	0.07	Indonesia	0.77	Poland
1.07	Belarus	0.48	Iran	0.62	Portugal
1.60	Belgium	1.61	Ireland	0.72	Romania
0.70	Benin	2.35	Israel	0.88	Russian Federation
	Bhutan	2.21	Italy	0.04	Rwanda
1.67	Bolivia	0.04	Jamaica		Saudi Arabia
	Botswana	2.80	Japan	0.01	Senegal
0.81	Brazil	0.26	Jordan	1.13	Singapore
0.57	Bulgaria	0.32	Kazakhstan	1.05	Slovak Republic
0.19	Burkina Faso		Kenya	1.46	Slovenia
0.31	Burundi	2.82	Korea, South	0.70	South Africa
	Cameroon	0.16	Kuwait	0.90	Spain
1.66	Canada	0.20	Kyrgyz Republic	0.19	Sri Lanka
0.25	Central African Republic	0.43	Latvia		Sudan
0.68	Chile		Lebanon	3.76	Sweden
0.66	China	0.22	Libya	2.60	Switzerland
0.12	Colombia	0.70	Lithuania	0.20	Syria
0.21	Costa Rica	0.31	Macedonia		Tanzania
1.03	Croatia	0.18	Madagascar	0.13	Thailand
0.84	Cuba		Malawi	0.48	Togo
1.20	Czech Republic	0.24	Malaysia		Trinidad and Tobago
1.95	Denmark		Mali	0.30	Tunisia
	Dominican Republic	0.32	Mauritius	0.45	Turkey
0.02	Ecuador	0.33	Mexico	0.57	Uganda
0.22	Egypt	0.90	Moldova		Ukraine
2.20	El Salvador		Mongolia	1.95	United Kingdom
0.57	Estonia		Morocco	2.63	United States
	Ethiopia		Mozambique		Uruguay
	Fiji		Nepal		Uzbekistan
2.76	Finland	2.08	Netherlands	0.49	Venezuela
2.25	France	1.04	New Zealand		Vietnam
0.01	Gabon		Nicaragua		Zambia
2.41	Germany		Niger		Zimbabwe
	Ghana	0.09	Nigeria		

Social and Institutional Capacity

Variable name

Scientific and Technical Articles per Million Population

Variable code	Variable number
ARTPOP	36

Units
Articles/Million Population

Reference year
1995

Median	Minimum	Maximum
109.36	1.84	395.6

Source
National Science Board, *Science and Engineering Indicators - 1998*. Arlington, VA: National Science Foundation (NSF), 1998.

Logic
The rate at which a country's scientific establishment publishes articles in the natural and earth sciences is correlated with its capacity to respond to environmental problems.

	Albania	97.15	Greece	218.44	Norway
	Algeria		Guatemala		Pakistan
26.92	Argentina		Haiti		Panama
	Armenia		Honduras		Papua New Guinea
281.29	Australia	95.83	Hungary		Paraguay
150.49	Austria	197.47	Iceland		Peru
	Azerbaijan	6.27	India		Philippines
	Bangladesh		Indonesia	82.70	Poland
	Belarus		Iran	50.73	Portugal
183.98	Belgium	106.95	Ireland		Romania
	Benin	395.60	Israel	93.62	Russian Federation
	Bhutan	124.19	Italy		Rwanda
	Bolivia		Jamaica		Saudi Arabia
	Botswana	174.76	Japan		Senegal
10.24	Brazil		Jordan	178.88	Singapore
72.59	Bulgaria		Kazakhstan	108.32	Slovak Republic
	Burkina Faso	2.87	Kenya	124.12	Slovenia
	Burundi	53.15	Korea, South	26.95	South Africa
	Cameroon		Kuwait	131.80	Spain
314.58	Canada		Kyrgyz Republic		Sri Lanka
	Central African Republic		Latvia		Sudan
21.96	Chile		Lebanon	318.18	Sweden
4.29	China		Libya	390.88	Switzerland
	Colombia		Lithuania		Syria
	Costa Rica		Macedonia		Tanzania
66.77	Croatia		Madagascar		Thailand
	Cuba		Malawi		Togo
110.41	Czech Republic		Malaysia		Trinidad and Tobago
271.01	Denmark		Mali		Tunisia
	Dominican Republic		Mauritius	11.82	Turkey
	Ecuador	10.02	Mexico		Uganda
14.32	Egypt		Moldova	44.70	Ukraine
	El Salvador		Mongolia	250.36	United Kingdom
	Estonia		Morocco		United States
	Ethiopia		Mozambique		Uruguay
	Fiji		Nepal		Uzbekistan
242.96	Finland	254.10	Netherlands		Venezuela
224.70	France	271.56	New Zealand		Vietnam
	Gabon		Nicaragua		Zambia
212.94	Germany		Niger		Zimbabwe
	Ghana	1.84	Nigeria		

Social and Institutional Capacity

Variable name

IUCN Member Organizations per Million Population

Variable code
IUCN

Variable number
37

Units
Organizations/Million Population

Reference year
2000

Median	Minimum	Maximum
0.42	0.01	7.85

Source
Membership List, IUCN-The World Conservation Union, 1 August 2000 (updated with new data on 10-Nov-00)

Logic
IUCN is the oldest international environmental membership organization, currently with over 900 members (governmental and NGO) worldwide, often including the most significant environmental NGOs in each country.

	Country		Country		Country
	Albania	0.59	Greece	1.41	Norway
0.12	Algeria	1.26	Guatemala	0.18	Pakistan
0.68	Argentina		Haiti	3.34	Panama
	Armenia	1.02	Honduras	0.26	Papua New Guinea
2.19	Australia	0.39	Hungary	0.95	Paraguay
0.78	Austria	7.85	Iceland	0.37	Peru
	Azerbaijan	0.02	India	0.05	Philippines
0.13	Bangladesh	0.01	Indonesia	0.21	Poland
	Belarus	0.02	Iran	0.30	Portugal
0.70	Belgium	0.86	Ireland	0.13	Romania
	Benin	0.86	Israel	0.05	Russian Federation
	Bhutan	0.33	Italy		Rwanda
1.22	Bolivia	1.27	Jamaica	0.19	Saudi Arabia
6.27	Botswana	0.17	Japan	0.55	Senegal
0.10	Brazil	2.38	Jordan	1.33	Singapore
0.23	Bulgaria	0.24	Kazakhstan	0.57	Slovak Republic
0.33	Burkina Faso	0.30	Kenya	0.52	Slovenia
	Burundi	0.14	Korea, South	0.65	South Africa
0.09	Cameroon	1.40	Kuwait	0.71	Spain
1.08	Canada	0.23	Kyrgyz Republic	0.70	Sri Lanka
	Central African Republic	0.37	Latvia	0.04	Sudan
0.23	Chile	2.74	Lebanon	0.82	Sweden
0.01	China	0.23	Libya	1.17	Switzerland
0.34	Colombia	0.54	Lithuania	0.08	Syria
2.62	Costa Rica	0.52	Macedonia	0.12	Tanzania
0.66	Croatia	0.09	Madagascar	0.05	Thailand
	Cuba	0.32	Malawi	0.28	Togo
0.49	Czech Republic	0.34	Malaysia		Trinidad and Tobago
1.36	Denmark	0.68	Mali	0.61	Tunisia
0.42	Dominican Republic	1.89	Mauritius	0.07	Turkey
1.46	Ecuador	0.11	Mexico	0.30	Uganda
0.05	Egypt	0.46	Moldova	0.06	Ukraine
1.37	El Salvador	0.45	Mongolia	0.76	United Kingdom
1.27	Estonia	0.25	Morocco	0.21	United States
0.02	Ethiopia	0.21	Mozambique	1.61	Uruguay
2.75	Fiji	0.48	Nepal	0.05	Uzbekistan
1.00	Finland	1.34	Netherlands	0.36	Venezuela
0.56	France	2.08	New Zealand	0.04	Vietnam
	Gabon	0.52	Nicaragua	0.97	Zambia
0.21	Germany	0.26	Niger	2.03	Zimbabwe
0.20	Ghana	0.03	Nigeria		

Social and Institutional Capacity

Variable name

Civil and Political Liberties

Variable code   **Variable number**
CIVLIB   38

Units
Index Ranging from 1 (High Levels of Liberties) to 7 (Low Levels)

Reference year
2000

Median   **Minimum**   **Maximum**
3   1   7

Source
Source: Freedom House, *Freedom in the World 1999-2000*,
New York: Freedom House, 2000,
http://www.freedomhouse.org/research/freeworld/2000/table5.htm

Logic
In countries that guarantee freedom of expression, rights to organize,
rule of law, economic rights, and multi-party elections, there is more likely to be a
vigorous public debate about values and issues relevant to environmental quality,
and legal safeguards that encourage innovation.

4.50	Albania	2.00	Greece	1.00	Norway
5.50	Algeria	3.50	Guatemala	6.00	Pakistan
2.50	Argentina	5.00	Haiti	1.50	Panama
4.00	Armenia	3.00	Honduras	2.50	Papua New Guinea
1.00	Australia	1.50	Hungary	3.50	Paraguay
1.00	Austria	1.00	Iceland	4.50	Peru
5.00	Azerbaijan	2.50	India	2.50	Philippines
3.50	Bangladesh	4.00	Indonesia	1.50	Poland
6.00	Belarus	6.00	Iran	1.00	Portugal
1.50	Belgium	1.00	Ireland	2.00	Romania
2.50	Benin	1.50	Israel	4.50	Russian Federation
6.50	Bhutan	1.50	Italy	6.50	Rwanda
2.00	Bolivia	2.00	Jamaica	7.00	Saudi Arabia
2.00	Botswana	1.50	Japan	4.00	Senegal
3.50	Brazil	4.00	Jordan	5.00	Singapore
2.50	Bulgaria	5.50	Kazakhstan	1.50	Slovak Republic
4.00	Burkina Faso	5.50	Kenya	1.50	Slovenia
6.00	Burundi	2.00	Korea, South	1.50	South Africa
6.50	Cameroon	4.50	Kuwait	1.50	Spain
1.00	Canada	5.00	Kyrgyz Republic	3.50	Sri Lanka
3.50	Central African Republic	1.50	Latvia	7.00	Sudan
2.00	Chile	5.50	Lebanon	1.00	Sweden
6.50	China	7.00	Libya	1.00	Switzerland
4.00	Colombia	1.50	Lithuania	7.00	Syria
1.50	Costa Rica	3.00	Macedonia	4.00	Tanzania
4.00	Croatia	3.00	Madagascar	2.50	Thailand
7.00	Cuba	3.00	Malawi	5.00	Togo
1.50	Czech Republic	5.00	Malaysia	1.50	Trinidad and Tobago
1.00	Denmark	3.00	Mali	5.50	Tunisia
2.50	Dominican Republic	1.50	Mauritius	4.50	Turkey
2.50	Ecuador	3.50	Mexico	5.00	Uganda
5.50	Egypt	3.00	Moldova	3.50	Ukraine
2.50	El Salvador	2.50	Mongolia	1.50	United Kingdom
1.50	Estonia	4.50	Morocco	1.00	United States
5.00	Ethiopia	3.50	Mozambique	1.50	Uruguay
2.50	Fiji	3.50	Nepal	6.50	Uzbekistan
1.00	Finland	1.00	Netherlands	4.00	Venezuela
1.50	France	1.00	New Zealand	7.00	Vietnam
4.50	Gabon	3.00	Nicaragua	4.50	Zambia
1.50	Germany	5.00	Niger	5.50	Zimbabwe
3.00	Ghana	3.50	Nigeria		

Social and Institutional Capacity

Variable name

Stringency and Consistency of Environmental Regulations

Variable code   **Variable number**
WEFSTR   39

Units
Survey Responses Ranging from 1 (Strongly Disagree) to 7 (Strongly Agree)

Reference year
2000

Median	Minimum	Maximum
3.86	2.35	6.45

Source
Michael E. Porter et al, *The Global Competitiveness Report 2000*, Oxford: Oxford University Press, 2000.

Logic
Stronger regulations prompt more effective action, other things equal.

Details
Average of responses to the following survey questions: "Air pollution regulations are among the world's most stringent"; "Water pollution regulations are among the world's most stringent"; "Environmental regulations are enforced consistently and fairly"; and "Environmental regulations are typically enacted ahead of most other countries."

	Albania	3.65	Greece	5.65	Norway
	Algeria		Guatemala		Pakistan
2.90	Argentina		Haiti		Panama
	Armenia		Honduras		Papua New Guinea
5.53	Australia	3.88	Hungary		Paraguay
6.30	Austria	4.63	Iceland	2.80	Peru
	Azerbaijan	2.78	India	2.55	Philippines
	Bangladesh	2.63	Indonesia	3.48	Poland
	Belarus		Iran	4.13	Portugal
5.20	Belgium	4.55	Ireland		Romania
	Benin	4.13	Israel	3.40	Russian Federation
	Bhutan	4.48	Italy		Rwanda
2.40	Bolivia		Jamaica		Saudi Arabia
	Botswana	5.60	Japan		Senegal
3.83	Brazil	3.65	Jordan	5.85	Singapore
3.20	Bulgaria		Kazakhstan	3.93	Slovak Republic
	Burkina Faso		Kenya		Slovenia
	Burundi	4.15	Korea, South	3.75	South Africa
	Cameroon		Kuwait	4.40	Spain
5.50	Canada		Kyrgyz Republic		Sri Lanka
	Central African Republic		Latvia		Sudan
3.85	Chile		Lebanon	6.10	Sweden
2.85	China		Libya	6.13	Switzerland
3.23	Colombia		Lithuania		Syria
3.90	Costa Rica		Macedonia		Tanzania
	Croatia		Madagascar	2.98	Thailand
	Cuba		Malawi		Togo
4.35	Czech Republic	3.83	Malaysia		Trinidad and Tobago
6.38	Denmark		Mali		Tunisia
	Dominican Republic	3.10	Mauritius	3.63	Turkey
3.00	Ecuador	3.53	Mexico		Uganda
3.35	Egypt		Moldova	2.98	Ukraine
2.35	El Salvador		Mongolia	5.40	United Kingdom
	Estonia		Morocco	5.88	United States
	Ethiopia		Mozambique		Uruguay
	Fiji		Nepal		Uzbekistan
6.38	Finland	6.08	Netherlands	2.88	Venezuela
5.30	France	5.35	New Zealand	2.63	Vietnam
	Gabon		Nicaragua		Zambia
6.45	Germany		Niger	2.70	Zimbabwe
	Ghana		Nigeria		

Social and Institutional Capacity

Variable name

Degree to which Environmental Regulations Promote Innovation

Variable code   **Variable number**
WEFINN   40

Units
Survey Responses Ranging from 1 (Strongly Disagree) to 7 (Strongly Agree)

Reference year
2000

Median	Minimum	Maximum
4.03	2.75	5.7

Source
Michael E. Porter et al, *The Global Competitiveness Report 2000*, Oxford: Oxford University Press, 2000.

Logic
Where regulations and management strategies prompt effective innovation, better results follow.

Details
Average of responses to the following survey questions: "Environmental regulations are flexible and offer many points for achieving compliance"; and "Environmental regulations are transparent and stable".

	Country		Country		Country
	Albania	4.00	Greece	5.00	Norway
	Algeria		Guatemala		Pakistan
3.60	Argentina		Haiti		Panama
	Armenia		Honduras		Papua New Guinea
4.80	Australia	3.95	Hungary		Paraguay
4.75	Austria	4.70	Iceland	3.70	Peru
	Azerbaijan	3.70	India	3.65	Philippines
	Bangladesh	3.60	Indonesia	3.70	Poland
	Belarus		Iran	4.15	Portugal
4.45	Belgium	4.75	Ireland		Romania
	Benin	4.15	Israel	4.00	Russian Federation
	Bhutan	3.85	Italy		Rwanda
3.35	Bolivia		Jamaica		Saudi Arabia
	Botswana	4.85	Japan		Senegal
3.95	Brazil	4.25	Jordan	5.55	Singapore
3.60	Bulgaria		Kazakhstan	4.10	Slovak Republic
	Burkina Faso		Kenya		Slovenia
	Burundi	4.00	Korea, South	4.25	South Africa
	Cameroon		Kuwait	4.25	Spain
4.90	Canada		Kyrgyz Republic		Sri Lanka
	Central African Republic		Latvia		Sudan
3.60	Chile		Lebanon	5.00	Sweden
3.90	China		Libya	5.15	Switzerland
3.80	Colombia		Lithuania		Syria
3.90	Costa Rica		Macedonia		Tanzania
	Croatia		Madagascar	3.60	Thailand
	Cuba		Malawi		Togo
4.05	Czech Republic	4.25	Malaysia		Trinidad and Tobago
5.10	Denmark		Mali		Tunisia
	Dominican Republic	3.70	Mauritius	4.15	Turkey
2.75	Ecuador	3.85	Mexico		Uganda
4.00	Egypt		Moldova	3.40	Ukraine
2.85	El Salvador		Mongolia	4.90	United Kingdom
	Estonia		Morocco	4.65	United States
	Ethiopia		Mozambique		Uruguay
	Fiji		Nepal		Uzbekistan
5.70	Finland	5.00	Netherlands	3.50	Venezuela
4.30	France	4.80	New Zealand	3.35	Vietnam
	Gabon		Nicaragua		Zambia
4.60	Germany		Niger	3.75	Zimbabwe
	Ghana		Nigeria		

Social and Institutional Capacity

Variable name

Percentage of Land Area Under Protected Status

Variable code   **Variable number**
PRAREA   41

Units
Percent Land Area

Reference year
1997

Median	Minimum	Maximum
6.47	0.00	43.08

Source
World Resources Institute, *World Resources 2000-01*, Washington, DC: World Resources Institute, 2000.

Logic
The percentage of land area dedicated to protected areas represents a proxy for the investment by the country in biodiversity conservation.

2.76	Albania
2.47	Algeria
1.70	Argentina
7.59	Armenia
6.99	Australia
28.33	Austria
5.49	Azerbaijan
0.75	Bangladesh
4.16	Belarus
2.59	Belgium
7.03	Benin
21.23	Bhutan
14.39	Bolivia
18.52	Botswana
4.20	Brazil
4.44	Bulgaria
10.44	Burkina Faso
5.61	Burundi
4.51	Cameroon
9.99	Canada
8.20	Central African Republic
18.88	Chile
6.44	China
9.01	Colombia
13.75	Costa Rica
6.70	Croatia
17.37	Cuba
15.83	Czech Republic
32.24	Denmark
31.48	Dominican Republic
43.08	Ecuador
0.80	Egypt
0.25	El Salvador
12.00	Estonia
5.52	Ethiopia
1.03	Fiji
5.99	Finland
11.66	France
2.81	Gabon
26.95	Germany
4.85	Ghana

2.24	Greece
16.83	Guatemala
0.35	Haiti
9.94	Honduras
6.81	Hungary
9.70	Iceland
4.80	India
9.66	Indonesia
5.12	Iran
0.86	Ireland
14.92	Israel
7.30	Italy
0.14	Jamaica
6.77	Japan
3.35	Jordan
2.75	Kazakhstan
6.16	Kenya
6.91	Korea, South
1.52	Kuwait
3.59	Kyrgyz Republic
12.48	Latvia
0.34	Lebanon
0.10	Libya
9.96	Lithuania
7.08	Macedonia
1.92	Madagascar
11.25	Malawi
4.51	Malaysia
3.71	Mali
6.01	Mauritius
2.39	Mexico
1.18	Moldova
10.30	Mongolia
0.71	Morocco
6.09	Mozambique
7.77	Nepal
6.71	Netherlands
23.59	New Zealand
7.44	Nicaragua
7.65	Niger
3.32	Nigeria

6.76	Norway
4.83	Pakistan
19.09	Panama
0.02	Papua New Guinea
3.53	Paraguay
2.70	Peru
4.87	Philippines
9.56	Poland
6.50	Portugal
4.66	Romania
3.06	Russian Federation
14.68	Rwanda
2.31	Saudi Arabia
11.32	Senegal
4.43	Singapore
21.76	Slovak Republic
5.70	Slovenia
5.39	South Africa
8.44	Spain
13.28	Sri Lanka
3.64	Sudan
9.01	Sweden
18.03	Switzerland
0.00	Syria
15.64	Tanzania
13.09	Thailand
7.87	Togo
3.04	Trinidad and Tobago
0.29	Tunisia
1.39	Turkey
9.57	Uganda
1.55	Ukraine
20.46	United Kingdom
13.39	United States
0.26	Uruguay
2.05	Uzbekistan
36.25	Venezuela
3.05	Vietnam
8.56	Zambia
7.93	Zimbabwe

Social and Institutional Capacity

Variable name

Number of Sectoral EIA Guidelines

Source

IIED, WRI and IUCN. *A Directory of Impact Assessment Guidelines (Second Edition)*, London: International Institute for Environment and Development (IIED), 1998.

Variable code	**Variable number**
EIA	42

Units

Number of Guidelines

Reference year

1998

Logic

Environmental Impact Assessment represents an important tool for promoting sound environmental management.

Median	**Minimum**	**Maximum**
0	0	13

0.00	Albania	1.00	Greece	0.00	Norway
0.00	Algeria	0.00	Guatemala	8.00	Pakistan
6.00	Argentina	0.00	Haiti	0.00	Panama
0.00	Armenia	0.00	Honduras	0.00	Papua New Guinea
1.00	Australia	0.00	Hungary	4.00	Paraguay
1.00	Austria	0.00	Iceland	6.00	Peru
0.00	Azerbaijan	9.00	India	1.00	Philippines
3.00	Bangladesh	5.00	Indonesia	0.00	Poland
0.00	Belarus	0.00	Iran	7.00	Portugal
9.00	Belgium	2.00	Ireland	0.00	Romania
0.00	Benin	0.00	Israel	2.00	Russian Federation
0.00	Bhutan	4.00	Italy	0.00	Rwanda
7.00	Bolivia	0.00	Jamaica	0.00	Saudi Arabia
0.00	Botswana	0.00	Japan	0.00	Senegal
2.00	Brazil	0.00	Jordan	1.00	Singapore
0.00	Bulgaria	0.00	Kazakhstan	8.00	Slovak Republic
0.00	Burkina Faso	1.00	Kenya	0.00	Slovenia
0.00	Burundi	0.00	Korea, South	8.00	South Africa
0.00	Cameroon	2.00	Kuwait	6.00	Spain
9.00	Canada	0.00	Kyrgyz Republic	2.00	Sri Lanka
0.00	Central African Republic	0.00	Latvia	0.00	Sudan
9.00	Chile	0.00	Lebanon	3.00	Sweden
1.00	China	0.00	Libya	6.00	Switzerland
2.00	Colombia	0.00	Lithuania	0.00	Syria
8.00	Costa Rica	0.00	Macedonia	1.00	Tanzania
0.00	Croatia	0.00	Madagascar	7.00	Thailand
0.00	Cuba	2.00	Malawi	0.00	Togo
1.00	Czech Republic	13.00	Malaysia	0.00	Trinidad and Tobago
1.00	Denmark	0.00	Mali	0.00	Tunisia
0.00	Dominican Republic	0.00	Mauritius	0.00	Turkey
1.00	Ecuador	2.00	Mexico	0.00	Uganda
11.00	Egypt	0.00	Moldova	0.00	Ukraine
0.00	El Salvador	0.00	Mongolia	9.00	United Kingdom
0.00	Estonia	0.00	Morocco	9.00	United States
0.00	Ethiopia	1.00	Mozambique	0.00	Uruguay
0.00	Fiji	6.00	Nepal	0.00	Uzbekistan
5.00	Finland	3.00	Netherlands	2.00	Venezuela
7.00	France	3.00	New Zealand	2.00	Vietnam
0.00	Gabon	0.00	Nicaragua	0.00	Zambia
3.00	Germany	1.00	Niger	9.00	Zimbabwe
1.00	Ghana	1.00	Nigeria		

Social and Institutional Capacity

Variable name

ISO 14001 Certified Companies per Million Dollars GDP

Variable code   **Variable number**
ISO14   43

Units
Number of ISO 14001 Certified Companies/Million US Dollars GDP

Reference year
2000

Median	Minimum	Maximum
0.05	0.03	30.8

Source
ISO14001/EMAS registered companies, ISO World, International Standards Organisation, available at http://www.ecology.or.jp/isoworld/english/analy14k.htm, visited November 2000.

Logic
ISO 14001 specifies standards for corporate environmental management. The commitment to ISO 14001 certification serves as a proxy for the degree to which industries are instituting management practices that reduce waste and resource consumption.

0.00	Albania	0.00	Greece	0.00	Norway
0.00	Algeria	0.33	Guatemala	0.10	Pakistan
2.58	Argentina	0.00	Haiti	0.00	Panama
0.00	Armenia	1.69	Honduras	0.00	Papua New Guinea
13.45	Australia	14.43	Hungary	0.55	Paraguay
0.00	Austria	0.00	Iceland	1.08	Peru
0.00	Azerbaijan	1.56	India	2.46	Philippines
0.00	Bangladesh	1.59	Indonesia	1.89	Poland
0.00	Belarus	0.45	Iran	0.00	Portugal
0.00	Belgium	0.00	Ireland	0.08	Romania
0.00	Benin	4.47	Israel	0.03	Russian Federation
0.00	Bhutan	0.00	Italy	0.00	Rwanda
0.00	Bolivia	0.00	Jamaica	0.31	Saudi Arabia
0.00	Botswana	14.77	Japan	0.00	Senegal
2.75	Brazil	5.17	Jordan	27.25	Singapore
0.00	Bulgaria	0.00	Kazakhstan	7.06	Slovak Republic
0.00	Burkina Faso	0.00	Kenya	8.39	Slovenia
0.00	Burundi	8.01	Korea, South	3.39	South Africa
0.00	Cameroon	0.00	Kuwait	0.00	Spain
4.21	Canada	0.00	Kyrgyz Republic	0.39	Sri Lanka
0.00	Central African Republic	0.65	Latvia	0.00	Sudan
0.78	Chile	4.52	Lebanon	0.00	Sweden
0.94	China	0.00	Libya	30.80	Switzerland
0.95	Colombia	0.42	Lithuania	0.56	Syria
8.22	Costa Rica	0.00	Macedonia	0.00	Tanzania
2.62	Croatia	0.00	Madagascar	9.40	Thailand
0.00	Cuba	0.00	Malawi	0.00	Togo
7.85	Czech Republic	12.95	Malaysia	1.10	Trinidad and Tobago
0.00	Denmark	0.00	Mali	0.23	Tunisia
0.31	Dominican Republic	3.42	Mauritius	1.80	Turkey
0.32	Ecuador	2.31	Mexico	0.00	Uganda
4.09	Egypt	0.00	Moldova	0.06	Ukraine
0.00	El Salvador	0.00	Mongolia	0.00	United Kingdom
3.31	Estonia	0.63	Morocco	1.22	United States
0.00	Ethiopia	0.00	Mozambique	3.73	Uruguay
3.25	Fiji	0.00	Nepal	0.00	Uzbekistan
0.00	Finland	0.00	Netherlands	0.62	Venezuela
0.00	France	10.33	New Zealand	0.80	Vietnam
0.00	Gabon	0.00	Nicaragua	3.84	Zambia
0.00	Germany	0.00	Niger	1.14	Zimbabwe
0.00	Ghana	0.14	Nigeria		

Social and Institutional Capacity

Variable name

Dow Jones Sustainability Group Index

Variable code	Variable number
DJSGI	44

Units
Percentage

Reference year
2000

Median	Minimum	Maximum
10.73	0.00	46.34

Source
"Assessment of the Country Allocation of the Dow Jones Sustainability Group Index", SAM Sustainability Group, 2001.

Logic
The Dow Jones Sustainability Group Index tracks a group of companies that have been rated as the top 10% in terms of sustainability. Firms that are already in the Dow Jones Global Index are eligible to enter the Sustainability Group Index. Countries in which a higher percentage of eligible firms meet the requirements have a private sector that is contributing more vigorously to environmental sustainability.

Details
For each country, the number of companies in the Sustainability Index was divided by the number of companies in the Global Index.

	Albania	10.34	Greece		Norway	
	Algeria		Guatemala		Pakistan	
	Argentina		Haiti		Panama	
	Armenia		Honduras		Papua New Guinea	
25.00	Australia		Hungary		Paraguay	
33.33	Austria		Iceland		Peru	
	Azerbaijan		India	7.69	Philippines	
	Bangladesh	0.00	Indonesia		Poland	
	Belarus		Iran		Portugal	
22.22	Belgium	11.11	Ireland		Romania	
	Benin		Israel		Russian Federation	
	Bhutan	9.52	Italy		Rwanda	
	Bolivia		Jamaica		Saudi Arabia	
	Botswana	7.10	Japan		Senegal	
11.76	Brazil		Jordan	6.98	Singapore	
	Bulgaria		Kazakhstan		Slovak Republic	
	Burkina Faso		Kenya		Slovenia	
	Burundi	10.00	Korea, South	2.86	South Africa	
	Cameroon		Kuwait	17.39	Spain	
17.20	Canada		Kyrgyz Republic		Sri Lanka	
	Central African Republic		Latvia		Sudan	
10.00	Chile		Lebanon	40.74	Sweden	
	China		Libya	42.31	Switzerland	
	Colombia		Lithuania		Syria	
	Costa Rica		Macedonia		Tanzania	
	Croatia		Madagascar	20.00	Thailand	
	Cuba		Malawi		Togo	
	Czech Republic		Malaysia		Trinidad and Tobago	
28.57	Denmark		Mali		Tunisia	
	Dominican Republic		Mauritius		Turkey	
	Ecuador	20.59	Mexico		Uganda	
	Egypt		Moldova		Ukraine	
	El Salvador		Mongolia	19.08	United Kingdom	
	Estonia		Morocco	7.56	United States	
	Ethiopia		Mozambique		Uruguay	
	Fiji		Nepal		Uzbekistan	
33.33	Finland		Netherlands		Venezuela	
8.57	France	25.00	New Zealand		Vietnam	
	Gabon		Nicaragua		Zambia	
46.34	Germany		Niger		Zimbabwe	
	Ghana		Nigeria			

Social and Institutional Capacity

Variable name

Average Innovest EcoValue '21 Rating of Firms

Variable code   **Variable number**
ECOVAL   45

Units
Scale ranging from -3 (low) to 3 (high)

Reference year
2001

Median   **Minimum**   **Maximum**
0.83   -3   2.34

Source
Innovest Strategic Value Advisors

Logic
The Innnovest EcoValue '21 rating measures environmental performance at the firm level.

Details
Within each country, EcoValue levels were weighted by market capitalization share and then averaged to get a value for the individual country, based on the location of company headquarters.

	Country			Country			Country
	Albania			Greece		2.00	Norway
	Algeria			Guatemala			Pakistan
	Argentina			Haiti			Panama
	Armenia			Honduras			Papua New Guinea
-3.00	Australia			Hungary			Paraguay
	Austria			Iceland			Peru
	Azerbaijan			India			Philippines
	Bangladesh			Indonesia			Poland
	Belarus			Iran			Portugal
-0.82	Belgium		2.00	Ireland			Romania
	Benin			Israel			Russian Federation
	Bhutan		-2.01	Italy			Rwanda
	Bolivia			Jamaica			Saudi Arabia
	Botswana		1.69	Japan			Senegal
	Brazil			Jordan		-2.97	Singapore
	Bulgaria			Kazakhstan			Slovak Republic
	Burkina Faso			Kenya			Slovenia
	Burundi			Korea, South			South Africa
	Cameroon			Kuwait		-0.96	Spain
1.36	Canada			Kyrgyz Republic			Sri Lanka
	Central African Republic			Latvia			Sudan
	Chile			Lebanon		2.34	Sweden
	China			Libya		1.40	Switzerland
	Colombia			Lithuania			Syria
	Costa Rica			Macedonia			Tanzania
	Croatia			Madagascar			Thailand
	Cuba			Malawi			Togo
	Czech Republic		-3.00	Malaysia			Trinidad and Tobago
2.15	Denmark			Mali			Tunisia
	Dominican Republic			Mauritius			Turkey
	Ecuador		-3.00	Mexico			Uganda
	Egypt			Moldova			Ukraine
	El Salvador			Mongolia		1.07	United Kingdom
	Estonia			Morocco		0.33	United States
	Ethiopia			Mozambique			Uruguay
	Fiji			Nepal			Uzbekistan
1.95	Finland		1.30	Netherlands			Venezuela
-0.48	France			New Zealand			Vietnam
	Gabon			Nicaragua			Zambia
0.59	Germany			Niger			Zimbabwe
	Ghana			Nigeria			

Social and Institutional Capacity

Variable name

World Business Council on Sustainable Development Members (per million dollars GDP)

Variable code	Variable number
WBCSD	46

Units

Members per Million Dollars GDP

Reference year

2001

Median	Minimum	Maximum
0.00	0.00	1148.45

Source

World Business Council on Sustainable Development, "List of Members," http://www.wbcsd.org/memlist2.htm, visited 6 January 2001.

Logic

The WBCSD is a prominent private-sector organization promoting the principles of sustainable development and encouraging high standards of environmental management within firms.

0.00	Albania	0.00	Greece	0.00	Norway		
208.68	Algeria	0.00	Guatemala	0.00	Pakistan		
0.00	Argentina	0.00	Haiti	0.00	Panama		
0.00	Armenia	0.00	Honduras	0.00	Papua New Guinea		
133.62	Australia	0.00	Hungary	233.23	Paraguay		
0.00	Austria	0.00	Iceland	0.00	Peru		
0.00	Azerbaijan	0.00	India	0.00	Philippines		
0.00	Bangladesh	0.00	Indonesia	0.00	Poland		
0.00	Belarus	0.00	Iran	204.06	Portugal		
0.00	Belgium	0.00	Ireland	0.00	Romania		
0.00	Benin	0.00	Israel	619.20	Russian Federation		
0.00	Bhutan	145.74	Italy	0.00	Rwanda		
0.00	Bolivia	0.00	Jamaica	0.00	Saudi Arabia		
0.00	Botswana	859.94	Japan	0.00	Senegal		
452.85	Brazil	0.00	Jordan	0.00	Singapore		
0.00	Bulgaria	0.00	Kazakhstan	0.00	Slovak Republic		
0.00	Burkina Faso	0.00	Kenya	0.00	Slovenia		
0.00	Burundi	222.59	Korea, South	117.82	South Africa		
0.00	Cameroon	0.00	Kuwait	0.00	Spain		
254.43	Canada	0.00	Kyrgyz Republic	0.00	Sri Lanka		
0.00	Central African Republic	0.00	Latvia	0.00	Sudan		
113.81	Chile	0.00	Lebanon	0.00	Sweden		
322.03	China	0.00	Libya	470.36	Switzerland		
0.00	Colombia	0.00	Lithuania	0.00	Syria		
167.04	Costa Rica	0.00	Macedonia	0.00	Tanzania		
148.17	Croatia	0.00	Madagascar	183.30	Thailand		
0.00	Cuba	0.00	Malawi	0.00	Togo		
161.79	Czech Republic	0.00	Malaysia	0.00	Trinidad and Tobago		
123.88	Denmark	0.00	Mali	0.00	Tunisia		
0.00	Dominican Republic	0.00	Mauritius	0.00	Turkey		
0.00	Ecuador	389.41	Mexico	0.00	Uganda		
0.00	Egypt	0.00	Moldova	0.00	Ukraine		
0.00	El Salvador	0.00	Mongolia	688.42	United Kingdom		
0.00	Estonia	0.00	Morocco	1,148.45	United States		
0.00	Ethiopia	0.00	Mozambique	0.00	Uruguay		
0.00	Fiji	0.00	Nepal	0.00	Uzbekistan		
191.87	Finland	270.56	Netherlands	0.00	Venezuela		
283.35	France	57.84	New Zealand	0.00	Vietnam		
0.00	Gabon	0.00	Nicaragua	0.00	Zambia		
360.86	Germany	0.00	Niger	0.00	Zimbabwe		
0.00	Ghana	0.00	Nigeria				

Social and Institutional Capacity

Variable name

Levels of Environmental Competitiveness

Variable code
WEFCOM

Variable number
47

Units
Survey Responses Ranging from 1 (Strongly Disagree) to 7 (Strongly Agree)

Reference year
2000

Median	Minimum	Maximum
4.35	3.20	5.90

Source
Michael E. Porter et al, *The Global Competitiveness Report 2000*, Oxford: Oxford University Press, 2000.

Logic
In countries where compliance with environmental standards is seen as beneficial to the economic interests of firms, the prospects for environmental sustainability are enhanced.

Details
Response to the statement "Complying with environmental standards has a positive influence on long-term competitiveness by prompting companies to improve products and processes."

	Albania	4.00	Greece	5.40	Norway
	Algeria		Guatemala		Pakistan
4.10	Argentina		Haiti		Panama
	Armenia		Honduras		Papua New Guinea
5.30	Australia	4.10	Hungary		Paraguay
5.40	Austria	5.10	Iceland	4.00	Peru
	Azerbaijan	3.90	India	3.70	Philippines
	Bangladesh	3.60	Indonesia	3.80	Poland
	Belarus		Iran	4.50	Portugal
5.20	Belgium	4.70	Ireland		Romania
	Benin	4.40	Israel	3.50	Russian Federation
	Bhutan	4.30	Italy		Rwanda
3.50	Bolivia		Jamaica		Saudi Arabia
	Botswana	5.40	Japan		Senegal
4.60	Brazil	4.20	Jordan	5.90	Singapore
3.60	Bulgaria		Kazakhstan	4.30	Slovak Republic
	Burkina Faso		Kenya		Slovenia
	Burundi	4.60	Korea, South	4.40	South Africa
	Cameroon		Kuwait	5.00	Spain
5.10	Canada		Kyrgyz Republic		Sri Lanka
	Central African Republic		Latvia		Sudan
4.20	Chile		Lebanon	5.40	Sweden
4.30	China		Libya	5.60	Switzerland
3.90	Colombia		Lithuania		Syria
4.50	Costa Rica		Macedonia		Tanzania
	Croatia		Madagascar	3.50	Thailand
	Cuba		Malawi		Togo
4.70	Czech Republic	4.20	Malaysia		Trinidad and Tobago
5.70	Denmark		Mali		Tunisia
	Dominican Republic	3.70	Mauritius	4.20	Turkey
4.20	Ecuador	4.40	Mexico		Uganda
4.30	Egypt		Moldova	3.20	Ukraine
3.30	El Salvador		Mongolia	5.00	United Kingdom
	Estonia		Morocco	5.00	United States
	Ethiopia		Mozambique		Uruguay
	Fiji		Nepal		Uzbekistan
5.80	Finland	5.60	Netherlands	3.60	Venezuela
4.90	France	5.00	New Zealand	3.50	Vietnam
	Gabon		Nicaragua		Zambia
5.50	Germany		Niger	4.10	Zimbabwe
	Ghana		Nigeria		

Social and Institutional Capacity

Variable name

Availability of Sustainable Development Information at the National Level

Variable code	Variable number
SDINFO	48

Units
Index Ranging from 1 (Low Levels of SD Information) to 4 (High Levels)

Reference year
1997

Median	Minimum	Maximum
2.57	1.50	4.00

Source
United Nations Department of Economic and Social Affairs web site, http://www.un.org/esa/agenda21/natlinfo/agenda21/issue/inst.htm#info, visited December 1999.

Logic
Agenda 21 represents a major effort to frame the sustainable development agenda, and therefore the quality of information related to Agenda 21 chapters has a direct bearing on decision-makers' abilities to pursue sustainability.

Details
In their reports to Rio+5 in 1997, countries rated themselves on the availability of information pertaining to chapters of Agenda 21 (from 1 for low levels of information, to 4 for high levels). We averaged the scores for seven key chapters, included Chapter 9 on protection of the atmosphere, Chapter 14 on sustainable agriculture and rural development, Chapter 15 on conservation of biological diversity, Chapter 18 on freshwater resources, Chapter 19 on toxic chemicals, Chapter 21 on solid wastes, and Chapter 40 on information for decision-making.

3.00	Albania			Ghana			Niger	
2.29	Algeria		2.29	Greece			Nigeria	
	Argentina			Guatemala		3.14	Norway	
	Armenia		1.57	Haiti			Pakistan	
	Australia			Honduras		2.00	Panama	
3.57	Austria		2.43	Hungary			Papua New Guinea	
	Azerbaijan		3.00	Iceland		1.71	Paraguay	
1.50	Bangladesh			India			Peru	
	Belarus			Indonesia		1.83	Philippines	
	Belgium			Iran		2.17	Poland	
2.50	Benin		3.43	Ireland		2.71	Portugal	
	Bhutan		3.29	Israel			Romania	
2.00	Bolivia			Italy		2.14	Russian Federation	
	Botswana			Jamaica			Rwanda	
2.00	Brazil		2.86	Japan			Saudi Arabia	
	Bulgaria			Jordan			Senegal	
	Burkina Faso			Kazakhstan			Singapore	
	Burundi			Kenya		3.50	Slovak Republic	
	Cameroon			Korea, South		2.86	Slovenia	
2.50	Canada			Kuwait		2.00	South Africa	
	Central African Republic			Kyrgyz Republic		3.57	Spain	
	Chile			Latvia			Sri Lanka	
	China		2.29	Lebanon			Sudan	
2.00	Colombia			Libya			Sweden	
2.86	Costa Rica		3.00	Lithuania		3.00	Switzerland	
	Croatia		2.71	Macedonia		2.00	Syria	
2.83	Cuba			Madagascar			Tanzania	
	Czech Republic		2.00	Malawi		2.71	Thailand	
3.00	Denmark		2.43	Malaysia			Togo	
	Dominican Republic			Mali			Trinidad and Tobago	
2.57	Ecuador		2.14	Mauritius		2.14	Tunisia	
2.71	Egypt		2.14	Mexico		2.29	Turkey	
	El Salvador			Moldova		2.00	Uganda	
2.57	Estonia		2.57	Mongolia		3.00	Ukraine	
	Ethiopia			Morocco			United Kingdom	
2.17	Fiji			Mozambique		4.00	United States	
3.14	Finland		3.43	Nepal			Uruguay	
2.83	France		3.57	Netherlands		2.00	Uzbekistan	
	Gabon		1.86	New Zealand			Venezuela	
3.14	Germany		3.29	Nicaragua		3.29	Vietnam	

Variable name

Environmental Strategies and Action Plans

Variable code   **Variable number**
PLANS   49

Units
Number of Strategies and Action Plans

Reference year
1992-1996

Median   **Minimum**   **Maximum**
2   0   7

Source
Sustainable Development Information Service, World Resources Institute, May 1996, http://www.wri.org/wdces/, site visited October 2000.

Logic
Environmental Strategies, Action Plans and Assessments provide valuable information for sustainable development decision making.

Details
Countries received one credit for each strategy, action plan or assessment produced from 1992-96 in the following categories: climate change, environmental synopsis, environmental profiles, environmental strategies, forestry, OECD, state of the environment, and other.

2.00	Albania	0.00	Greece	6.00	Norway
0.00	Algeria	2.00	Guatemala	4.00	Pakistan
3.00	Argentina	0.00	Haiti	0.00	Panama
0.00	Armenia	1.00	Honduras	1.00	Papua New Guinea
3.00	Australia	3.00	Hungary	0.00	Paraguay
3.00	Austria	2.00	Iceland	2.00	Peru
1.00	Azerbaijan	5.00	India	2.00	Philippines
1.00	Bangladesh	7.00	Indonesia	4.00	Poland
0.00	Belarus	1.00	Iran	3.00	Portugal
0.00	Belgium	2.00	Ireland	0.00	Romania
3.00	Benin	2.00	Israel	2.00	Russian Federation
1.00	Bhutan	3.00	Italy	2.00	Rwanda
4.00	Bolivia	4.00	Jamaica	0.00	Saudi Arabia
1.00	Botswana	2.00	Japan	2.00	Senegal
1.00	Brazil	0.00	Jordan	3.00	Singapore
0.00	Bulgaria	1.00	Kazakhstan	2.00	Slovak Republic
2.00	Burkina Faso	3.00	Kenya	1.00	Slovenia
2.00	Burundi	0.00	Korea, South	3.00	South Africa
1.00	Cameroon	0.00	Kuwait	0.00	Spain
2.00	Canada	0.00	Kyrgyz Republic	5.00	Sri Lanka
2.00	Central African Republic	3.00	Latvia	0.00	Sudan
3.00	Chile	0.00	Lebanon	1.00	Sweden
6.00	China	0.00	Libya	2.00	Switzerland
5.00	Colombia	1.00	Lithuania	0.00	Syria
0.00	Costa Rica	1.00	Macedonia	5.00	Tanzania
1.00	Croatia	0.00	Madagascar	2.00	Thailand
1.00	Cuba	2.00	Malawi	2.00	Togo
4.00	Czech Republic	4.00	Malaysia	1.00	Trinidad and Tobago
2.00	Denmark	1.00	Mali	2.00	Tunisia
1.00	Dominican Republic	1.00	Mauritius	1.00	Turkey
5.00	Ecuador	4.00	Mexico	5.00	Uganda
4.00	Egypt	4.00	Moldova	2.00	Ukraine
3.00	El Salvador	3.00	Mongolia	4.00	United Kingdom
3.00	Estonia	0.00	Morocco	4.00	United States
3.00	Ethiopia	3.00	Mozambique	2.00	Uruguay
4.00	Fiji	2.00	Nepal	1.00	Uzbekistan
4.00	Finland	6.00	Netherlands	0.00	Venezuela
4.00	France	1.00	New Zealand	0.00	Vietnam
2.00	Gabon	1.00	Nicaragua	2.00	Zambia
2.00	Germany	3.00	Niger	3.00	Zimbabwe
2.00	Ghana	2.00	Nigeria		

Social and Institutional Capacity

Variable name

Number of ESI Variables Missing from a Subset of All Variables

Variable code   **Variable number**
ESIMIS   50

Units
Percentage

Reference year
2001

Median	Minimum	Maximum
12	0	18

Source
2001 Environmental Sustainability Index data set

Logic
The more ESI variables a country is missing, the poorer the availability of environmental information in general.

Details
We counted the number of missing variables from the set of variables for which it could reasonably be expected that any country could have coverage if it wanted to: GMS_EC, GMS_PH, GMS_SS, NO2, PRTBRD, PRTMAM, SO2, TSP, GMS_DO, BODWAT, CARSKM, COALKM, FOREST, FERTHA, GR2050, NOXKM, NUKE, PESTHA, PRESS, SO2KM, TFR, VOCKM, DISINT, DISRES, U5MORT, CALOR, WATSUP, CIVLIB, ENEFF, ARTPOP, EIA, GASPR, ISO14, IUCN, PLANS, PRAREA, RDEXP, RDPERS, RENEWP, SDINFO, CFC, CO2_EM, CO2HIS, SO2EXP, CITES, EIONUM, FOOT, FSC, GEF, MONFUN, VIENNA.

14.00	Albania	8.00	Greece	2.00	Norway
14.00	Algeria	12.00	Guatemala	9.00	Pakistan
3.00	Argentina	18.00	Haiti	12.00	Panama
17.00	Armenia	14.00	Honduras	16.00	Papua New Guinea
7.00	Australia	0.00	Hungary	14.00	Paraguay
6.00	Austria	12.00	Iceland	15.00	Peru
17.00	Azerbaijan	5.00	India	5.00	Philippines
11.00	Bangladesh	6.00	Indonesia	2.00	Poland
12.00	Belarus	9.00	Iran	2.00	Portugal
8.00	Belgium	10.00	Ireland	7.00	Romania
17.00	Benin	11.00	Israel	4.00	Russian Federation
17.00	Bhutan	8.00	Italy	18.00	Rwanda
15.00	Bolivia	14.00	Jamaica	18.00	Saudi Arabia
15.00	Botswana	3.00	Japan	12.00	Senegal
4.00	Brazil	11.00	Jordan	12.00	Singapore
5.00	Bulgaria	16.00	Kazakhstan	6.00	Slovak Republic
15.00	Burkina Faso	14.00	Kenya	8.00	Slovenia
17.00	Burundi	2.00	Korea, South	6.00	South Africa
15.00	Cameroon	17.00	Kuwait	7.00	Spain
1.00	Canada	16.00	Kyrgyz Republic	13.00	Sri Lanka
17.00	Central African Republic	10.00	Latvia	13.00	Sudan
5.00	Chile	17.00	Lebanon	5.00	Sweden
5.00	China	16.00	Libya	3.00	Switzerland
6.00	Colombia	5.00	Lithuania	14.00	Syria
7.00	Costa Rica	13.00	Macedonia	15.00	Tanzania
11.00	Croatia	14.00	Madagascar	5.00	Thailand
8.00	Cuba	16.00	Malawi	16.00	Togo
8.00	Czech Republic	4.00	Malaysia	16.00	Trinidad and Tobago
7.00	Denmark	13.00	Mali	12.00	Tunisia
17.00	Dominican Republic	14.00	Mauritius	6.00	Turkey
9.00	Ecuador	4.00	Mexico	14.00	Uganda
8.00	Egypt	12.00	Moldova	11.00	Ukraine
10.00	El Salvador	12.00	Mongolia	2.00	United Kingdom
11.00	Estonia	12.00	Morocco	5.00	United States
16.00	Ethiopia	16.00	Mozambique	12.00	Uruguay
17.00	Fiji	14.00	Nepal	16.00	Uzbekistan
1.00	Finland	1.00	Netherlands	8.00	Venezuela
4.00	France	4.00	New Zealand	16.00	Vietnam
17.00	Gabon	12.00	Nicaragua	16.00	Zambia
6.00	Germany	17.00	Niger	12.00	Zimbabwe
12.00	Ghana	14.00	Nigeria		

Social and Institutional Capacity

Variable name

Energy Efficiency (total energy consumption per unit GDP)

Variable code   **Variable number**
ENEFF   51

Units
Billion Btu/Million Dollars GDP

Reference year
1998

Median	**Minimum**	**Maximum**
15.37	2.76	101.19

Source
US Energy Information Agency, http://www.eia.doe.gov/emeu/international/contents.html
site visited September 2000.

Logic
The more eco-efficient an economy is, the higher its resource productivity
and the less energy it needs to produce goods and services.

	Albania	12.95	Greece	12.17	Norway
18.64	Algeria	11.52	Guatemala	30.70	Pakistan
12.22	Argentina	8.78	Haiti	18.70	Panama
19.26	Armenia	8.97	Honduras	10.57	Papua New Guinea
11.46	Australia	32.29	Hungary	15.32	Paraguay
7.09	Austria	14.49	Iceland	10.81	Peru
101.19	Azerbaijan	28.13	India	19.74	Philippines
13.15	Bangladesh	22.96	Indonesia	45.05	Poland
39.21	Belarus	26.89	Iran	11.77	Portugal
11.83	Belgium	6.85	Ireland	58.39	Romania
	Benin	9.96	Israel	74.19	Russian Federation
15.29	Bhutan	6.66	Italy	6.13	Rwanda
18.41	Bolivia	35.58	Jamaica	35.11	Saudi Arabia
9.33	Botswana	6.55	Japan	7.87	Senegal
14.01	Brazil	34.52	Jordan	20.41	Singapore
60.71	Bulgaria	76.93	Kazakhstan	63.95	Slovak Republic
2.76	Burkina Faso	15.41	Kenya	11.26	Slovenia
6.93	Burundi	17.91	Korea, South	37.92	South Africa
6.74	Cameroon	30.81	Kuwait	8.73	Spain
17.54	Canada	66.03	Kyrgyz Republic	13.70	Sri Lanka
	Central African Republic	25.01	Latvia	5.36	Sudan
16.63	Chile	41.21	Lebanon	9.14	Sweden
39.10	China	23.64	Libya	5.19	Switzerland
23.98	Colombia	54.92	Lithuania	22.36	Syria
16.13	Costa Rica	67.74	Macedonia	8.53	Tanzania
	Croatia	6.69	Madagascar	19.29	Thailand
39.37	Cuba	9.36	Malawi	8.51	Togo
56.22	Czech Republic	22.88	Malaysia	76.60	Trinidad and Tobago
4.84	Denmark	4.15	Mali	16.63	Tunisia
18.68	Dominican Republic	9.11	Mauritius	13.85	Turkey
27.57	Ecuador	17.72	Mexico	3.40	Uganda
31.03	Egypt	47.38	Moldova	96.53	Ukraine
13.75	El Salvador	44.24	Mongolia	8.59	United Kingdom
16.09	Estonia	12.82	Morocco	13.41	United States
3.93	Ethiopia	12.59	Mozambique	12.86	Uruguay
13.10	Fiji	7.38	Nepal	88.73	Uzbekistan
8.37	Finland	11.01	Netherlands	44.11	Venezuela
7.39	France	15.09	New Zealand	64.57	Vietnam
11.04	Gabon	36.46	Nicaragua	28.31	Zambia
7.28	Germany	6.40	Niger	22.34	Zimbabwe
13.20	Ghana	23.66	Nigeria		

Social and Institutional Capacity

Variable name

Renewable Energy Production as a Percentage of Total Energy Consumption

Variable code | **Variable number**
RENEWP | 52

Units

Renewable Energy Production as a Percent of Total Energy Consumption

Reference year

1998

Median	Minimum	Maximum
533.9	0	8

Source

US Energy Information Agency, http://www.eia.doe.gov/emeu/international/contents.html, site visited September 2000.

Logic

The higher the proportion of hydroelectric and renewable energy sources, the lower the reliance on more environmentally damaging sources such as fossil fuel energy.

Details

Hydroelectric, biomass, geothermal, solar and wind electric power production is calculated as a percentage of total energy consumption. Some countries exceed 100 percent because they are net exporters of renewable energy. A logarithmic transformation of this variable was used in calculating the ESI.

64.39	Albania	3.16	Greece	64.03	Norway
0.04	Algeria	19.36	Guatemala	12.88	Pakistan
13.75	Argentina	15.12	Haiti	24.26	Panama
17.38	Armenia	26.53	Honduras	12.55	Papua New Guinea
4.57	Australia	0.15	Hungary	533.89	Paraguay
29.81	Austria	64.83	Iceland	27.21	Peru
3.18	Azerbaijan	6.41	India	18.90	Philippines
1.72	Bangladesh	3.22	Indonesia	1.41	Poland
0.02	Belarus	1.68	Iran	15.19	Portugal
0.55	Belgium	2.32	Ireland	9.84	Romania
0.00	Benin	0.05	Israel	6.02	Russian Federation
265.37	Bhutan	6.80	Italy	13.16	Rwanda
12.34	Bolivia	3.20	Jamaica	0.00	Saudi Arabia
0.00	Botswana	5.72	Japan	0.00	Senegal
38.38	Brazil	0.15	Jordan	0.00	Singapore
3.31	Bulgaria	3.30	Kazakhstan	5.03	Slovak Republic
7.46	Burkina Faso	28.99	Kenya	11.66	Slovenia
18.93	Burundi	0.64	Korea, South	0.37	South Africa
38.32	Cameroon	0.00	Kuwait	7.62	Spain
29.29	Canada	50.45	Kyrgyz Republic	23.92	Sri Lanka
17.66	Central African Republic	20.05	Latvia	14.92	Sudan
17.42	Chile	4.13	Lebanon	35.06	Sweden
6.22	China	0.00	Libya	29.37	Switzerland
25.71	Colombia	2.08	Lithuania	12.59	Syria
48.04	Costa Rica	8.32	Macedonia	27.81	Tanzania
14.51	Croatia	23.33	Madagascar	3.15	Thailand
3.95	Cuba	43.57	Malawi	0.50	Togo
1.49	Czech Republic	1.80	Malaysia	0.08	Trinidad and Tobago
4.81	Denmark	19.46	Mali	0.14	Tunisia
12.20	Dominican Republic	2.91	Mauritius	15.18	Turkey
19.53	Ecuador	6.18	Mexico	38.22	Uganda
6.89	Egypt	2.59	Moldova	1.92	Ukraine
23.93	El Salvador	0.00	Mongolia	1.20	United Kingdom
0.02	Estonia	5.52	Morocco	4.51	United States
35.10	Ethiopia	34.23	Mozambique	63.55	Uruguay
26.11	Fiji	31.67	Nepal	3.64	Uzbekistan
19.11	Finland	1.11	Netherlands	20.74	Venezuela
6.50	France	37.24	New Zealand	25.36	Vietnam
12.79	Gabon	27.29	Nicaragua	83.66	Zambia
2.04	Germany	0.00	Niger	6.89	Zimbabwe
55.36	Ghana	6.11	Nigeria		

Social and Institutional Capacity

Variable name

Price of Premium Gasoline

Variable code
GASPR

Variable number
53

Units
US Dollars/Liter

Reference year
1998 (last quarter)

Median	Minimum	Maximum
0.53	0.08	1.21

Source

Gesellschaft fuer Technische Zusammenarbeit (GTZ), *Fuel Prices and Taxation*, Frankfurt: GTZ, May 1999.

Logic

Unsubsidized gasoline prices are an indicator that appropriate price signals are being sent and that environmental externalities have been internalized. Artificially low prices encourage wasteful consumption and thus air pollution and greenhouse gas emissions.

0.86	Albania	0.65	Greece	1.21	Norway
0.31	Algeria	0.41	Guatemala	0.46	Pakistan
0.94	Argentina	0.59	Haiti	0.41	Panama
0.49	Armenia	0.50	Honduras	0.41	Papua New Guinea
0.46	Australia	0.72	Hungary	0.47	Paraguay
1.04	Austria	1.12	Iceland	0.55	Peru
0.46	Azerbaijan	0.56	India	0.34	Philippines
0.47	Bangladesh	0.16	Indonesia	0.54	Poland
0.34	Belarus	0.08	Iran	1.02	Portugal
1.12	Belgium	1.02	Ireland	0.53	Romania
0.39	Benin	0.86	Israel	0.28	Russian Federation
0.59	Bhutan	1.19	Italy	0.72	Rwanda
0.53	Bolivia	0.37	Jamaica	0.16	Saudi Arabia
0.31	Botswana	1.02	Japan	0.71	Senegal
0.80	Brazil	0.42	Jordan	0.72	Singapore
0.66	Bulgaria	0.30	Kazakhstan	0.61	Slovak Republic
0.68	Burkina Faso	0.70	Kenya	0.66	Slovenia
0.72	Burundi	0.93	Korea, South	0.43	South Africa
0.64	Cameroon	0.17	Kuwait	0.84	Spain
0.41	Canada	0.47	Kyrgyz Republic	0.84	Sri Lanka
0.81	Central African Republic	0.55	Latvia	0.33	Sudan
0.49	Chile	0.35	Lebanon	1.09	Sweden
0.28	China	0.22	Libya	0.86	Switzerland
0.24	Colombia	0.51	Lithuania	0.45	Syria
0.41	Costa Rica	0.70	Macedonia	0.63	Tanzania
0.67	Croatia	0.47	Madagascar	0.30	Thailand
0.50	Cuba	0.51	Malawi	0.42	Togo
0.72	Czech Republic	0.28	Malaysia	0.39	Trinidad and Tobago
1.05	Denmark	0.77	Mali	0.60	Tunisia
0.40	Dominican Republic		Mauritius	0.78	Turkey
0.38	Ecuador	0.36	Mexico	0.86	Uganda
0.29	Egypt	0.45	Moldova	0.49	Ukraine
0.54	El Salvador	0.23	Mongolia	1.11	United Kingdom
0.45	Estonia	0.79	Morocco	0.32	United States
0.36	Ethiopia	0.55	Mozambique	0.90	Uruguay
0.50	Fiji	0.59	Nepal	0.11	Uzbekistan
1.17	Finland	1.14	Netherlands	0.14	Venezuela
1.11	France	0.64	New Zealand	0.35	Vietnam
0.63	Gabon	0.47	Nicaragua	0.53	Zambia
0.96	Germany	0.76	Niger	0.26	Zimbabwe
0.32	Ghana	0.13	Nigeria		

Social and Institutional Capacity

Variable name

Subsidies for Energy or Materials Usage

Variable code   **Variable number**
WEFSUB   54

Units
Survey Responses Ranging from 1 (Strongly Disagree) to 7 (Strongly Agree)

Reference year
2000

Median   **Minimum**   **Maximum**
4.45   2.60   6.20

Source
Michael E. Porter et al, *The Global Competitveness Report 2000*,
Oxford: Oxford University Press, 2000.

Logic
Subsidies encourage wasteful consumption of energy and materials.

Details
Response to the statement "No government subsidies for energy or materials usage are present."

	Albania	4.20	Greece	4.50	Norway		
	Algeria		Guatemala		Pakistan		
5.10	Argentina		Haiti		Panama		
	Armenia		Honduras		Papua New Guinea		
5.20	Australia	4.40	Hungary		Paraguay		
5.10	Austria	5.10	Iceland	5.50	Peru		
	Azerbaijan	3.00	India	4.20	Philippines		
	Bangladesh	2.90	Indonesia	3.70	Poland		
	Belarus		Iran	4.40	Portugal		
4.90	Belgium	5.20	Ireland		Romania		
	Benin	5.10	Israel	4.20	Russian Federation		
	Bhutan	4.40	Italy		Rwanda		
5.40	Bolivia		Jamaica		Saudi Arabia		
	Botswana	4.60	Japan		Senegal		
4.60	Brazil	4.40	Jordan	5.30	Singapore		
3.50	Bulgaria		Kazakhstan	3.50	Slovak Republic		
	Burkina Faso		Kenya		Slovenia		
	Burundi	4.10	Korea, South	4.60	South Africa		
	Cameroon		Kuwait	4.40	Spain		
4.80	Canada		Kyrgyz Republic		Sri Lanka		
	Central African Republic		Latvia		Sudan		
5.10	Chile		Lebanon	4.70	Sweden		
4.00	China		Libya	5.10	Switzerland		
4.40	Colombia		Lithuania		Syria		
4.20	Costa Rica		Macedonia		Tanzania		
	Croatia		Madagascar	3.80	Thailand		
	Cuba		Malawi		Togo		
4.40	Czech Republic	4.40	Malaysia		Trinidad and Tobago		
5.10	Denmark		Mali		Tunisia		
	Dominican Republic	4.50	Mauritius	4.60	Turkey		
2.60	Ecuador	4.10	Mexico		Uganda		
4.00	Egypt		Moldova	3.50	Ukraine		
3.90	El Salvador		Mongolia	5.20	United Kingdom		
	Estonia		Morocco	4.60	United States		
	Ethiopia		Mozambique		Uruguay		
	Fiji		Nepal		Uzbekistan		
6.20	Finland	5.30	Netherlands	3.60	Venezuela		
5.00	France	6.10	New Zealand	3.70	Vietnam		
	Gabon		Nicaragua		Zambia		
4.60	Germany		Niger	3.60	Zimbabwe		
	Ghana		Nigeria				

Social and Institutional Capacity

Variable name

Reducing Corruption

Source
Dataset from "Aggregating Governance Indicators" and "Governance Matters", Kaufmann D., Kraay A. and Zoido-Lobaton P., May 2000, World Bank.

Variable code	Variable number
GRAFT	55

Units
Standardized Scale in which high values represent low levels of corruption.

Logic
Corruption contributes to lax enforcement of environmental regulations and an ability on the part of producers and consumers to evade responsibility for the environmental harms they cause.

Reference year
2000

Median	Minimum	Maximum
-0.28	-1.57	2.13

-0.99	Albania	0.82	Greece	1.69	Norway		
-0.88	Algeria	-0.82	Guatemala	-0.77	Pakistan		
-0.27	Argentina	-0.53	Haiti	-0.46	Panama		
-0.80	Armenia	-0.94	Honduras	-0.85	Papua New Guinea		
1.60	Australia	0.61	Hungary	-0.96	Paraguay		
1.46	Austria	1.83	Iceland	-0.20	Peru		
-1.00	Azerbaijan	-0.31	India	-0.23	Philippines		
-0.29	Bangladesh	-0.80	Indonesia	0.49	Poland		
-0.65	Belarus	-0.85	Iran	1.22	Portugal		
0.67	Belgium	1.57	Ireland	-0.46	Romania		
-0.78	Benin	1.28	Israel	-0.62	Russian Federation		
	Bhutan	0.80	Italy		Rwanda		
-0.44	Bolivia	-0.12	Jamaica	-0.58	Saudi Arabia		
0.54	Botswana	0.72	Japan	-0.24	Senegal		
0.06	Brazil	0.14	Jordan	1.95	Singapore		
-0.56	Bulgaria	-0.87	Kazakhstan	0.03	Slovak Republic		
-0.37	Burkina Faso	-0.65	Kenya	1.02	Slovenia		
	Burundi	0.16	Korea, South	0.30	South Africa		
-1.10	Cameroon	0.62	Kuwait	1.21	Spain		
2.06	Canada	-0.76	Kyrgyz Republic	-0.12	Sri Lanka		
	Central African Republic	-0.26	Latvia	-1.02	Sudan		
1.03	Chile	-0.40	Lebanon	2.09	Sweden		
-0.29	China	-0.88	Libya	2.07	Switzerland		
-0.49	Colombia	0.03	Lithuania	-0.79	Syria		
0.58	Costa Rica	-0.52	Macedonia	-0.92	Tanzania		
-0.46	Croatia	-0.47	Madagascar	-0.16	Thailand		
0.27	Cuba	-0.19	Malawi	-0.24	Togo		
0.38	Czech Republic	0.63	Malaysia	0.51	Trinidad and Tobago		
2.13	Denmark	-0.48	Mali	0.02	Tunisia		
-0.77	Dominican Republic	0.34	Mauritius	-0.35	Turkey		
-0.82	Ecuador	-0.28	Mexico	-0.47	Uganda		
-0.27	Egypt	-0.39	Moldova	-0.89	Ukraine		
-0.35	El Salvador	-0.15	Mongolia	1.71	United Kingdom		
0.59	Estonia	0.13	Morocco	1.41	United States		
-0.44	Ethiopia	-0.53	Mozambique	0.43	Uruguay		
0.81	Fiji		Nepal	-0.96	Uzbekistan		
2.08	Finland	2.03	Netherlands	-0.72	Venezuela		
1.28	France	2.07	New Zealand	-0.33	Vietnam		
-1.02	Gabon	-0.84	Nicaragua	-0.61	Zambia		
1.62	Germany	-1.57	Niger	-0.32	Zimbabwe		
-0.30	Ghana	-0.95	Nigeria				

Global Stewardship

Variable name

Number of Memberships in Environmental Intergovernmental Organizations

Variable code
EIONUM

Variable number
56

Units
Number of Memberships

Reference year
1998

Median	Minimum	Maximum
12	2	35

Source
Organizational Memberships from Yearbook of International Organizations. Digital data set provided by Center for International Development and Conflict Management, University of Maryland.

Logic
Countries contribute to global environmental governance by participating in intergovernmental environmental organizations.

Details
100 intergovernmental organizations were coded as "environmental" by CIESIN. (List available upon request)

6.00	Albania
14.00	Algeria
15.00	Argentina
4.00	Armenia
19.00	Australia
20.00	Austria
5.00	Azerbaijan
7.00	Bangladesh
5.00	Belarus
26.00	Belgium
10.00	Benin
2.00	Bhutan
15.00	Bolivia
6.00	Botswana
20.00	Brazil
11.00	Bulgaria
9.00	Burkina Faso
5.00	Burundi
18.00	Cameroon
18.00	Canada
7.00	Central African Republic
10.00	Chile
12.00	China
16.00	Colombia
12.00	Costa Rica
9.00	Croatia
13.00	Cuba
12.00	Czech Republic
26.00	Denmark
10.00	Dominican Republic
17.00	Ecuador
21.00	Egypt
10.00	El Salvador
8.00	Estonia
9.00	Ethiopia
8.00	Fiji
25.00	Finland
35.00	France
13.00	Gabon
13.00	Ghana

23.00	Greece
13.00	Guatemala
8.00	Haiti
9.00	Honduras
15.00	Hungary
	Iceland
23.00	India
15.00	Indonesia
11.00	Iran
19.00	Ireland
12.00	Israel
26.00	Italy
10.00	Jamaica
24.00	Japan
11.00	Jordan
5.00	Kazakhstan
17.00	Kenya
16.00	Korea, South
10.00	Kuwait
3.00	Kyrgyz Republic
8.00	Latvia
10.00	Lebanon
10.00	Libya
8.00	Lithuania
6.00	Macedonia
9.00	Madagascar
12.00	Malawi
16.00	Malaysia
12.00	Mali
10.00	Mauritius
15.00	Mexico
5.00	Moldova
5.00	Mongolia
18.00	Morocco
6.00	Mozambique
6.00	Nepal
30.00	Netherlands
12.00	New Zealand
12.00	Nicaragua
10.00	Niger
17.00	Nigeria

26.00	Norway
14.00	Pakistan
14.00	Panama
11.00	Papua New Guinea
9.00	Paraguay
15.00	Peru
14.00	Philippines
16.00	Poland
21.00	Portugal
13.00	Romania
22.00	Russian Federation
5.00	Rwanda
8.00	Saudi Arabia
14.00	Senegal
8.00	Singapore
12.00	Slovak Republic
11.00	Slovenia
13.00	South Africa
27.00	Spain
14.00	Sri Lanka
15.00	Sudan
27.00	Sweden
24.00	Switzerland
15.00	Syria
16.00	Tanzania
16.00	Thailand
13.00	Togo
12.00	Trinidad and Tobago
16.00	Tunisia
14.00	Turkey
13.00	Uganda
8.00	Ukraine
28.00	United Kingdom
23.00	United States
11.00	Uruguay
5.00	Uzbekistan
16.00	Venezuela
8.00	Vietnam
10.00	Zambia
11.00	Zimbabwe

Global Stewardship

Variable name

Percentage of CITES Reporting Requirements Met

Variable code
CITES

Variable number
57

Units
Percent of Requirements Met

Reference year
2000

Median	Minimum	Maximum
75	0	100

Source
Convention on International Trade in Endangered Species of Wild Fauna and Flora, Report on National Reports Required Under Article VIII, Paragraph 7(a), of the Convention, Eleventh Meeting of the Conference of the Parties, Gigiri, Kenya, April 2000, available at http://www.unep-wcmc.org/CITES/eng/cop/11/docs/19.pdf, site visited November 2000.

Logic
Preparing and submitting national reports is a fundamental responsibility under CITES. The degree to which a country fulfills this responsibility is an indication of how seriously it takes its commitment to protection of endangered species.

0.00	Albania	100.00	Greece	87.00	Norway
60.00	Algeria	89.50	Guatemala	78.30	Pakistan
88.90	Argentina	0.00	Haiti	81.00	Panama
0.00	Armenia	21.40	Honduras	73.90	Papua New Guinea
100.00	Australia	85.70	Hungary	68.20	Paraguay
100.00	Austria	0.00	Iceland	75.00	Peru
0.00	Azerbaijan	100.00	India	83.30	Philippines
70.60	Bangladesh	95.00	Indonesia	88.90	Poland
50.00	Belarus	69.60	Iran	72.20	Portugal
100.00	Belgium	0.00	Ireland	40.00	Romania
26.70	Benin	52.60	Israel	78.30	Russian Federation
0.00	Bhutan	100.00	Italy	16.70	Rwanda
60.00	Bolivia	50.00	Jamaica	0.00	Saudi Arabia
90.50	Botswana	89.50	Japan	81.80	Senegal
54.20	Brazil	35.00	Jordan	100.00	Singapore
62.50	Bulgaria	0.00	Kazakhstan	100.00	Slovak Republic
55.60	Burkina Faso	65.00	Kenya	0.00	Slovenia
27.30	Burundi	100.00	Korea, South	95.80	South Africa
72.20	Cameroon	0.00	Kuwait	100.00	Spain
95.80	Canada	0.00	Kyrgyz Republic	70.00	Sri Lanka
47.40	Central African Republic	100.00	Latvia	56.30	Sudan
75.00	Chile	0.00	Lebanon	100.00	Sweden
100.00	China	0.00	Libya	100.00	Switzerland
83.30	Colombia	0.00	Lithuania	0.00	Syria
83.30	Costa Rica	0.00	Macedonia	84.20	Tanzania
0.00	Croatia	87.50	Madagascar	68.80	Thailand
88.90	Cuba	77.80	Malawi	75.00	Togo
100.00	Czech Republic	85.70	Malaysia	66.70	Trinidad and Tobago
95.50	Denmark	100.00	Mali	100.00	Tunisia
100.00	Dominican Republic	87.50	Mauritius	66.70	Turkey
70.80	Ecuador	87.50	Mexico	50.00	Uganda
19.00	Egypt	0.00	Moldova	0.00	Ukraine
33.30	El Salvador	100.00	Mongolia	100.00	United Kingdom
85.70	Estonia	60.90	Morocco	87.50	United States
90.00	Ethiopia	77.80	Mozambique	62.50	Uruguay
0.00	Fiji	75.00	Nepal	50.00	Uzbekistan
82.60	Finland	100.00	Netherlands	76.20	Venezuela
100.00	France	100.00	New Zealand	40.00	Vietnam
70.00	Gabon	90.90	Nicaragua	72.20	Zambia
100.00	Germany	50.00	Niger	88.90	Zimbabwe
87.00	Ghana	45.80	Nigeria		

Global Stewardship

Variable name

Levels of Participation in the Vienna Convention/ Montreal Protocol

Variable code **Variable number**
VIENNA 58

Units
Index Ranging from 0 (no participation) to 3 (high levels of participation)

Reference year
2000

Median	Minimum	Maximum
2.50	0	3

Source
United Nations Environment Program, The Ozone Secretariat,
http://www.unep.org/ozone/ratif.htm, site visited November 2000.

Logic
The number of protocols and amendments that a country has acceded to or ratified under the Vienna Convention is an indication of its commitment to fight ozone depletion.

Details
The index assigned values as follows. Countries received a score of zero if they were not signatory to the Vienna Convention. They received a score of 1 if they had ratified the Montreal Protocol only. They received a score of 2 if they ratified the above plus the London Amendment. They received a score of 2.5 if they ratified the above plus the Copenhagen Amendment. They received a score of 3 if they ratified the above plus the Montreal Amendment.

1.00	Albania	2.50	Greece	3.00	Norway
2.50	Algeria	1.00	Guatemala	2.50	Pakistan
2.50	Argentina	2.00	Haiti	3.00	Panama
1.00	Armenia	1.00	Honduras	2.00	Papua New Guinea
3.00	Australia	3.00	Hungary	2.00	Paraguay
3.00	Austria	3.00	Iceland	2.50	Peru
3.00	Azerbaijan	2.00	India	2.00	Philippines
2.00	Bangladesh	2.50	Indonesia	3.00	Poland
2.00	Belarus	2.50	Iran	2.50	Portugal
2.50	Belgium	2.50	Ireland	2.00	Romania
2.50	Benin	2.50	Israel	2.00	Russian Federation
0.00	Bhutan	2.50	Italy	0.00	Rwanda
3.00	Bolivia	2.50	Jamaica	2.50	Saudi Arabia
2.50	Botswana	2.50	Japan	3.00	Senegal
2.50	Brazil	3.00	Jordan	3.00	Singapore
3.00	Bulgaria	1.00	Kazakhstan	3.00	Slovak Republic
2.50	Burkina Faso	2.50	Kenya	3.00	Slovenia
1.00	Burundi	3.00	Korea, South	2.00	South Africa
2.50	Cameroon	2.50	Kuwait	3.00	Spain
3.00	Canada	1.00	Kyrgyz Republic	3.00	Sri Lanka
1.00	Central African Republic	2.50	Latvia	1.00	Sudan
3.00	Chile	3.00	Lebanon	3.00	Sweden
2.00	China	1.00	Libya	2.50	Switzerland
2.50	Colombia	2.50	Lithuania	3.00	Syria
2.50	Costa Rica	3.00	Macedonia	2.00	Tanzania
3.00	Croatia	1.00	Madagascar	2.50	Thailand
2.50	Cuba	2.50	Malawi	2.50	Togo
3.00	Czech Republic	2.50	Malaysia	3.00	Trinidad and Tobago
2.50	Denmark	2.00	Mali	3.00	Tunisia
1.00	Dominican Republic	2.50	Mauritius	2.50	Turkey
2.50	Ecuador	2.50	Mexico	3.00	Uganda
3.00	Egypt	1.00	Moldova	2.00	Ukraine
1.00	El Salvador	2.50	Mongolia	2.50	United Kingdom
2.50	Estonia	2.50	Morocco	2.50	United States
1.00	Ethiopia	2.50	Mozambique	3.00	Uruguay
2.50	Fiji	2.00	Nepal	2.50	Uzbekistan
2.50	Finland	3.00	Netherlands	2.50	Venezuela
2.50	France	3.00	New Zealand	2.50	Vietnam
1.00	Gabon	2.50	Nicaragua	2.00	Zambia
3.00	Germany	3.00	Niger	2.50	Zimbabwe
2.00	Ghana	1.00	Nigeria		

Global Stewardship

Variable name

Compliance with Environmental Agreements

Variable code   **Variable number**
WEFAGR   59

Units
Survey Responses Ranging from 1 (Strongly Disagree) to 7 (Strongly Agree)

Reference year
2000

Median	Minimum	Maximum
4.40	3	6.70

Source
Michael E. Porter et al, *The Global Competitiveness Report 2000*, Oxford: Oxford University Press, 2000.

Logic
Where compliance is a high priority, other things equal, global obligations are more effectively honored.

Details
Response to the statement: "Compliance with international environmental agreements is a high priority."

	Albania	4.30	Greece	6.10	Norway
	Algeria		Guatemala		Pakistan
3.40	Argentina		Haiti		Panama
	Armenia		Honduras		Papua New Guinea
5.20	Australia	4.60	Hungary		Paraguay
6.30	Austria	4.60	Iceland	3.10	Peru
	Azerbaijan	3.80	India	3.50	Philippines
	Bangladesh	3.60	Indonesia	3.60	Poland
	Belarus		Iran	4.70	Portugal
5.00	Belgium	4.90	Ireland		Romania
	Benin	4.40	Israel	3.70	Russian Federation
	Bhutan	4.60	Italy		Rwanda
3.10	Bolivia		Jamaica		Saudi Arabia
	Botswana	5.60	Japan		Senegal
4.20	Brazil	4.20	Jordan	5.70	Singapore
4.20	Bulgaria		Kazakhstan	4.00	Slovak Republic
	Burkina Faso		Kenya		Slovenia
	Burundi	4.40	Korea, South	4.20	South Africa
	Cameroon		Kuwait	4.50	Spain
5.60	Canada		Kyrgyz Republic		Sri Lanka
	Central African Republic		Latvia		Sudan
4.30	Chile		Lebanon	6.40	Sweden
4.80	China		Libya	6.00	Switzerland
4.50	Colombia		Lithuania		Syria
4.60	Costa Rica		Macedonia		Tanzania
	Croatia		Madagascar	3.60	Thailand
	Cuba		Malawi		Togo
4.90	Czech Republic	4.30	Malaysia		Trinidad and Tobago
6.70	Denmark		Mali		Tunisia
	Dominican Republic	4.10	Mauritius	4.10	Turkey
3.40	Ecuador	4.10	Mexico		Uganda
4.10	Egypt		Moldova	3.70	Ukraine
3.00	El Salvador		Mongolia	5.50	United Kingdom
	Estonia		Morocco	5.30	United States
	Ethiopia		Mozambique		Uruguay
	Fiji		Nepal		Uzbekistan
6.40	Finland	6.30	Netherlands	3.00	Venezuela
5.50	France	5.50	New Zealand	4.20	Vietnam
	Gabon		Nicaragua		Zambia
6.10	Germany		Niger	3.30	Zimbabwe
	Ghana		Nigeria		

Global Stewardship

Variable name

Montreal Protocol Multilateral Fund Participation

Variable code
MONFUN

Variable number
60

Units
Standardized Scale (Z score)

Reference year
2000

Median	Minimum	Maximum
0.10	-0.30	10.61

Source
Report of the Twelfth Meeting of the Sub-Committee on Monitoring, Evaluation and Finance UNEP/OzL.Pro/ExCom /32/3, and MontrealProtocol Unit (MPU), SEED/UNDP.

Logic
Managing global environmental problems requires active participation, both from donors and from funding recipients who implement projects. The Montreal Protocol Multilateral Fund is a major organized effort to finance reductions in production and consumption of ozone-depleting substances.

Details
This score combines payments (contributions to the Montreal Protocol Multilateral Fund and bilateral payments credited under the terms of the Fund) and receipts by countries from the Fund to implement CFC abatement projects. To make payments and receipts comparable, the two were first standardized, and countries were assigned the higher of the two possible Z scores. Payments were normalized by share of United Nations budget, and receipts were normalized by share of total Fund payments. Covers payments and receipts during 1991-1999.

-0.30	Albania	1.62	Greece	1.35	Norway
-0.30	Algeria	0.47	Guatemala	-0.29	Pakistan
1.41	Argentina	-0.30	Haiti	1.74	Panama
-0.30	Armenia	-0.30	Honduras	-0.30	Papua New Guinea
1.54	Australia	2.15	Hungary	0.40	Paraguay
1.26	Austria	1.40	Iceland	0.80	Peru
5.97	Azerbaijan	-0.19	India	0.46	Philippines
-0.27	Bangladesh	-0.10	Indonesia	1.30	Poland
-0.30	Belarus	-0.26	Iran	0.55	Portugal
1.42	Belgium	1.29	Ireland	-0.30	Romania
-0.30	Benin	0.87	Israel	-0.30	Russian Federation
-0.30	Bhutan	0.51	Italy	-0.30	Rwanda
-0.07	Bolivia	2.07	Jamaica	-0.30	Saudi Arabia
-0.20	Botswana	0.86	Japan	-0.30	Senegal
0.54	Brazil	-0.25	Jordan	0.10	Singapore
10.61	Bulgaria	-0.30	Kazakhstan	5.10	Slovak Republic
-0.30	Burkina Faso	-0.21	Kenya	-0.16	Slovenia
-0.14	Burundi	-0.30	Korea, South	1.09	South Africa
-0.30	Cameroon		Kuwait	1.23	Spain
1.75	Canada	-0.30	Kyrgyz Republic	0.30	Sri Lanka
0.03	Central African Republic	-0.08	Latvia	-0.30	Sudan
-0.28	Chile	0.63	Lebanon	1.69	Sweden
0.03	China	-0.30	Libya	1.44	Switzerland
1.44	Colombia	7.58	Lithuania	0.16	Syria
2.51	Costa Rica	-0.30	Macedonia	-0.24	Tanzania
-0.30	Croatia	-0.30	Madagascar	1.10	Thailand
0.09	Cuba	-0.07	Malawi	-0.30	Togo
4.93	Czech Republic	6.42	Malaysia	2.68	Trinidad and Tobago
1.51	Denmark	-0.30	Mali	-0.30	Tunisia
1.25	Dominican Republic	4.03	Mauritius	-0.30	Turkey
-0.21	Ecuador	1.08	Mexico	-0.28	Uganda
1.37	Egypt	0.19	Moldova	0.26	Ukraine
0.74	El Salvador	-0.30	Mongolia	0.90	United Kingdom
-0.01	Estonia	0.57	Morocco	1.21	United States
-0.30	Ethiopia	-0.24	Mozambique	3.32	Uruguay
-0.30	Fiji	-0.30	Nepal	0.58	Uzbekistan
1.64	Finland	1.20	Netherlands	1.57	Venezuela
0.86	France	1.67	New Zealand	-0.21	Vietnam
1.62	Gabon	-0.30	Nicaragua	-0.18	Zambia
0.97	Germany	-0.29	Niger	-0.29	Zimbabwe
0.21	Ghana	0.03	Nigeria		

Global Stewardship

Variable name

Global Environmental Facility Participation

Variable code　　**Variable number**
GEF　　　　　　　　61

Units
Standardized Scale (Z score)

Reference year
2000

Median　　**Minimum**　　**Maximum**
-0.05　　　　-0.17　　　　6.01

Source

"GEF Projects – Allocations and Disbursements" www.gefweb.org/Allocations_Disbursements.pdf and "GEF Council December 8-10, 1999, Agenda Item 10, DRAFT ANNUAL REPORT 1999 VOLUME II: FINANCIAL STATEMENT" www.gefweb.org/COUNCIL/GEF_C14/gef_c14_8.pdf

Logic

Managing global environmental problems requires active participation, from both donors and recipients. The GEF represents the most significant global-scale effort to support world-wide environmental protection efforts.

Details

This score combines payments and receipts. To make payments and receipts comparable, the two were first standardized, and countries were assigned the higher of the two possible Z scores. Payments were normalized by share of United Nations budget, and receipts were normalized by share of total GEF payments. Covers payments and receipts during through the entire Phase I period and through October 30, 2000 of Phase 2.

-0.17	Albania		0.14	Greece		0.96	Norway
-0.12	Algeria		-0.05	Guatemala		1.92	Pakistan
-0.11	Argentina		-0.17	Haiti		0.15	Panama
-0.08	Armenia		-0.03	Honduras		0.46	Papua New Guinea
0.26	Australia		0.09	Hungary		-0.17	Paraguay
0.30	Austria		-0.17	Iceland		-0.04	Peru
0.21	Azerbaijan		0.45	India		0.07	Philippines
6.01	Bangladesh		-0.17	Indonesia		0.03	Poland
0.13	Belarus		-0.17	Iran		0.12	Portugal
0.47	Belgium		0.06	Ireland		-0.11	Romania
0.07	Benin		-0.17	Israel		-0.13	Russian Federation
2.80	Bhutan		0.29	Italy		-0.17	Rwanda
0.19	Bolivia		0.52	Jamaica		-0.17	Saudi Arabia
-0.17	Botswana		0.27	Japan		-0.08	Senegal
-0.11	Brazil		0.81	Jordan		-0.17	Singapore
0.33	Bulgaria		-0.17	Kazakhstan		3.36	Slovak Republic
-0.15	Burkina Faso		-0.16	Kenya		1.13	Slovenia
-0.17	Burundi		-0.17	Korea, South		-0.10	South Africa
-0.05	Cameroon		-0.17	Kuwait		-0.02	Spain
0.53	Canada		-0.17	Kyrgyz Republic		-0.03	Sri Lanka
-0.06	Central African Republic		0.12	Latvia		-0.17	Sudan
-0.14	Chile		0.13	Lebanon		1.02	Sweden
-0.05	China		-0.17	Libya		0.64	Switzerland
-0.17	Colombia		1.05	Lithuania		-0.17	Syria
1.47	Costa Rica		-0.17	Macedonia		-0.17	Tanzania
-0.17	Croatia		-0.02	Madagascar		-0.12	Thailand
-0.08	Cuba		-0.17	Malawi		-0.17	Togo
0.98	Czech Republic		-0.17	Malaysia		-0.17	Trinidad and Tobago
0.95	Denmark		-0.10	Mali		-0.10	Tunisia
-0.02	Dominican Republic		1.40	Mauritius		0.11	Turkey
0.17	Ecuador		-0.01	Mexico		-0.02	Uganda
0.72	Egypt		-0.17	Moldova		-0.15	Ukraine
-0.12	El Salvador		0.56	Mongolia		0.41	United Kingdom
-0.17	Estonia		-0.17	Morocco		0.19	United States
-0.17	Ethiopia		-0.08	Mozambique		0.62	Uruguay
-0.17	Fiji		-0.11	Nepal		-0.16	Uzbekistan
0.71	Finland		0.79	Netherlands		-0.16	Venezuela
0.31	France		0.39	New Zealand		-0.14	Vietnam
-0.17	Gabon		0.02	Nicaragua		-0.17	Zambia
0.37	Germany		-0.15	Niger		0.19	Zimbabwe
0.02	Ghana		-0.17	Nigeria			

Global Stewardship

Variable name

FSC Accredited Forest Area as a Percentage of Total Forest Area

Variable code FSC **Variable number** 62

Units
FSC Forest Area as Percent of Total Forest Area

Reference year
2000

Median 0 **Minimum** 0 **Maximum** 36

Source
Forest Stewardship Council, personal communication.

Logic
This variable measures the extent to which a country seeks sustainable forestry practices.

Details
A logarithmic transformation of this variable was used in calculating the ESI.

0.00	Albania	0.00	Greece	0.00	Norway
0.00	Algeria	3.00	Guatemala	0.00	Pakistan
0.00	Argentina	0.00	Haiti	0.00	Panama
0.00	Armenia	0.50	Honduras	0.01	Papua New Guinea
0.00	Australia	0.00	Hungary	0.00	Paraguay
0.00	Austria	0.00	Iceland	0.00	Peru
0.00	Azerbaijan	0.00	India	0.00	Philippines
0.00	Bangladesh	0.10	Indonesia	30.00	Poland
0.00	Belarus	0.00	Iran	0.00	Portugal
0.30	Belgium	0.00	Ireland	0.00	Romania
0.00	Benin	0.00	Israel	0.00	Russian Federation
0.00	Bhutan	0.10	Italy	0.00	Rwanda
1.30	Bolivia	0.00	Jamaica	0.00	Saudi Arabia
0.00	Botswana	0.00	Japan	0.00	Senegal
0.10	Brazil	0.00	Jordan	0.00	Singapore
0.00	Bulgaria	0.00	Kazakhstan	0.00	Slovak Republic
0.00	Burkina Faso	0.00	Kenya	0.00	Slovenia
0.00	Burundi	0.00	Korea, South	9.00	South Africa
0.00	Cameroon	0.00	Kuwait	0.00	Spain
0.08	Canada	0.00	Kyrgyz Republic	0.11	Sri Lanka
0.00	Central African Republic	0.00	Latvia	0.00	Sudan
0.00	Chile	0.00	Lebanon	33.00	Sweden
0.00	China	0.00	Libya	4.00	Switzerland
0.00	Colombia	0.00	Lithuania	0.00	Syria
3.00	Costa Rica	0.00	Macedonia	0.00	Tanzania
1.58	Croatia	0.00	Madagascar	0.00	Thailand
0.00	Cuba	0.00	Malawi	0.00	Togo
4.00	Czech Republic	0.40	Malaysia	0.00	Trinidad and Tobago
0.00	Denmark	0.00	Mali	0.00	Tunisia
0.00	Dominican Republic	0.00	Mauritius	0.00	Turkey
0.00	Ecuador	0.30	Mexico	0.00	Uganda
0.00	Egypt	0.00	Moldova	0.38	Ukraine
0.00	El Salvador	0.00	Mongolia	36.00	United Kingdom
0.00	Estonia	0.00	Morocco	0.80	United States
0.00	Ethiopia	0.00	Mozambique	0.00	Uruguay
0.00	Fiji	0.00	Nepal	0.00	Uzbekistan
0.00	Finland	20.00	Netherlands	0.00	Venezuela
0.01	France	0.03	New Zealand	0.00	Vietnam
0.00	Gabon	0.00	Nicaragua	0.00	Zambia
0.80	Germany	0.00	Niger	0.80	Zimbabwe
0.00	Ghana	0.00	Nigeria		

Global Stewardship

Variable name

Ecological Footprint "Deficit"

Variable code   **Variable number**

FOOT   63

Units

Area Units (hectares of biologically productive space with world-average productivity)/person

Reference year

1996

Median   **Minimum**   **Maximum**

-0.53   -12.21   31.72

Source

World Wide Fund for Nature (WWF), *Living Planet Report 2000*, Gland, Switzerland: WWF, 2000.

Logic

The ecological footprint is a measure of a country's impact on global environmental resources. A negative number (deficit) indicates that a country requires more land area than it actually has in order to support its economy, and a positive number means that it has a surplus of biologically productive land.

Details

The amount by which the ecological footprint of the country's population exceeds the biological capacity of the space available to that population.

Value	Country	Value	Country	Value	Country
-0.48	Albania	-3.83	Greece	0.01	Norway
-1.21	Algeria	0.36	Guatemala	-0.40	Pakistan
1.31	Argentina	-0.48	Haiti	1.82	Panama
-0.47	Armenia	0.83	Honduras	30.20	Papua New Guinea
0.93	Australia	-1.94	Hungary	2.68	Paraguay
-1.30	Austria		Iceland	7.90	Peru
-1.54	Azerbaijan	-0.32	India	-0.54	Philippines
-0.52	Bangladesh	1.70	Indonesia	-3.05	Poland
-1.80	Belarus	-1.71	Iran	-2.76	Portugal
	Belgium	-2.72	Ireland	-1.10	Romania
0.58	Benin	-4.64	Israel	-1.26	Russian Federation
1.82	Bhutan	-3.59	Italy	-0.48	Rwanda
11.96	Bolivia	-1.95	Jamaica	-5.74	Saudi Arabia
0.24	Botswana	-5.08	Japan	-0.11	Senegal
8.96	Brazil	-1.50	Jordan	-12.21	Singapore
-1.80	Bulgaria	-2.40	Kazakhstan	-1.92	Slovak Republic
-0.11	Burkina Faso	-0.59	Kenya	-2.77	Slovenia
-0.25	Burundi	-4.86	Korea, South	-2.65	South Africa
3.35	Cameroon	-9.67	Kuwait	-2.98	Spain
3.50	Canada	-0.37	Kyrgyz Republic	-0.43	Sri Lanka
13.38	Central African Republic	0.33	Latvia	0.62	Sudan
-1.38	Chile	-2.50	Lebanon	0.48	Sweden
-0.96	China	-3.78	Libya	-4.33	Switzerland
3.76	Colombia	-1.04	Lithuania	-1.46	Syria
-0.60	Costa Rica	-2.05	Macedonia	0.33	Tanzania
-0.17	Croatia	2.00	Madagascar	-1.35	Thailand
-0.98	Cuba	-0.10	Malawi	0.00	Togo
-3.37	Czech Republic	0.29	Malaysia	-1.66	Trinidad and Tobago
-4.19	Denmark	0.41	Mali	-1.05	Tunisia
-0.34	Dominican Republic	-0.23	Mauritius	-1.24	Turkey
1.74	Ecuador	-1.04	Mexico	0.13	Uganda
-1.06	Egypt	-0.77	Moldova	-2.49	Ukraine
-0.87	El Salvador	1.37	Mongolia	-4.46	United Kingdom
-3.10	Estonia	-0.57	Morocco	-6.66	United States
-0.18	Ethiopia	0.35	Mozambique	0.22	Uruguay
	Fiji	-0.07	Nepal	-1.70	Uzbekistan
1.32	Finland	-3.35	Netherlands	3.01	Venezuela
	France	6.26	New Zealand	-0.30	Vietnam
31.72	Gabon	2.96	Nicaragua	3.03	Zambia
-3.01	Germany	-0.56	Niger	-0.77	Zimbabwe
0.08	Ghana	-0.43	Nigeria		

Global Stewardship

Variable name

CO2 Emissions (total times per capita)

Variable code	**Variable number**
CO2_EM	64

Units
MetricTons

Reference year
1997

Median	**Minimum**	**Maximum**
5780.36	0.61	8163271.04

Source
Carbon Dioxide Information Analysis Center, available at http://cdiac.esd.ornl.gov/

Logic
Carbon dioxide is the most significant greenhouse gas. This variable combines total and per capita emissions, reflecting two ways to measure global responsibility.

Details
The indicator was obtained by multiplying the Total CO2 emissions from fossil-fuels (thousand metric tons of C) with the Per capita CO2 emissions (metric tons of carbon). A logarithmic transformation of this variable was used in calculating the ESI.

60.48	Albania	45,816.16	Greece	77,574.00	Norway
22,856.24	Algeria	419.40	Guatemala	4,605.84	Pakistan
39,510.45	Argentina	18.95	Haiti	1,608.53	Panama
159.18	Armenia	214.51	Honduras	100.35	Papua New Guinea
406,642.56	Australia	24,763.44	Hungary	208.20	Paraguay
33,776.28	Austria	1,195.48	Iceland	2,635.05	Peru
9,910.02	Azerbaijan	81,170.71	India	5,669.72	Philippines
334.15	Bangladesh	20,832.96	Indonesia	235,670.11	Poland
27,146.34	Belarus	95,873.70	Iran	18,525.14	Portugal
78,193.06	Belgium	27,025.92	Ireland	38,207.00	Romania
8.12	Benin	41,445.46	Israel	1,035,132.40	Russian Federation
5.35	Bhutan	215,966.62	Italy	2.70	Rwanda
1,121.76	Bolivia	3,396.48	Jamaica	270,857.68	Saudi Arabia
540.44	Botswana	793,571.64	Japan	85.50	Senegal
37,759.68	Brazil	2,413.53	Jordan	139,998.51	Singapore
21,523.20	Bulgaria	68,280.84	Kazakhstan	18,786.02	Slovak Republic
5.30	Burkina Faso	108.78	Kenya	8,464.54	Slovenia
0.61	Burundi	297,587.55	Korea, South	192,966.36	South Africa
32.35	Cameroon	107,585.64	Kuwait	111,861.12	Spain
591,793.80	Canada	674.88	Kyrgyz Republic	230.56	Sri Lanka
1.32	Central African Republic	2,001.60	Latvia	39.52	Sudan
17,313.56	Chile	5,891.00	Lebanon	18,899.70	Sweden
685,326.00	China	24,935.34	Libya	16,875.04	Switzerland
8,533.46	Colombia	4,389.43	Lithuania	11,319.57	Syria
487.80	Costa Rica	4,287.99	Macedonia	13.46	Tanzania
6,020.40	Croatia	3.28	Madagascar	54,142.40	Thailand
4,249.48	Cuba	3.96	Malawi	10.95	Togo
108,858.75	Czech Republic	60,707.00	Malaysia	28,116.55	Trinidad and Tobago
45,063.40	Denmark	1.31	Mali	2,225.09	Tunisia
1,624.05	Dominican Republic	190.65	Mauritius	45,935.70	Turkey
2,552.54	Ecuador	105,961.84	Mexico	2.92	Uganda
13,721.34	Egypt	1,842.75	Moldova	197,841.19	Ukraine
372.25	El Salvador	1,745.49	Mongolia	342,451.36	United Kingdom
18,337.56	Estonia	2,784.64	Morocco	8,163,271.04	United States
5.17	Ethiopia	6.06	Mozambique	661.05	Uruguay
53.56	Fiji	11.06	Nepal	33,523.20	Uzbekistan
45,614.86	Finland	125,244.48	Netherlands	115,074.00	Venezuela
147,676.02	France	19,191.92	New Zealand	1,739.25	Vietnam
727.20	Gabon	149.58	Nicaragua	53.60	Zambia
629,798.28	Germany	9.06	Niger	2,185.92	Zimbabwe
66.36	Ghana	4,935.70	Nigeria		

Global Stewardship

Variable name

Historic Cumulative CO2 Emissions

Variable code
CO2HIS

Variable number
65

Units
Metric Tons

Reference year
1997

Median	Minimum	Maximum
16688.55	120.56	2961127.75

Source
Carbon Dioxide Information Analysis Center, available at http://cdiac.esd.ornl.gov/

Logic
Given the long atmospheric lifetime of CO2, historic emissions represent an important factor in climate change.

Details
Historic carbon-dioxide emissions data were utilized, applying an annual decay rate of .9926, which is consistent with the estimate that 80 percent of any given carbon-dioxid emission remains in the atomsphere after 30 years. A logarithmic transformation of this variable was used in calculating the ESI.

1,008.45	Albania
60,924.54	Algeria
77,116.71	Argentina
1,546.22	Armenia
173,397.53	Australia
33,683.03	Austria
17,231.66	Azerbaijan
13,184.42	Bangladesh
34,313.61	Belarus
58,960.83	Belgium
541.73	Benin
231.83	Bhutan
6,761.45	Bolivia
1,751.77	Botswana
165,229.50	Brazil
28,401.88	Bulgaria
527.05	Burkina Faso
120.56	Burundi
1,348.58	Cameroon
267,360.26	Canada
129.53	Central African Republic
30,130.43	Chile
1,962,527.17	China
39,137.02	Colombia
2,864.11	Costa Rica
10,645.19	Croatia
16,621.52	Cuba
68,734.85	Czech Republic
34,223.47	Denmark
7,501.50	Dominican Republic
13,506.01	Ecuador
63,080.36	Egypt
3,002.19	El Salvador
10,458.04	Estonia
1,207.16	Ethiopia
438.15	Fiji
31,674.15	Finland
196,669.68	France
1,681.15	Gabon
471,169.60	Germany
2,583.21	Ghana

47,768.68	Greece
4,206.87	Guatemala
662.88	Haiti
2,464.83	Honduras
32,957.78	Hungary
1,191.47	Iceland
573,969.71	India
142,539.06	Indonesia
168,386.62	Iran
19,876.32	Ireland
32,291.61	Israel
229,549.39	Italy
5,819.15	Jamaica
657,751.13	Japan
8,442.42	Jordan
72,319.25	Kazakhstan
4,011.08	Kenya
243,100.00	Korea, South
28,706.76	Kuwait
3,610.49	Kyrgyz Republic
4,790.81	Latvia
9,368.47	Lebanon
25,310.18	Libya
8,261.24	Lithuania
6,233.58	Macedonia
668.92	Madagascar
432.39	Malawi
75,572.63	Malaysia
265.02	Mali
937.48	Mauritius
204,256.06	Mexico
5,992.58	Moldova
4,309.27	Mongolia
19,383.48	Morocco
627.58	Mozambique
1,154.62	Nepal
91,134.14	Netherlands
16,755.58	New Zealand
1,710.41	Nicaragua
608.94	Niger
75,148.68	Nigeria

55,844.63	Norway
53,568.92	Pakistan
3,568.59	Panama
1,321.14	Papua New Guinea
2,160.88	Paraguay
15,100.14	Peru
41,303.25	Philippines
196,634.79	Poland
28,783.64	Portugal
63,170.06	Romania
809,347.32	Russian Federation
270.40	Rwanda
81,328.47	Saudi Arabia
1,892.38	Senegal
42,257.25	Singapore
21,389.96	Slovak Republic
8,266.72	Slovenia
168,507.98	South Africa
136,735.34	Spain
4,259.69	Sri Lanka
2,072.70	Sudan
28,337.90	Sweden
22,946.15	Switzerland
28,867.57	Syria
1,552.85	Tanzania
120,899.06	Thailand
524.13	Togo
13,470.81	Trinidad and Tobago
10,333.83	Tunisia
114,672.43	Turkey
604.62	Uganda
208,956.66	Ukraine
297,179.93	United Kingdom
2,961,127.75	United States
3,115.83	Uruguay
57,024.21	Uzbekistan
108,769.35	Venezuela
25,247.68	Vietnam
1,420.11	Zambia
10,267.46	Zimbabwe

Global Stewardship

Variable name

CFC Consumption (total times per capita)

Variable code	Variable number
CFC	66

Units

Ozone Depletion Potential (ODP) tons (Metric Tons x ODP)

Reference year

MRYA 1996-98

Median	Minimum	Maximum
3096.17	0	2096731.55

Source

UNEP, Production and Consumption of Ozone Depleting Substances, 1986-1998, October 1999.

Logic

Emissions of CFCs contribute to the breakdown of the Earth's protective ozone layer and to global climate change. This variable combines total and per capita emission, reflecting the long atmospheric lifetime of CFCs.

Details

The indicator was obtained by multiplying the Total CFCs emissions (metric tons times ozone depletion potential) with the Per capita CFCs emissions (obtained by dividing the total CFCs emissions by the population in 1997). A logarithmic transformation of this variable was used in calculating the ESI.

	Albania			Greece	58.24	Norway
81,627.89	Algeria	2,225.37	Guatemala	11,091.52	Pakistan
31,916.38	Argentina		Haiti	43,976.07	Panama
	Armenia	1,638.72	Honduras	288.08	Papua New Guinea
0.22	Australia	0.10	Hungary	2,509.55	Paraguay
	Austria	0.00	Iceland	4,388.27	Peru
5,286.64	Azerbaijan	46,502.34	India	105,641.32	Philippines
5,643.89	Bangladesh	88,310.73	Indonesia	2,451.70	Poland
6,331.14	Belarus	480,228.61	Iran		Portugal
	Belgium		Ireland	15,021.65	Romania
34.82	Benin	0.00	Israel	817,386.43	Russian Federation
	Bhutan		Italy		Rwanda
272.19	Bolivia	15,736.64	Jamaica	142,831.18	Saudi Arabia
31.81	Botswana	101.31	Japan	1,867.71	Senegal
588,838.63	Brazil	119,897.02	Jordan	84.33	Singapore
0.00	Bulgaria		Kazakhstan	0.19	Slovak Republic
124.44	Burkina Faso	2,214.78	Kenya	0.00	Slovenia
643.81	Burundi	1,858,868.33	Korea, South	619.83	South Africa
4,855.01	Cameroon	135,805.16	Kuwait		Spain
58.29	Canada		Kyrgyz Republic	3,420.18	Sri Lanka
0.00	Central African Republic	214.94	Latvia	3,378.16	Sudan
37,241.22	Chile	71,790.14	Lebanon		Sweden
2,096,731.55	China	80,339.88	Libya	231.85	Switzerland
37,414.36	Colombia	2,919.55	Lithuania	279,497.02	Syria
11,103.16	Costa Rica	1,997.95	Macedonia	1,125.00	Tanzania
1,649.37	Croatia	739.80	Madagascar	239,571.46	Thailand
39,953.99	Cuba	322.74	Malawi		Togo
11.75	Czech Republic	259,617.88	Malaysia	19,060.25	Trinidad and Tobago
	Denmark	1,180.63	Mali	67,931.19	Tunisia
11,944.58	Dominican Republic	1,342.64	Mauritius	236,217.77	Turkey
6,197.71	Ecuador	128,672.29	Mexico	6.05	Uganda
36,637.74	Egypt	365.59	Moldova	23,739.77	Ukraine
6,433.23	El Salvador	157.67	Mongolia		United Kingdom
3,385.93	Estonia	29,193.18	Morocco	23,385.16	United States
24.80	Ethiopia	26.24	Mozambique	11,525.63	Uruguay
249.25	Fiji	37.69	Nepal	121.02	Uzbekistan
	Finland		Netherlands	602,347.63	Venezuela
	France	0.00	New Zealand	3,272.79	Vietnam
126.65	Gabon	292.60	Nicaragua	97.96	Zambia
	Germany	356.53	Niger	16,872.89	Zimbabwe
134.00	Ghana	218,257.67	Nigeria		

Global Stewardship

Variable name

S02 Exports

Variable code | **Variable number**
SO2EXP | 67

Units
100 Metric Tons

Reference year
1997 (Asia) and 1998 (Europe)

Median | **Minimum** | **Maximum**
538 | 4.12 | 12300

Source
International Institute for Applied Systems Analysis, RAINS-ASIA and Co-operative Programme for monitoring and evaluation of the long range transmission of air pollutants in Europe (EMEP).

Logic
The transport of sulphur emissions across national boundaries contributes to poor air quality and acid rain in receiving countries.

307.00	Albania	2,029.00	Greece	98.00	Norway
	Algeria		Guatemala	420.00	Pakistan
	Argentina		Haiti		Panama
12.00	Armenia		Honduras		Papua New Guinea
	Australia	2,348.00	Hungary		Paraguay
175.00	Austria	110.00	Iceland		Peru
	Azerbaijan	3,400.00	India	723.00	Philippines
238.00	Bangladesh	1,320.00	Indonesia	5,849.00	Poland
628.00	Belarus		Iran	1,349.00	Portugal
832.00	Belgium	565.00	Ireland	2,768.00	Romania
	Benin		Israel	4,148.00	Russian Federation
4.12	Bhutan	3,876.00	Italy		Rwanda
	Bolivia		Jamaica		Saudi Arabia
	Botswana	1,420.00	Japan		Senegal
	Brazil		Jordan	642.00	Singapore
4,974.00	Bulgaria		Kazakhstan	746.00	Slovak Republic
	Burkina Faso		Kenya	538.00	Slovenia
	Burundi	438.00	Korea, South		South Africa
	Cameroon		Kuwait	5,201.00	Spain
	Canada		Kyrgyz Republic	81.50	Sri Lanka
	Central African Republic	155.00	Latvia		Sudan
	Chile		Lebanon	144.00	Sweden
12,300.00	China		Libya	94.00	Switzerland
	Colombia	363.00	Lithuania		Syria
	Costa Rica	71.00	Macedonia		Tanzania
367.00	Croatia		Madagascar		Thailand
	Cuba		Malawi		Togo
1,762.00	Czech Republic	401.00	Malaysia		Trinidad and Tobago
326.00	Denmark		Mali		Tunisia
	Dominican Republic		Mauritius	3,465.00	Turkey
	Ecuador		Mexico		Uganda
	Egypt	143.00	Moldova	3,560.00	Ukraine
	El Salvador	69.00	Mongolia	5,591.00	United Kingdom
496.00	Estonia		Morocco		United States
	Ethiopia		Mozambique		Uruguay
	Fiji	188.00	Nepal		Uzbekistan
245.00	Finland	425.00	Netherlands		Venezuela
2,537.00	France		New Zealand	201.00	Vietnam
	Gabon		Nicaragua		Zambia
4,448.00	Germany		Niger		Zimbabwe
	Ghana		Nigeria		